MONOGRAPHS ON INDUSTRIAL CHEMISTRY
EDITED BY SIR EDWARD THORPE, C.B., LL.D., F.R.S.

THE ELECTRIC FURNACE

THE
ELECTRIC FURNACE

BY

J. N. PRING, M.B.E., D.Sc.

RESEARCH DEPARTMENT, ROYAL ARSENAL, WOOLWICH
FORMERLY READER IN ELECTRO-CHEMISTRY AT THE UNIVERSITY OF MANCHESTER

WITH ILLUSTRATIONS

LONGMANS, GREEN AND CO.

39 PATERNOSTER ROW, LONDON

FOURTH AVENUE & 30TH STREET, NEW YORK

BOMBAY, CALCUTTA, AND MADRAS

1921

AUTHOR'S PREFACE

THE very high range of temperatures attainable in electric furnaces has opened up a large field of chemistry which was not previously accessible. In the case of processes in which the electric current merely acts as a heating agent and is dissociated from any electrolytic action, products such as artificial graphite, the highly refractory metals in a compact form, and a large series of compounds of carbon with metals, have been isolated for the first time. In many cases these products have received an industrial application which has already ranged them in the forefront of the world's manufacturing processes.

With other substances, notably certain classes of steel and iron and ferro-alloys, aluminium and sodium, the agency of electricity applied to chemical processes, has led to a very great simplification and cheapening of the manufacture, while with yet another class of products, such as copper and zinc, these methods have enabled materials to be prepared of a degree of purity unattainable by any other means.

The electric furnace, in facilitating the production of a range of temperatures well beyond that attainable by any other known method, has thus inaugurated a new department of chemical industry. On account of the rapid progress of this branch of industrial electro-chemistry, it has until recently been generally overlooked that an extension of electrical methods of heating of no less importance has taken place in another direction, viz. in chemical and metallurgical operations, including those with both ferrous and non-ferrous metals for the production of heat at temperatures which are well within the range of ordinary fuel-heating methods. Though

v

many of these systems are old-established and developed to a high degree of efficiency, there is a progressive tendency for the extension of electrical methods of heating to chemical processes at all ranges of temperature.

The position in this country with reference to electro-chemical developments is that a large share of the pioneering work, particularly during the nineteenth century, has been conducted here. As notable instances, there may be mentioned the electrolytic refining of copper which originated in South Wales, the manufacture of aluminium alloys by Cowles in Staffordshire, the electrical production of steel by Siemens, the inception of the induction furnace by Ferranti, and the synthesis of nitric acid in the high-tension arc by McDougal & Howles in Manchester.

In nearly all cases, however, the successful commercial development of these processes was first undertaken abroad to be later reintroduced, in many cases, at home.

The conservative attitude in this country has, in this connexion, for a long time been fostered by a number of misconceptions which are still widely held. The oldest of these relates to the subject of power supply and arises from the fact that the earlier electro-chemical processes became established in districts where cheap water-power is available. It is frequently assumed from this that the large supplies of electrical energy needed in electro-chemical processes cannot be economically generated from steam power and that this class of industries must be confined to large water-power centres. However, the position with regard to the generation of electrical energy on a large scale from steam-power has undergone a very rapid development in the last fifteen years.

It is further recalled that with the best steam-driven turbo-generators which have been designed up to the present, an efficiency of only 18 to 20 per cent is obtained in the transformation of the heat value of the fuel to electrical energy, and that, in consequence, this electricity cannot be applied to produce heat in a furnace in competition with direct fuel-heating.

The fallacies of this conclusion are as follow :—

1. In fuel-heated chemical processes, a large proportion of the heat is, at the best, carried away by the waste flue gases and there are many cases, particularly where the heat applied has to penetrate through the walls of crucibles or muffles, where the efficiency in the application of the heat is considerably less than the 18 per cent realised with turbogenerators, while with a system of internal electric heating, the electrical energy is applied as heat in the body of the charge with an efficiency, after allowing for radiation losses, which may amount to 80 per cent or upwards.

2. There are comparatively few pyro-chemical processes in which the cost of heat, even if applied electrically from steam-generated power, is the main factor in determining the economics of the process. The consumption of power per unit value of product varies very widely with different substances, but in very few cases is it so high as to outweigh the consideration of such factors as the availability of raw materials and vicinity of subsidiary industries. In considering the substitution of fuel-heating methods by electrical heating, a disadvantage in the cost of electrical heat is frequently more than counter-balanced by such advantages as those of greater cleanliness and less contamination of products and better regulation and control of temperature.

The electro-chemical industry is of comparatively recent origin, and its whole scope has rapidly expanded with the general developments in electrical engineering, and particularly with the cheapening in the cost of power production resulting from such advances.

In 1887, at the time of the installation of the Cowles plant for the manufacture of aluminium alloys at Milton, in Staffordshire, the 500 h.p. dynamo which was specially constructed for this occasion was, on account of its size, regarded as a notable triumph in electrical engineering, whereas to-day 30,000 h.p. generators are not uncommon at large power stations.

The position of electro-chemical processes in the metallurgical industry has been described by Prof. H. C. H. Carpenter[1] as follows:—

[1] "Nature," 1919, **104**, 243.

"Viewing the industry to-day, it is manifest that there is a notable trend towards the substitution of furnace or pyro-metallurgy by hydro and electro-metallurgy. Even where furnace operations still hold the field, attempts are being continually made towards the substitution of fuel heat by electric heat. . . . The great importance of this tendency is that it permits of a more complete beneficiation of any given ore, and, indeed, brings a far wider range of raw materials within the scope of economic exploitation than otherwise would be the case. In the refining of metals, 'electric heat' is tending in some cases to supplement, in others to supplant, fuel heat. It is widely held that the quality of steel which can be produced in the electric furnace is superior to that obtained in the open-hearth furnace. This is due to the fact that, owing to the high temperature employed, more refractory basic fluxes can be used which permit of a greater removal of sulphur and phosphorus, with a consequent improvement in the properties of the refined steel. Moreover, in the electric furnace, the charge is decidedly less contaminated with gases. For high-grade materials, such as high-speed cutting tools, where quality is of paramount importance, the electric furnace seems to have a field all its own."

In addition to their metallurgical applications, the directions in which electro-chemical processes have recently proved to be of national and vital importance are in the production from the atmosphere of synthetic nitrogen compounds which form the basis of all modern explosives, and are needed in rapidly increasing amounts as fertilisers. A further modern development of considerable promise is in the production of organic compounds from carbides.

The most noteworthy branches of the electro-chemical and electro-metallurgical industry have, within the last decade, been described in a number of publications. Nevertheless, the present rapid progress of these enterprises demands a frequent revision and extension of the literature of this subject as contained in text-books. No apology is needed, therefore, for the attempt, at this stage, to introduce an additional contribution to the general technical discussion of the position and prospects of high-temperature industrial chemistry. In compiling this work, the author is indebted to the proprietors of a large number of scientific journals, publications, and industrial companies, for permission and facilities kindly extended to reproduce data and illustrations. Material has also been utilised from earlier publications of the writer by the Manchester University Press. For the preparation of a number of the diagrams, the author is much indebted to Mr. M. Francis, and for assistance in reading the proofs to Mr. W. H. Watson.

CONTENTS

LIST OF PLATES

xii

SECTION I.

HISTORICAL.

As different methods of producing electric currents in increasing magnitudes were discovered, one of the earliest applications made, in every case, was the study of their effects on chemical change both by electrolytic action and, on account of the high temperatures attainable, by their electrothermal effects. Thus Davy,[1] in 1810, using a voltaic pile of 1000 plates, isolated the alkali metals and aluminium by electrolysis, and conducted experiments on the fusion of iron wire. Pepys, in 1815,[2] by means of an electrically heated iron rod demonstrated the cementation of iron by absorption of carbon. The experiment was conducted by bending a piece of soft iron, cutting a longitudinal groove at the bend, and filling with diamond dust (cf. Fig. 1). The wire was then mounted between wider metal poles, and after covering with talc to protect from oxidation was heated by the passage of an electric current, when the diamond was absorbed and the iron converted into steel.

FIG. 1.

H. Wilde, in 1886, in Manchester, by means of a current from one of the earliest types of magneto-electric machines, was able to melt a bar of platinum 6 mm. thick and 2 feet long. William von Siemens, in 1878, designed several forms of arc furnaces which contained all the important features of modern types. Two of these designs are illustrated in Figs. 2 and 3. In Fig. 2, the metallic charge in the crucible was connected to one pole of the current supply and an adjustable water-cooled electrode, entering through the roof of the furnace, was connected to the second pole. Fig. 3 shows a type in which the electrode is of carbon and adjusted to the distance necessary to maintain an arc with the desired current by a magnetic solenoid S operated by the main current passing through a surrounding

---

[1] "Phil. Trans.," 1810, 16.   [2] *Ibid.*, 1815, 371.

I

coil.   The crucible T is insulated by being surrounded with carbon packed in a wider container.   Siemens also designed a furnace using horizontal electrodes (Fig. 4).   In this the charge was heated by downward radiation from the arc.   The negative electrode A consisted of a copper tube closed at one end and cooled by water circulation, while the positive electrode B was a hollow carbon rod, through which a stream of gas could be admitted to the furnace.

In some cases the arc was deflected downwards electro-magnetically.   With these apparatus Siemens was able to reduce iron-ore and melt steel to the amount of 20 lb. in one hour and platinum

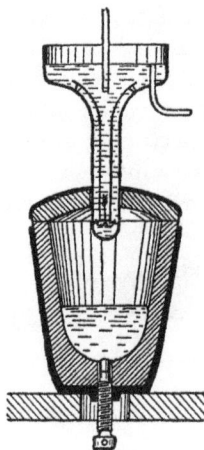

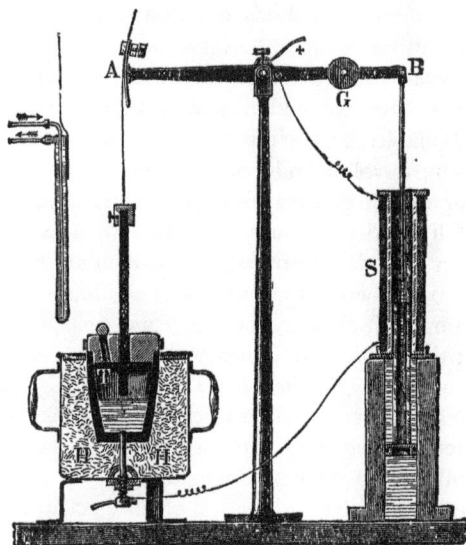

FIG. 2.   FIG. 3.

to the amount of 9 lb. in a quarter of an hour.   The commercial application of this and similar devices was at the time only precluded by the high cost of electric current.

In 1887 Héroult, in France, and Hall, in America, successfully devised an electrolytic furnace for the production of aluminium.

Despretz, in 1849, carried out experiments in which high temperatures produced electrically were utilised.   The earlier work forms the subject of a paper entitled "The Fusion and Volatilisation of some Refractory Bodies: Notes on some experiments carried out with the triple aid of the voltaic pile, the sun, and the blowpipe ".[1]

[1] " Comptes rendus," 1849, **28**, 755.

In later experiments an apparatus is described which consists of a tube of sugar-charcoal about ¼ inch wide and ½ inch long, closed by two charcoal plugs which, with its contents, was raised to a high temperature by the passage of an electric current. In further experiments, use was made of a small retort of sugar-charcoal within which the arc was formed with the aid of a carbon rod, the retort itself serving as positive electrode.

Fig. 4.

BERTHELOT'S ARC FURNACE.

The application of the electric arc in an enclosed vessel which enabled reaction with the surrounding gases to be studied was first

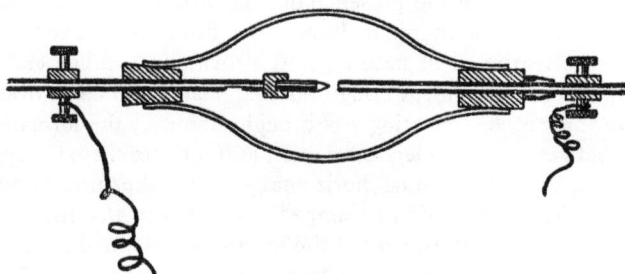

Fig. 5.

made by Berthelot in 1862 who employed an apparatus as illustrated in Fig. 5. Two carbon electrodes were supported by the plugs closing the tubular openings at opposite ends of a pear-shaped vessel. The electrodes were made hollow to enable the passage of gases, and an arc could be formed in the centre of the vessel between the carbon poles. By using an atmosphere of hydrogen, the synthesis of acetylene was demonstrated.

Cowles' Process for Aluminium Alloys.

One of the earliest electric furnace processes to be brought into commercial operation was that of the Cowles Brothers which was installed at Milton, Staffordshire, in 1886.[1] This process resulted as a development of experiments which had been carried out on the production of zinc by smelting the ore in an electric furnace (cf. p. 294). The process applied at Milton consisted in the reduction of

Fig. 6.

alumina by carbon in the presence of iron or other metals, resulting in the formation of aluminium alloys. The furnaces were rectangular in shape, constructed of fireclay, and provided at either end with an inclined cast-iron pipe, through which the electrode was introduced as shown in Fig. 6. Heating was brought about by the formation of an arc between carbon electrodes at E, and the current was supplied by two copper bars running horizontally across the furnace and by flexible cables which could be clamped to the bars in any desired position (cf. Fig. 7). Adjustment of the electrodes was made by a screw mechanism. A number of furnaces were arranged side by side but only one was in operation at a time, the other furnaces being meanwhile either charged or left to cool. Each electrode consisted of a bundle of from seven to nine carbons (each $2\frac{1}{2}$ ins. diam.) held to-

[1] " Industries," 1888, **5**, 237.

gether by a cylindrical head of metal which was connected to the flexible cables leading in the current (cf. Fig. 7). The bed of the furnace was formed of granular charcoal, which had been saturated with a weak solution of lime in order to increase its insulating properties. The furnace was prepared by inserting the electrodes, forming a rectangular space by temporarily inserting partitions of sheet iron, filling in the charge on the inside of the partition and charcoal around the outside. The charge consisted of a mixture of corundum (Al_2O_3), metal, and charcoal. A current of 5000 amps. at 60 volts was

FIG. 7.

passed and one and a half hours were required for the reduction of each charge.

To commence the heating, a few pieces of carbon were thrown into the furnace, so as to bridge across between the electrodes. The whole charge was then covered with a top layer of charcoal, and the furnace was covered with a cast-iron case provided with a hole in the centre, through which the gases generated during the process escaped. The reduced aluminium and alloying metal were volatilised, and the ascending vapours became condensed in the upper and cooler layer of charcoal. Combination occurred here and the liquid alloy flowed to the base of the furnace and was run off through a tap-hole (cf. Fig. 8). The electrical energy required for the production of 1 lb. of contained aluminium varied from fifteen to thirty h.p. hours according to the grade of alloy. The daily output at Milton amounted to about 20 cwt. of ferro-aluminium or aluminium bronze containing 15 to 17 per cent aluminium. The current was obtained from a Crompton dynamo of 300 kw. capacity, which was specially constructed for this work, and its capacity constituted at that time a record in electric generators.

It may be noted that the production of aluminium metal itself

Fig. 8.

by reduction of the oxide with carbon in the above process is precluded by the high reactivity of aluminium with carbon at the temperature of reaction, to give aluminium carbide.[1]

A furnace similar to that at Milton was about the same time installed at Lockport in the United States, but after the inauguration, in 1886, of the pioneer work of Hall and of Héroult on their electrolytic processes for producing aluminium, both plants suspended operation.

Moissan's Experiments.

Moissan, commencing in 1892, carried out a series of experimental researches [2] with the high temperatures obtained by means of an electric arc furnace. The main reactions investigated were the reduction of refractory oxides by carbon, the formation of carbides and allotropic changes of carbon. Metals such as chromium, tungsten, molybdenum, uranium and titanium, were thus obtained in a fused state for the first time and an investigation of their

---

[1] Cf. Pring, "Trans. Chem. Soc.," 1905, **87**, 1530.
[2] "The Electric Furnace," by H. Moissan, transl. by A. T. de Mouilpied. London, Ed. Arnold.

physical and chemical properties was made possible. The method adopted, as illustrated in Fig. 9, consists in placing a powerful arc in a cavity of minimum size in a limestone block and at a certain distance above the substance to be heated. In this way, actual contact with the carbon vapour from the arc is avoided and at the same time the thermal action of the current is separated from any electrolytic effect. The current generally used by Moissan was about 450 amps. at 60 volts. This type of furnace, which is still eminently suitable for laboratory experiments, consists of two slabs of lime carefully cut and superposed. The lower slab has a long groove in which the electrodes rest, and in the middle is a small cavity which serves to hold a small carbon crucible containing the substance to be treated. The intense heat of the current soon melts the lime above the arc, so that a small dome is obtained which reflects the heat on to the crucible. The electrodes are easily rendered movable by means of two adjustable supports or, better, by using two sliders which rest on a bed-plate. The freedom of motion of the electrodes enables an adjustment of the length of the arc to be conveniently made. The electrodes were made by Moissan

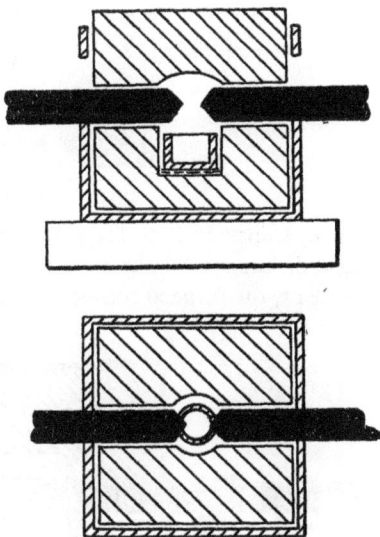

FIG. 9.

of cylinders of carbon as free from mineral matter as possible. For this purpose, retort carbon was powdered, treated with acids to free it from contained iron, washed, calcined, and finally rendered coherent by means of pitch. The cylinders formed by means of a high and regular pressure, were carefully dried, and baked at a high temperature. The arc at first is less than 1 cm. long but at the end of the experiment the length usually increases to from 2 to 2·5 cm. If the furnace be filled with a good conducting metallic vapour (e.g. aluminium), the electrodes may be 5 to 6 cm. apart. The length of the arc will thus be regulated according to the readings of the voltmeter and ammeter, so that an approximately constant resistance can be maintained. With a current of

8 *THE ELECTRIC FURNACE*

400 amps. at 80 volts, the experiment is completed in five or six minutes.

A Moissan laboratory type of furnace of 30 kw. capacity, showing resistances for regulating the current composed of water-cooled gun-metal tubes, and sliding electrode-holders for adjusting the length of the arc is shown in Plate I.

In a type of furnace employed by Moissan[1] for continuous heating of materials, a carbon tube is passed through the furnace

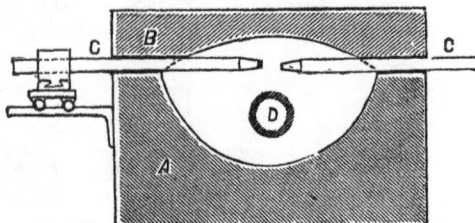

FIG. 10.

chamber transversely to the plane of the electrodes, and arranged to be at a distance of about 10 mm. below the arc. As shown in Figs. 10 and 11, the furnace consists of a chamber hollowed from limestone

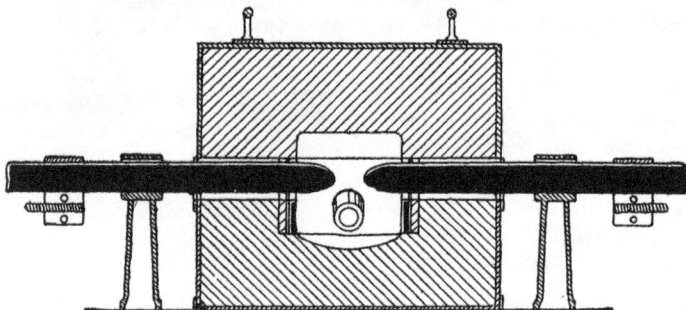

FIG. 11.

blocks A and B. The electrodes are arranged at C, and a carbon tube at D, inclined at an angle of 30° to facilitate the passage of the charge.

PRODUCTION OF CHROMIUM.

A furnace of the above type was developed by Chaplet for the production of metallic chromium, and brought into operation at the works of the Société Néo-Metallurgie in France. As shown in Fig. 12, the furnace is formed from limestone blocks made in two parts with an interior cavity lined with magnesia. The lower part is

[1] " Comptes Rendus," 1892, **115,** 1031; 1893, **117,** 679.

Plate I.—Moissan Furnace.

movable to enable replacement whenever the well or hollow *e* becomes filled with the product of the furnace operation. The upper part of the furnace contains an enclosure with an arched roof, while two inclined carbon tubes d_1 d_2 are brought through the side walls, and heated by a series of arcs formed between carbon poles arranged transversely to the plane of the diagram at C.

The arcs from the electrodes c' c' serve to prevent cooling of the product at the exit of the tubes. The charge was passed down the tubes d_1 d_2, and after exposure to the zone of very high temperature, issued at O into the trough *e*. As in the method first developed by

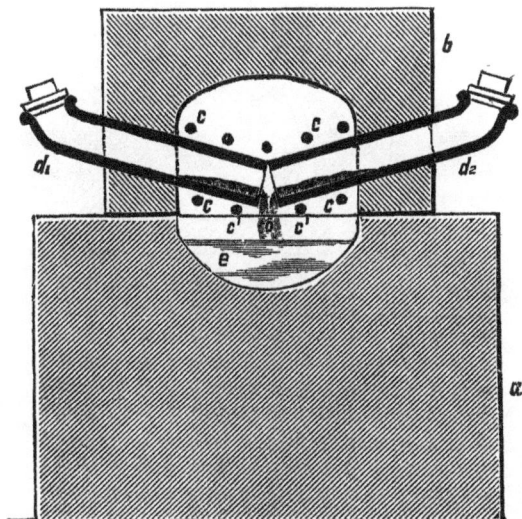

FIG. 12.

Moissan, chromium, containing from 8 to 12 per cent carbon as carbide, was first prepared by the reduction of the oxide by carbon, and subsequently refined to a carbon-free product by reheating with chromium oxide and lime.

MAGNETIC DEFLECTION OF ARC.

The principle of the deflection of the electric arc by the application of a magnetic field was first made use of in an electric furnace by Siemens (cf. p. 1).

In a furnace designed in 1886 by Rogerson, Statter and Stevenson,[1] an open horse-shoe electro-magnet is employed to deflect the arc as shown in Figs. 13 and 14.

[1] Eng. Pat. 10,600 of 1886.

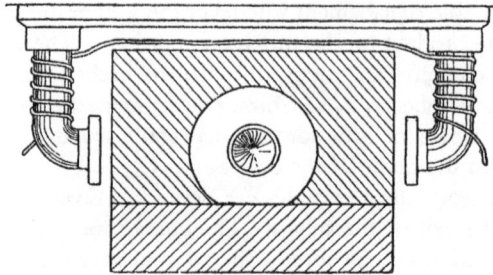

FIG. 13.

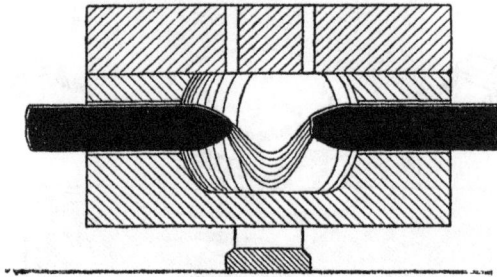

FIG. 14.

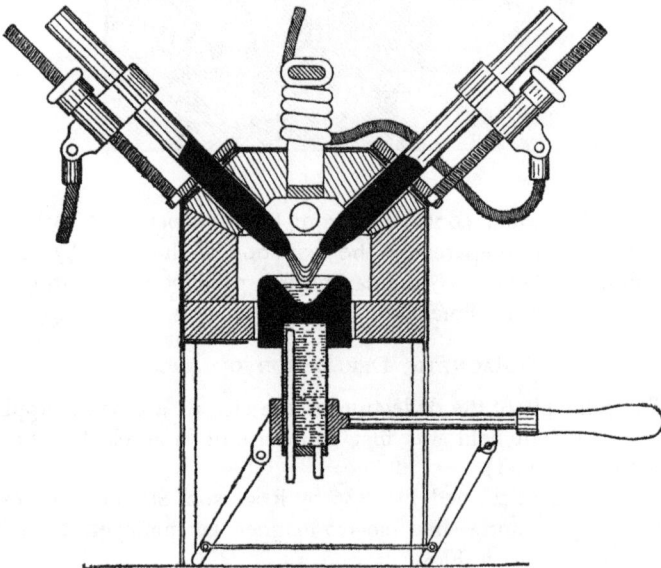

FIG. 15.

In a type of furnace designed by Schuen, a magnetic coil is, as shown in Figs. 15 and 16, placed vertically above the furnace, and deflects the arc downwards on to the charge contained in a crucible.

MANUFACTURE OF CALCIUM CARBIDE.

The next material of notable importance to the commercial production of which the electric furnace was applied is calcium carbide, which was introduced for use in the generation of acetylene.

Calcium carbide was first discovered in 1836 by Robert Davy and its properties studied by Wöhler in 1862, though these discoveries did not acquire any industrial significance until the year 1892, when Moissan, in Paris, and Willson, at Spray, North Carolina, U.S.A., in studying the reduction of lime by carbon, obtained this

FIG. 16.

compound on a larger scale as an electric furnace product. The material obtained by Willson attracted notice through its production, on being brought into contact with water, of a gas which burned with a sooty luminous flame, though some time elapsed before the identity of the furnace product was recognised. Moissan investigated the material in a more scientific manner, and pointed out its possibilities as an illuminant. Patents which were taken out later were granted in France to Bullier, Moissan's assistant, and in America and England to Willson. A large number of companies were floated for the manufacture of calcium carbide, water-powers were developed, and many works brought into operation in different countries. Manufacture was begun in the United Kingdom by a company which commenced operations at Foyers and leased power from the British Aluminium Company and, in America, at Spray,

N.C. In 1899 eighty-six carbide works were in operation or in the course of erection in Europe and a large period of over-production resulted, followed by the abandoning of the large majority of the processes.

The early set-back in the calcium carbide industry may be largely ascribed to the loss of public confidence resulting from the numerous

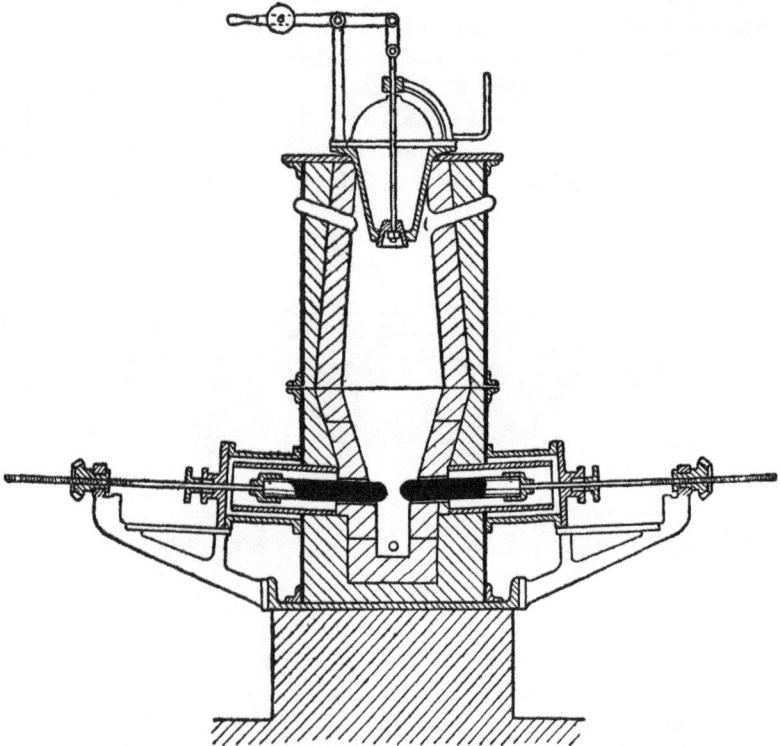

Fig. 17.

explosions which occurred through the presence of impurities such as phosphine in the gas.

In recent years, however, a revival in the manufacture has taken place and the output has steadily grown. The increasing demand has arisen through the extension of the consumption of carbide mainly for producing acetylene for use in oxy-acetylene welding, for flares in marine work, and as the starting product in the manufacture of cyanamide and other synthetic nitrogen products. An important application in this connexion which has been introduced during the war is in the synthetic formation of alcohol, acetic acid, and acetone.

ELECTRIC SMELTING OF IRON-ORE.

A new era in the growth of electric furnace enterprise was introduced in 1898, when Stassano, at the Arsenal at Turin, brought into operation a furnace for the electrical production of pig-iron from ore. The importance of this step consists in the endeavour to apply the electric furnace for the first time to the production of only moderate temperatures, in a process which had been developed to á high state of efficiency by the use of fuel heating, whereas

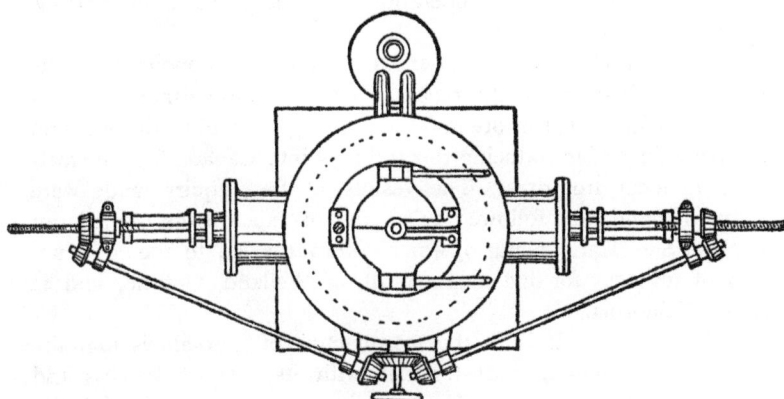

FIG. 18.

hitherto electric furnaces had only been developed industrially for products which could not be obtained by other means.

The earliest type of furnace to be introduced by Stassano consisted in form of the usual shaft furnace but with an arc formed between carbon poles near the base of the furnace taking the place of the tuyères (Fig. 17). The electrodes seen also in plan in Fig. 18 are surrounded by water-cooled jackets. The charge is admitted at the head of the shaft and the pig-iron and slag emptied by a tapping-hole in the base.

During the operation the arc was enclosed or "smothered" by the charge. It was found, however, that, with this type, the resistance of the charge to the passage of the current was too high and the resulting iron had too high a carbon content. The procedure was then modified by a system in which the arc was formed above the surface of the furnace charge and so operated by heat radiated downwards.

A furnace of this type of 500 h.p. capacity was installed at Darfo, Italy, in 1901. In place of the production of pig-iron, high-grade Italian ores were smelted together with charcoal and lime and a malleable iron or mild steel obtained. An alternating current of 2000 amps. at 170 volts was employed. The power expenditure amounted to 4000 kw. hours, and the consumption of electrodes 12 kg. per metric ton of iron or steel produced. The furnace was converted later to the production of steel from iron or scrap. The process of Stassano for the smelting of iron-ores was followed by other systems developed by Keller, Héroult, Harmet and others, and is now in successful operation on a large scale in Norway, Sweden and California.

In 1903 a Commission under Dr. Haanel was appointed by the Canadian Government to report on the electrothermic processes for the smelting of iron-ore and the making of steel in Europe, and the possibility of introducing this industry into Canada for the treatment of local iron-ores. The results of the inquiry made were favourable and were followed by a series of experiments carried out at Sault Ste. Marie, Ontario, which afterwards led to the construction of furnaces for this manufacture at Welland, Ontario, and at Baird, California.

Ferro-Alloys.—It is in the manufacture of ferro-alloys that the early electro-chemical industry met with its greatest success and most rapid development. The main incentive which led originally to the progress of this work was the decline in the calcium carbide industry which followed its early extension. Thus, in 1900, experiments were made in France[1] on the production of ferro-chromium, ferro-silicon, and other ferro-alloys. Carbide furnaces were found applicable for this manufacture, and the success obtained has finally led, in the case of many of these alloys, to the complete replacement of the older processes by electric furnace manufacture. Large works manufacturing ferro-alloys are now in operation mainly in Savoy, and Isère in the South of France and in the United States, Switzerland, and Scandinavia. The furnaces employed range in power up to 3000 h.p. As a rule, either one or two vertical carbon electrodes are employed, the system being classified as that of the " smothered " arc or resistance type.

[1] A. Keller, " The Application of the Electric Furnace in Metallurgy,'' " Journ. Iron and Steel Instit.," 1903 (i), 161.

Steel.—The electrical production of steel from pig-iron, scrap steel, etc., was originally very largely developed through the abandonment of many carbide furnaces and on account of the marked success obtained with ferro-alloy manufacture. In France, Héroult, and in Sweden, Kjellin, successfully applied many of these plants to the production of high-quality steel and this important industry was started about the year 1900.

At the present time a large number of different types of electric furnaces are almost exclusively used in the preparation of a large class of special alloy steels, the crucible process has been largely replaced for obtaining high-quality ordinary steels and the electric furnace has been extensively applied in conjunction with and supplementary to open-hearth and Bessemer treatment.

The use of the electric furnace has now extended itself very largely throughout the range of chemical and metallurgical practice. There are indeed comparatively few cases of processes involving the supplying of heat in which the application of electricity has not, in some instances, been applied.

SECTION II.

PRINCIPLES OF ELECTRIC FURNACES.

THE types of processes concerned in industrial electro-chemistry may be classified under the following headings :—

1. Those in which the current produces a definite electrolytic change, either in an aqueous or fused medium.

2. Processes involving solely electronic or ionic changes in gases.

3. Those in which some part is taken by ionic effects in conjunction with the production of a high temperature, such as in the union of oxygen and nitrogen brought about by means of a high-tension arc.

4. Purely thermal changes produced at temperatures which are unattainable by any other means than electrical.

5. The production of moderate temperatures, which are within the limits attainable with gas or coal-firing. In this case the electricity merely plays the part of a form of fuel.

Methods of Electrical Heating.—An electric furnace may be defined as an enclosure containing a channel or path of conducting material in which heat is generated by the passage of an electric current.

The current is, as a rule, introduced through terminals which are in electrical contact with the ends of the conducting path and the resistance offered to the passage of the current leads to a definite fall in the potential of the closed circuit. The heat generated is, in accordance with Joule's law, determined by this drop of potential and the current. The temperature attained is, apart from the influence of disturbing chemical changes, determined by the condition when the heat generated from the current is counterbalanced by that lost through conduction and radiation from the reaction zone of the furnace.

JOULE'S LAW.

The law which determines the transformation of electric energy into heat was expressed by Joule, who derived the following relationship between heat and electrical energy :—

$$Q = 0\text{·}239 \ ECt$$

where Q is the heat in gram-calories, E the fall of potential in volts between the furnace terminals, C the current in amperes and t the time in seconds. By applying to this the relation between current, voltage, and resistance given by Ohm's law, or $C = \dfrac{E}{R}$, we have—

$$Q = 0\text{·}239 \ C^2 Rt \qquad \text{or, } Q = 0\text{·}239 \ \frac{E^2}{R} \ t.$$

According to this, with any given resistance, the heating effect is seen to be proportional to the square of the current or to the square of the voltage.

The different systems in use for applying heat developed electrically to chemical processes may be classified under the headings of the arc, resistance, and induction methods.

ARC FURNACES.

These may be either of low-tension or high-tension type. In the former (*a*, Fig. 19), the two poles or electrodes of carbon arranged in juxtaposition are connected to a low potential electrical supply. On bringing the ends into contact, a flow of current occurs, and on withdrawing, a spark is formed accompanied by the production of ions, which enables the passage of a certain amount of current. Heat is thus generated, which leads to the production of vaporised carbon and further ionisation. A very high local temperature is obtained which finally leads to a constant production of carbon vapour and after separating the poles to a moderate distance, the passage of a continuous current is maintained. When operated with direct current, the positive carbon is hollowed out by the current due to the bombardment by a stream of electrons from the negative electrode (cf. Fig. 20). Through the heat thus generated the electrode becomes heated to the volatilisation temperature of carbon, when a crater is formed through evaporation. It should be noted that almost the whole heat is generated on the surface of the crater on the positive pole. An arc can also be formed by alternating current. As shown at *b*, Fig. 19, in place of two carbon electrodes,

2

an arc can be formed between a carbon pole and the base of the
furnace or the surface of a conducting charge.

The temperature of the hottest part of the positive carbon in
the electric arc has been estimated to be about 3700° C. (6700° F.).
When the very highest temperatures are required, the use of the

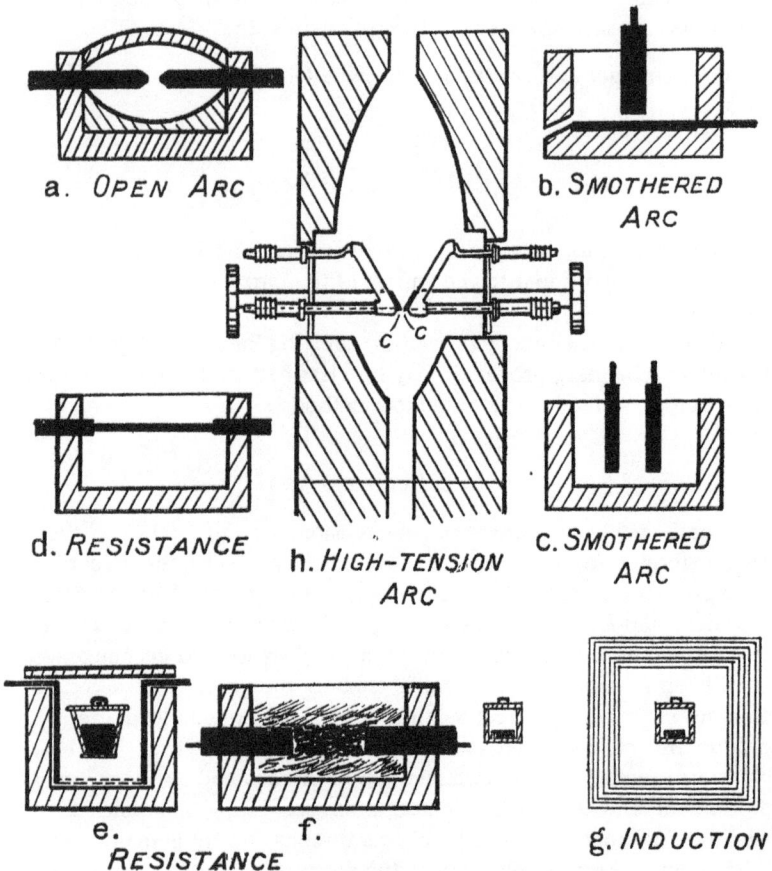

a. OPEN ARC

b. SMOTHERED ARC

d. RESISTANCE

h. HIGH-TENSION ARC

c. SMOTHERED ARC

e. RESISTANCE

f.

g. INDUCTION

FIG. 19.

electrical arc offers many advantages. It is much easier as a
rule to concentrate the expenditure of a large amount of power
in an arc than in any other form of electrical heating so that the
limit of temperature attained is generally higher. The arc may
be open and the heat radiated from it be thrown downwards upon
the material under treatment, as in the furnace of Moissan, or the

arc may be directly surrounded by the substance to be heated, and thus smothered. In the latter case, which really reverts to the resistance system, the heat can obviously be more economically employed, though serious contamination may result from the contact of the charge with the heated carbon vapour.

A minimum e.m.f. of 45 to 60 volts is necessary to maintain a low-tension arc, if with direct current, and 30 to 35 volts if alternating current is employed.

The arc furnace is, as a rule, employed when it is desired to produce the highest temperature in a necessarily limited zone of the charge and in order to avoid contamination where it is desired to heat by radiation rather than by contact of the charge with a heated resister.

FIG. 20.
(*From Ganot's " Natural Philosophy ".*)

ACTION OF MAGNETIC FIELD ON ARC.

Similarly to the case with all flexible conductors, an arc is displaced laterally under the influence of a magnetic field (cf. p. 9). With electrodes arranged vertically, and the current flowing downwards, i.e. the positive pole above, an arc is deflected outwards towards the right by a north magnetic pole which bears on it, and to the left by a south magnetic pole. If the magnet is strong, the arc may be extinguished. The deflection brought about by a magnetic field is utilised in some types of arc furnaces for directing the heat of the arc towards the charge. If the two poles of an electro-magnet are placed so that the path of current crosses the field between the poles, the arc is deflected to give a flame similar to that of a blow-pipe. A device of this nature is used for the purpose of electric welding.

Types of experimental furnaces which have been constructed for applying the magnetic deflection of an arc are shown in Figs. 13-16.

HIGH-TENSION ARC.

The high-tension arc (*h*, Fig. 19) is formed in air or other gases, generally between metal electrodes which at one point are in close proximity. An alternating current of about 5000 volts is usually employed to maintain the arc, though to start when cold a

2 *

considerably higher voltage is required and, in practice, is often applied by means of an induced current so as to produce a spark. A conducting path thus formed by ionisation enables a current to be maintained by the 5000 volt circuit.   The resistance of the gap between the electrodes rapidly falls on passing the current and in practice its flow is usually checked, either by introducing an outside reactance or choking resistance in the circuit or by arranging the electrodes to diverge in a vertical plane, which causes the arc, immediately after its formation, to ascend until the increased resistance of the path which is encountered leads to its rupture, followed by its re-formation at the narrowest part of the gap.

The main distinction between the high and low-tension arc is that in the former the current is conducted by ions of air or gas produced in the surrounding atmosphere and, on account of the smaller currents used at the high voltage, the heating effect at the surface of the electrodes is not great, whereas, with the low-tension arc, the very highly conducting path produced by carbon (or metal) vapour, enabling the passage of big currents at low voltages, causes the main heating effect to be developed at the surface of the anode.

Resistance Furnaces.

In the resistance principle the heat is produced by the passage of the electric current through a solid or liquid conductor.   In many types of furnaces, use is made of a combination of arc and resistance heating.

The conducting path which serves as the source of heat may be formed by a core of conducting material such as granular carbon, a tube or rod of carbon or metal, a metal wire or the material which is to undergo reaction in the furnace may itself form the conducting medium.

The loss of heat from the reaction zone of an electric furnace is checked by the provision of a surrounding casing of a refractory insulating material, generally consisting on the outside of a masonry, firebrick, or metal shell for the support and protection of the inner lining.   In many cases the unaltered charge itself forms the main heat insulation.

The resistance of the heating core is determined by its length, cross section, and specific resistance in accordance with the expression—

$$R = \theta \frac{l}{a}$$

where l is the length in cm., a the cross section in sq. cm., and θ the specific resistance.

The value of θ changes with the temperature according to a definite relation which varies with different materials. With metals, an increase of temperature generally produces an increase in resistance or the *temperature coefficient* is positive, while with carbon, metallic oxides, and most compounds the resistance diminishes with increase of temperature or the *temperature coefficient* is negative.

The value of θ for some of the materials more commonly used in electric furnace work and the temperature coefficient are given in the table below. The specific resistances are given in microhms (a microhm being one-millionth of an ohm), and the temperature coefficients are given as the relative increase of θ for one degree rise of temperature.

Material.	Specific Resistance. Microhms.	Temperature Coefficient.
A. *Metals and Alloys at 0°.*		
Silver	1·4 to 1·6	0·004
Copper	1·6	0·004
Aluminium	2·8 to 3·2	0·003
Nickel	8 to 11	0·005
Iron (with 0·1 per cent C.) . .	9 to 15	0·005
Steel (with 1·0 per cent C.) . .	15 to 50	0·003
Lead	20	0·004
Manganin	40	− 5 × 10⁻⁵
Constantan	44	− 2 × 10⁻⁵
" Kromore " {85 per cent Ni / 15 per cent Cr	119	5 × 10⁻⁷
" Mangaloy " {40 per cent Fe / 60 per cent {Ni / Mn	134	7 × 10⁻⁷
" Nichrome " {60 per cent Ni / 12 per cent Cr / 28 per cent Fe	137	2 × 10⁻⁷
B. *Non-metals at 0°.*		
Graphite	860 to 1450	− 0·001
Electrode Carbon	4200	− 0·0002
Retort Carbon	5000 to 6000	− 0·0002
Magnetite (Swedish) . . .	600,000	—
C. *Materials at Furnace Temperatures.*		*Temperature of Measurement*
Molten pig-iron	160	1300°
Graphite	690	100° − 2000°C.
Carbon (amorph.)	3600	100° − 2000°C.
Silicate slag	2 × 10⁶ to 5 × 10⁶	Molten
Firebricks (various) . . .	{20 × 10⁶ to 60 × 10⁶ / 60 × 10⁶ to 700 × 10⁶	1550° / 1500°

The values given only represent approximate averages, as the resistance of all materials varies largely through the influence of admixed chemical impurities and when in different physical conditions.

THE INDUCTION FURNACE.

In furnaces of this type (*g*, Fig. 19) which are mostly used in the manufacture of steel, the use of electrodes for introducing the current is eliminated entirely. The advantages obtained are that the tendency for the introduction of carbonic oxide and impurities from the electrodes is obviated, while access of gases generally to the metal is precluded and a product obtained which is particularly free from gas occlusion.

The general principle on which these furnaces work is the arrangement of the bath of molten metal in an annular ring crucible employing this as the secondary circuit of an induced current. A quadrangular iron core, formed of thin insulated sheets of soft iron, is placed in the centre of the circle and connected around the outside of the crucible. Insulated copper wire or a water-cooled copper tube is wrapped around the arm of the core inside the circle and serves as the primary coil for the alternating current. The current, when passing through this coil, excites a magnetic flux in the core and this flux induces an alternating electric current in the contents of the furnace chamber. The arrangement is consequently that of a step down transformer having a large number of primary turns and a single secondary turn, the secondary turn consisting of the metal in the furnace. The current in the metal is thus about equal to that in the primary circuit multiplied by the number of turns of wire in the primary coil and the voltage is reduced in the same ratio as the number of amperes are increased.

THE SILENT DISCHARGE.

A further type of electrical discharge which has been applied industrially to the production of chemical change is that of the " Dark " or " Silent " discharge. This phenomenon occurs between two adjacent poles on which a large difference of potential is maintained, by virtue of the free ions which are always present in smaller or larger amount in air and gases and which lead to a conductance of electricity between the electrodes. When the current passed exceeds a certain value, a large increase in the ionisation results from the effect of collisions and rise of temperature, which finally leads to a rush of current ending in a disruptive or high-tension discharge. When this occurs, the silent discharge changes over to an arc of the high-tension type.

The slow or silent discharge through gases may exert a chemical influence, thus, hydrogen and nitrogen are made to combine to form ammonia, hydrogen and cyanogen to form hydrocyanic acid, carbon monoxide and water to form formic acid, and oxygen to be transformed into ozone. The last of these reactions has been applied commercially on a large scale for the preparation of ozone.

SECTION III.

TYPES OF FURNACES USED IN EXPERIMENTAL AND LABORATORY WORK.

In addition to the arc furnace of Moissan and its modifications described in the previous section, the following noteworthy types of experimental furnaces have been designed:—

WIRE-WOUND FURNACES.

In these, heating is brought about by an electric current traversing a wire of a refractory metal or alloy, such as platinum,

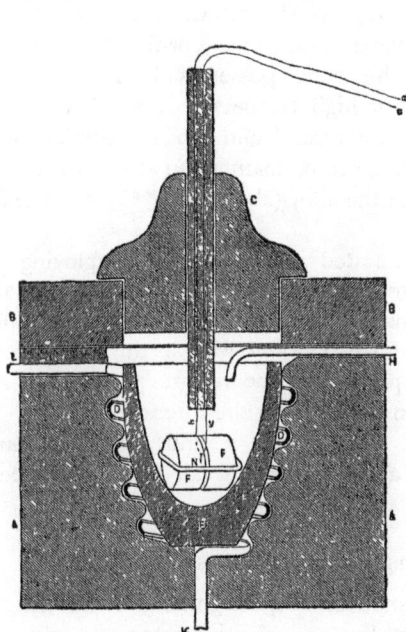   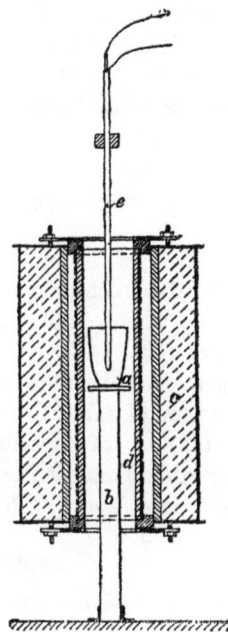

FIG. 21.   FIG. 22.

tungsten or "nichrome," which is coiled around a tube of porcelain, magnesia, alumina, or other refractory material, the tube being then jacketed with a thick layer of some heat-insulating material.

Furnaces of this type used in conjunction with a thermo-electric pyrometer admit of very delicate temperature adjustment, moreover the nature of the gas atmosphere in which the heating is carried out can be varied at will.

HERAÜS FURNACES.

Early types of furnaces which were designed by Heraüs and largely employed in laboratory work are shown in Figs. 21 and 22. The former consisted of a crucible furnace with a thick platinum wire built in the body of the furnace and the latter was constructed in the form of a tube wound with platinum foil so as to be suitable for connecting directly on to lighting circuits of normal voltages of 110 or 220. With these furnaces the material under treatment can be conveniently observed and the temperature accurately regulated and measured while any desired gas atmosphere can be maintained.

The length and diameter of the wire employed as the resister can be adjusted to suit the potential of the current supply. For lower temperatures, as required in drying ovens, iron wire may be successfully applied, and for higher temperatures, nickel, which has a melting-point of 1427° C., though this metal possesses the great disadvantage of crystallising rapidly at high temperatures, which quickly leads to fracture. The alloy "nichrome" and similar materials (cf. p. 21), if protected from oxidation, can be maintained at temperatures approaching the melting-point of the alloy (about 1450° C.) for several hours.

In the type of furnace illustrated in Fig. 22, by employing a cylindrical refractory tube of 9·4 cm. external, and 8·6 cm. internal diameter, wound with 80 turns of "nichrome" wire of 0·18 cm. diameter (Np. 15 S.W.G.) and surrounding with a thick layer of kieselguhr, a current of 10 amps. from a 200 volt supply, enables a temperature of 1200° C. to be maintained in the interior.

A cylinder of "alundum" or magnesia wound with tungsten wire or ribbon enables, in the absence of air, small crucibles to be heated to over 2000° C. (cf. p. 42).

CARBON RESISTANCE FURNACES.

Borchers Furnace.—A resistance furnace, designed by W. Borchers in 1891, consists, as shown in Fig. 23, of a refractory enclosure through the end walls of which massive carbon poles were introduced. A narrow carbon rod was inserted in the ends of the poles and served to lead the current across, thereby being raised to a high temperature while the material to be heated was arranged to sur-

round the carbon core. With this furnace, Borchers was able to
demonstrate the reduction of refractory oxides.

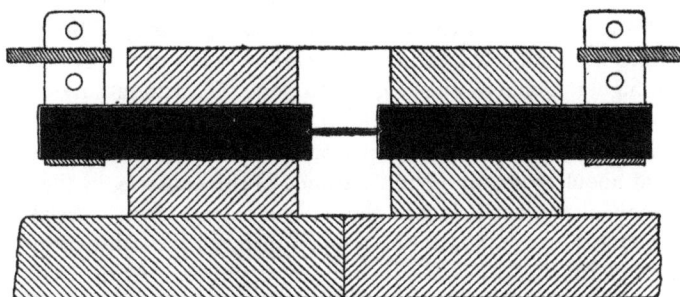

FIG. 23.

Hutton and Patterson Furnace.—For laboratory and experimental
purposes, very satisfactory furnaces can be constructed from carbon
tubes which become heated by the resistance they offer to the
passage of the current.   In a type designed by Hutton and Patterson [1] (Fig. 24), the inside of the tube forms the heating chamber, the
outside being jacketed with some material, which, in this case, must

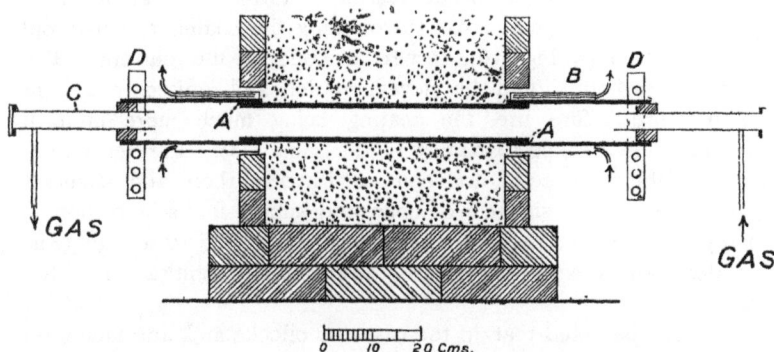

FIG. 24.

not only be a good heat insulator but must also protect the carbon
tube from oxidation.   Up to 2000° C., amorphous carborundum is
suitable, above this temperature, soot, carbonised cotton, or crushed
wood-charcoal is most effective.   Firm and durable connexions to
introduce the current are made by electro-coppering the ends of the

[1] "Trans. Faraday Soc.," 1905, **1**, 187.

carbon tube A A, and then inserting them for a short distance into a closely-fitting copper tube and attaching with soft solder. The copper tube is itself surrounded by a wider cylindrical metal jacket B and water circulated through the intervening space. The current terminals are connected to D. The carbon tubes can thus be taken to the highest temperatures, e.g. up to 3000° C., without any arcking developing at the contacts. Uniformity in the heating of the tube, particularly at the higher temperatures, tends to be automatically brought about to within a short distance of the contacts, by the negative temperature coefficient in the resistance of carbon. According to this factor any part at lower temperature receives, through its increased resistance, a larger expenditure of power. At temperatures above about 900° C. some form of optical pyrometer, such as that of Wanner or Féry, enables the progress of the heating to be followed with exactitude, observations of the interior being made through glass plates closing the ends of the tube C. The only disadvantage of this type of furnace lies in the low electrical resistance of the carbon tubes. Currents of several hundred amperes are required even for the smaller tubes, and with a tube of $2\frac{1}{2}$ inches internal diameter, and 2 feet length, a current of some 1000 amps. at 10 volts, is required to attain 1900° to 2000° C.

A different type of carbon resistance furnace is that in which granular carbon such as coke, surrounding the chamber or material to be heated (*e*, Fig. 19), serves to conduct the current. The granular carbon should be sieved, graded and ignited to a high temperature before use, the heating being much more uniform if the grains are approximately of the same size. Further, on account of the presence of undecomposed hydrocarbons and occluded gases, the initial resistance of coke and similar forms of carbon is many times higher than when it has been exposed to a high temperature which expels these gases and causes graphitisation of the carbon.

It may be noted that in the cases of oilcoke and anthracite, on calcining, an enormous reduction of the resistance (of the order of $10^6 : 1$) occurs, without appreciable graphitisation.

Instances of the use of granular carbon as a heating medium have been applied in the types of furnaces shown in Fig. 26 which consists of a crucible furnace, and Fig. 27, constructed as a muffle furnace.

Furnaces of the above type were widely adopted at the Berlin Porcelain Works for the production of temperatures up to about 1700° C. For very high temperature work, however, their application

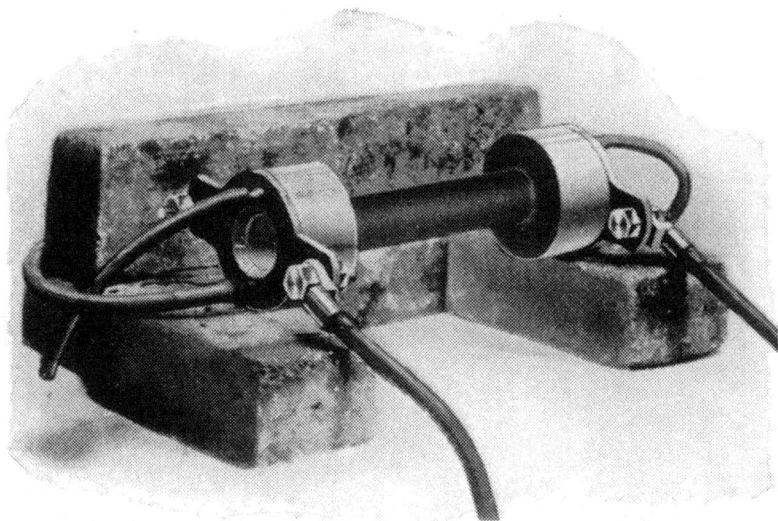

FIG. 25.—CARBON TUBE FURNACE.

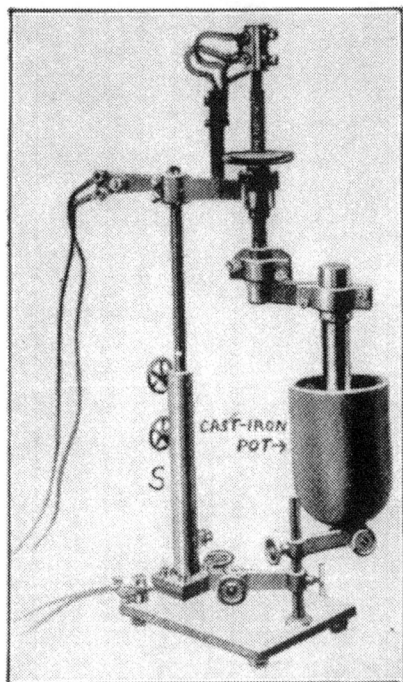

CAST-IRON POT→

FIG. 29.—ADJUSTABLE STAND FURNACE.

is limited by the impossibility of finding a suitable refractory material, which can be used in contact with carbon, for the construction of the walls of the heated chamber.

A carbon rod can be used for heating a surrounding charge, as shown at *d*, Fig. 19, and Fig. 28, or a core of granular carbon *f*, Fig. 19, a method which is used in large scale furnaces for the manufacture of carborundum and graphite.

A readily-controlled and convenient type of laboratory furnace can be made from a carbon tube attached by soldering, after electro-coppering the ends, to metal water-cooled holders. The tube, holders, and cables are mounted as shown in Fig. 25. The tube is surrounded by a suitable insulating material contained in a fire-brick enclosure. The disadvantage of the low resistance of the carbon, involving the application of very large currents, may be overcome by cutting a narrow spiral slot axially along the tube.[1]

In a type of furnace designed by W. Rosenhain and E. A. Coad-Pryor,[2] the device is adopted of forming a tubular heating element by superposing rings of graphite upon one another. On the passage of a current, heating is brought about by the resistance at the junction of the graphite surfaces.

Adjustable Stand Furnace.—A convenient type of electric furnace stand, which during use does not suffer from the heat of the furnace and can be applied to different methods of arc and resistance heating, is shown in Fig. 29. In this illustration, a cast-iron pot is used as the receptacle and is in metallic connection with one pole of the current. In place of this vessel a carbon block can be used as the base of the furnace and the walls can be built up of refractory bricks.

The vertical arm of the stand is bolted to the iron base plate, an ebonite or fibre sleeve being interposed at the junction to provide electrical insulation. Adjustment of the height over a large range is made by releasing the clamping screw in the centre of the vertical arm and sliding the top portion as desired. A close adjustment of the top electrode is made by the wheel screw gear placed above. By the use of the iron pot container with its charge as the lower electrode and inserting a carbon electrode in the top clamp, an arc furnace of either the " open " or " smothered " type is obtained. In a further modification a carbon rod can be heated through its electrical resistance by closely fitting into holes bored centrally in

---

[1] Cf. W. Arsem, " Trans. Amer. Electrochem Soc.," 1906, 9, 153.
[2] " J. Far. Soc.," 1919, 14, 264.

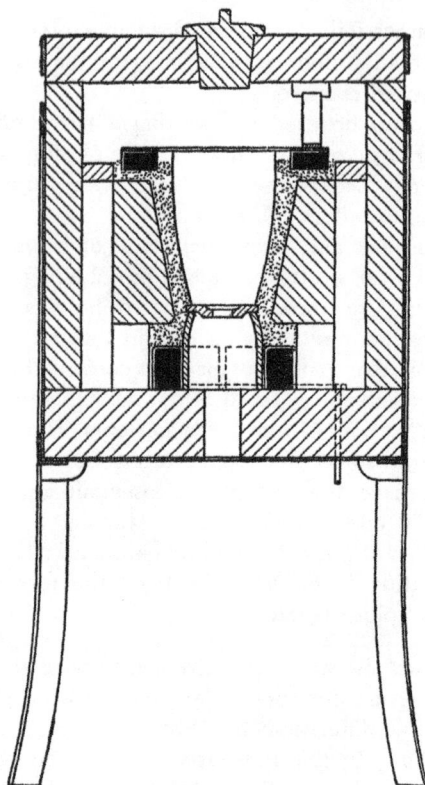

Fig. 26.

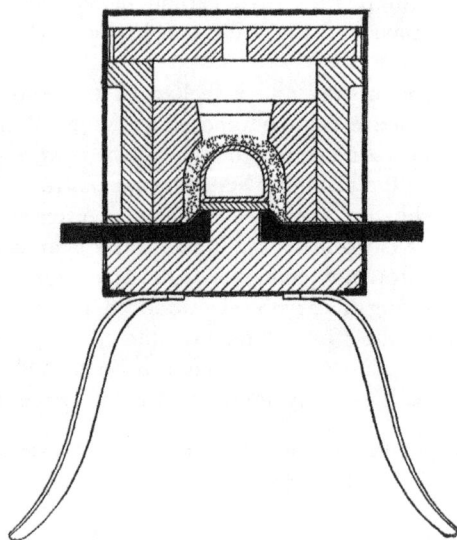

Fig. 27.

graphite blocks which are themselves held in the copper clamps, as shown in Fig. 28, or by electro-coppering the ends and soldering in water-cooled metal holders. This type of furnace stand is particularly convenient for the preparation of materials such as calcium carbide, carborundum and for the fusion of magnesia.

ELECTRIC FURNACES IN GLASS REACTION VESSELS.

A type of apparatus which was introduced for investigating the synthesis of hydrocarbons and for measurements of ionisation produced by carbon at high temperatures and in high vacua is shown

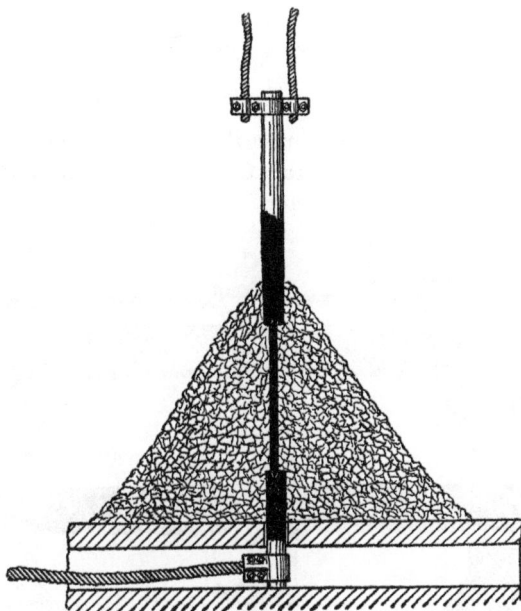

FIG. 28.

in Fig. 30. A carbon rod M, 6 to 10 cms. long and 0·4 to 0·5 cm. diameter, is mounted in graphite plugs *g g*, which are themselves inserted in holes bored in copper plugs stopping the ends of water-cooled copper tubes C C. The tubes and rod are mounted in a spherical glass containing vessel and an air-tight junction between the water-cooled tubes and the side arms of the glass vessel obtained by means of soft wax at W W.[1]

[1] " Direct Union of Carbon and Hydrogen at High Temperatures." Pring and Hutton, " Trans. Chem. Soc.," 1906, **89**, 1591 ; 1910, **97**, 498. " The Origin of Thermal Ionisation from Carbon," Pring, " Proc. Roy. Soc.," A., 1914, **89**, 344.

By the passage through a rod 0·5 cm. diameter of a current of 110 amps. at low gaseous pressures, a temperature of 2050° C.

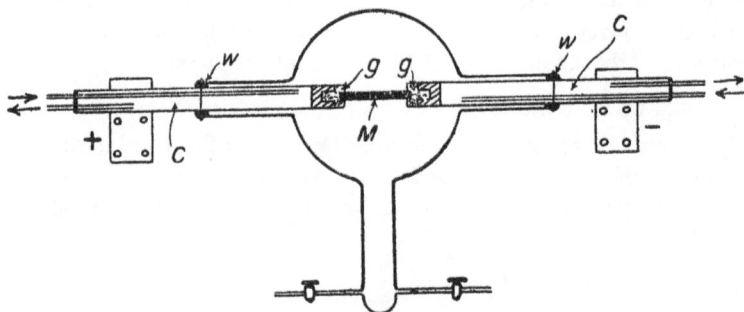

FIG. 30.

can be maintained for long intervals, while for short intervals the temperature can be taken to 2400° C. in this apparatus and measured by means of an optical pyrometer.

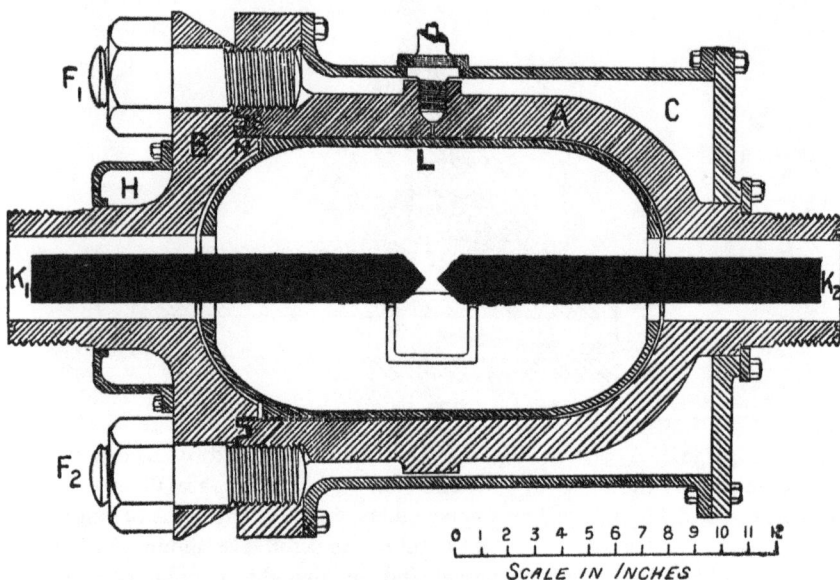

SCALE IN INCHES

FIG. 31.

ELECTRICAL FURNACE AT HIGH GASEOUS PRESSURES.

An investigation was made of the application of high gaseous pressures to chemical reactions in an electric furnace by Hutton and Petavel.[1]

[1] " Phil. Trans.," 1908, A., **207**, 421.

Heating was effected either by means of the " open " or " smothered " arc, or by the resistance of the charge. Pressures up to 200 atmospheres were used. The apparatus consists of a steel enclosure of about 20 litres capacity, provided with various fittings for the introduction or circulation of gas, gauges for the measurement of pressure, windows for observation and finally with insulated carbon holders leading the current to the inside of the furnace. The construction of the enclosure is seen in Fig. 31,

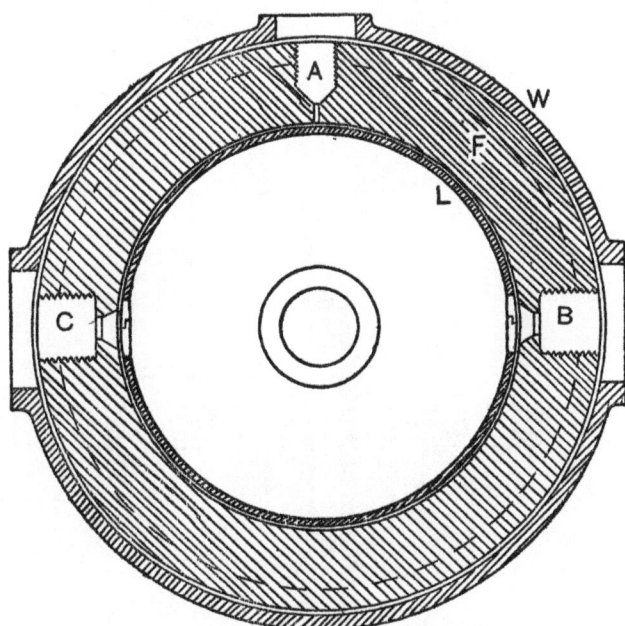

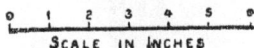

Fig. 32.

giving a sectional diagram. The shape of the interior is cylindrical, 10 inches diameter by 17 inches long, with hemispherical ends, one of which forms the cover B, and is held in place by ten $2\frac{1}{4}$-inch studs (F_1 F_2) which are fixed into a flange of the main forging. The cover is rendered gas-tight by a spigot-joint S packed with lead. The furnace body is surmounted by a cast-iron casing H, through which cooling water is circulated. The main forging A is surrounded by the cast-iron water jacket C. Both the hemispherical ends of the furnace have projections $K_1 K_2$, bored out to a distance of 3 inches. The carbon holders (Fig. 33) which move in these recesses are

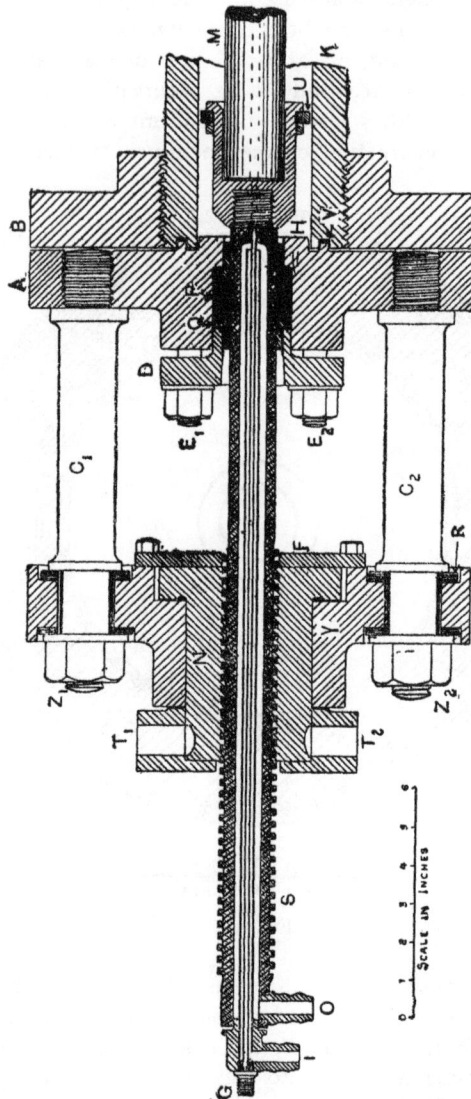

FIG. 33.

SCALE IN INCHES

thus protected from the direct heat or flame of the furnace. The length of the projections $K_1 K_2$ is sufficient to allow a feed of 8 inches. To obviate any risk of damage to the main forging by contact with the hot furnace material, a cast-iron lining L is always used.

The main forging is provided with three openings, as shown in Fig. 32, which is a transverse section through the centre of the furnace perpendicular to the axis of the carbon. F is the main forging, W the water jacket, and L the cast-iron lining. The aperture A serves to receive the valve through which the enclosure is filled with compressed gas. In most cases a window fitting is screwed into B and a pressure gauge into C.

The carbon-feeding mechanism is shown in Fig. 33. The feeding rod passes into the furnace through the insulated stuffing box F. This stuffing box serves the double purpose of making a gas-tight

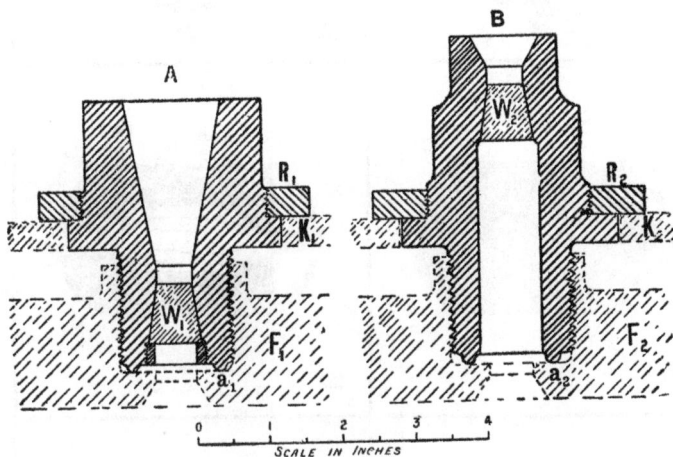

FIG. 34.

joint and providing insulation adequate for the relatively low e.m.f. required with this furnace. As packing, a mixture of asbestos and tallow is used, which, in itself, assists the insulation. The stuffing box is compressed by means of the ring D which presses in the gland but is electrically insulated from it by mica washers. The feeding rod is hollow and provided with water circulation.

The window (Fig. 34) consists of a glass or quartz cone W, which is forced into the gun-metal fitting after being surrounded with a thin film of cement, the shape of the glass tending to make the joint more perfect the higher the pressure. The joint between the fitting and the water jacket is made by means of a gun-metal ring R, which screws on the fitting itself. The side walls of the furnace enclosure

3

are shown at F and a gas-tight ring joint at *a*.   A mirror placed in
front of the window, and inclined at an angle of 45°, enables inspection without exposure to risk from fracture of the glass.   The design

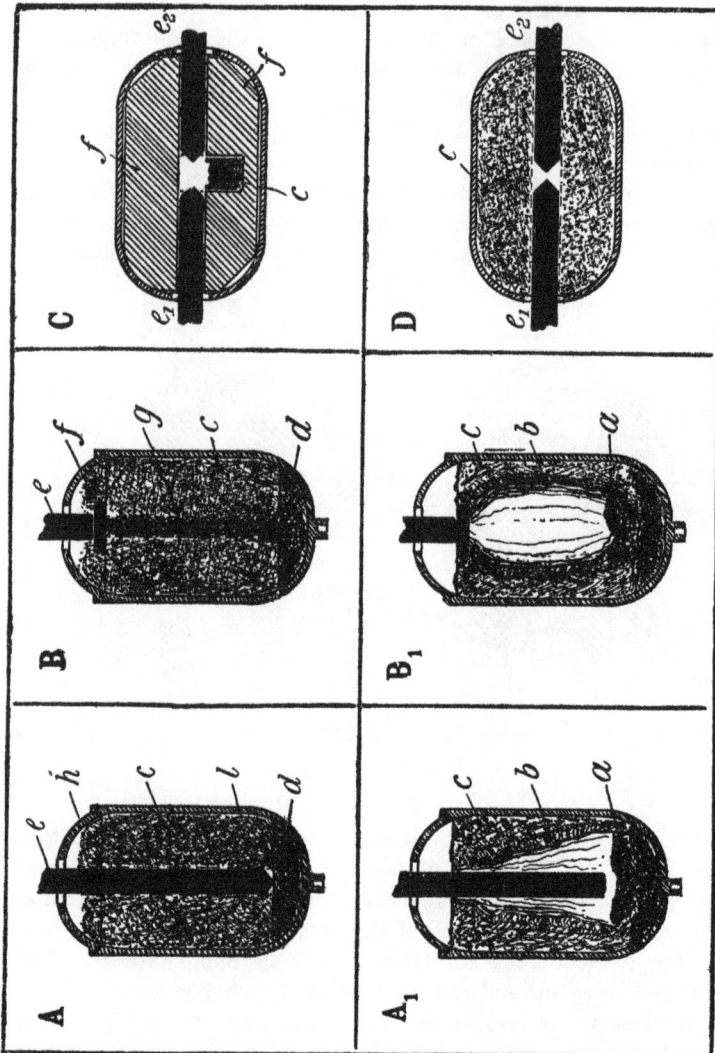

Fig. 35.

of window shown at A gives a clear view of the arc, but for very
large currents it is advisable to make use of the fitting B in which
the glass plug is more carefully protected from the source of heat.

Plate II.—Hutton and Petavel High-Pressure Furnace,

Fig. 35 shows sectional views of various furnace arrangements in the pressure cylinder.

A shows the smothered arc system before a run, l is the cast-iron liner in which the charge is placed, h the cover of same, e the vertical carbon electrode, 41 mm. diameter, d granular carbon bed forming the lower electrode, c the charge.

A_1 shows the smothered arc arrangement after the run in the case of the preparation of calcium carbide. a is an ingot of the fused product, b the fused and fritted material forming the walls of the cavity, and c the unacted-on material.

B shows the resistance type before the run, e is the carbon electrode, f the graphite end pieces leading to the core, g the resistance core of granular material or a carbon rod or other solid "resistor," c the charge, and d the granular carbon bed or other form of lower electrode.

B_1 is the resistance type after the run, the components of which are as in the case of A_1.

C shows the arrangement for horizontal arc radiation heating, $e_1 e_2$ are the electrodes, f the walls or jacket of heat insulating material, and c the charge in a carbon or other crucible.

D shows the smothered arc with two carbons $e_1 e_2$ embedded in the material c, which is used either in horizontal or vertical position.

A view of this furnace during an actual experiment arranged in a vertical position with compressed gas cylinder and various fittings is shown in Plate II.

Among the reactions which were studied with this furnace were the properties of the arc under high gaseous pressures, the formation of calcium carbide and carborundum, the fusion of silica and the direct reduction of alumina by carbon.

HIGH-PRESSURE FURNACE FOR 750 ATMOSPHERES.

A furnace designed by Sir J. E. Petavel and constructed by Chas. W. Cook for use with gaseous pressures up to 750 atmospheres and tested at 1500 atmospheres, hydraulic pressure, is illustrated in Fig. 36 which gives a vertical section and Fig. 37 giving a horizontal section through the line T T in Fig. 36.

The cylindrical body C and hexagonal bolt B are made of mild steel. The conical piece D is of hardened steel and, on being forced down by the bolt B, forms a gas-tight seal at E (Fig. 36) through a disc of soft copper embedded in the recessed ring. The gun-metal nozzle A through which the gas is introduced is secured by a collar which screws down on to the piece D. An annular space H

3 *

surrounded by a cast-iron cover N is provided for water circulation. A similar water-cooling channel not shown in the diagram is fitted in the walls of the furnace near the base.

The central cavity of the enclosure is $2\frac{1}{2}$ inches wide at its widest point. Electrical circuits are introduced into the enclosure by means of five plugs, T_1, T_2 . . . T_5, secured in the walls of the

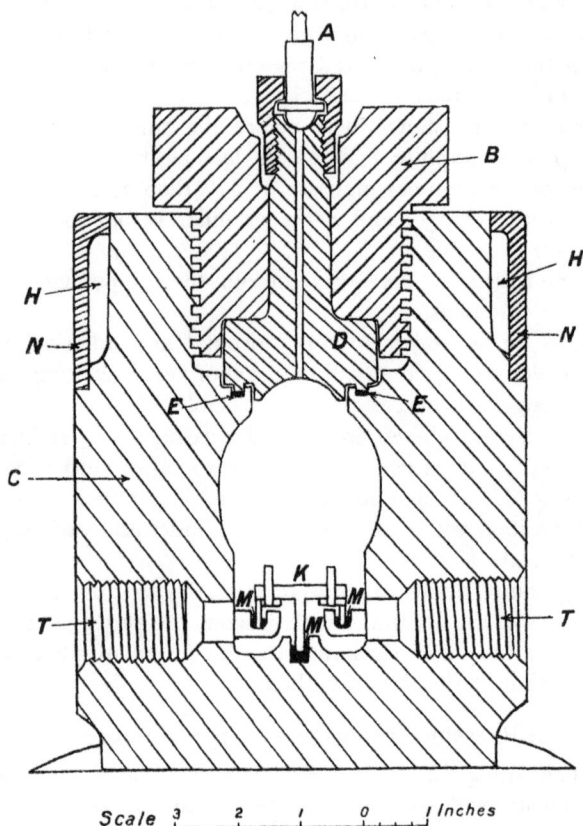

Scale 3 2 1 0 1 Inches

FIG. 36.

furnace as seen in Fig. 37. Each plug contains a central rod of silver steel insulated from the body of the plug by ebonite collars at E E. Hollowed recesses are provided near the widened terminal points as shown at M and serve to contain a quantity of mercury which enables electrical contact to be established with the apparatus in the enclosure. A further recess is provided in the centre of the base of the enclosure for containing mercury which is in metallic connexion with the body of the furnace and, by means of the terminal T_6, enables a sixth circuit to be provided. Electrical connexion between the terminals M and the apparatus employed in the enclosure are conveniently made by means of a circular ebonite support or table K fitted with five rods around the circumference and one in the centre which project downwards so as to fit in the mercury wells M. The apparatus is first mounted on the table and the different terminals such as those for the heat-

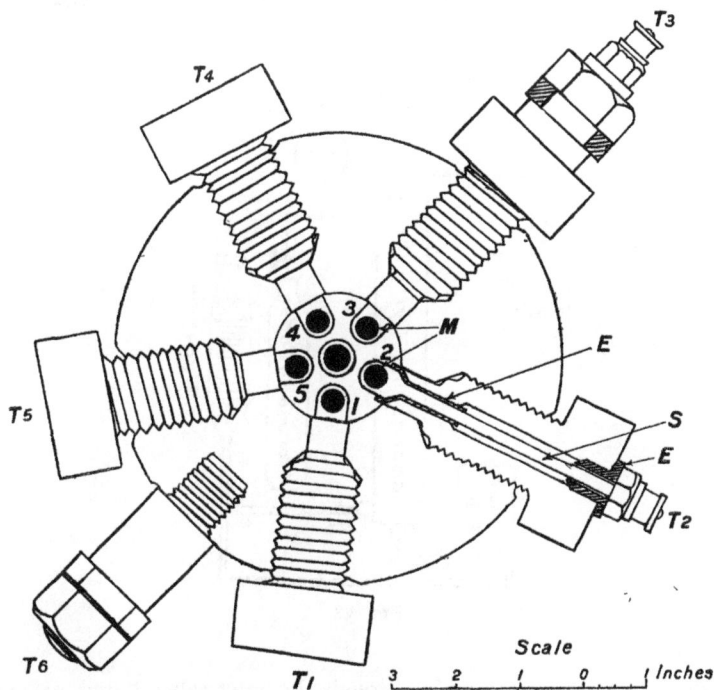

FIG. 37.

ing coil and thermo-junction circuits are connected to the poles of the table, which, with its attached apparatus, is then lowered into position in the furnace cavity.

The device employed for compressing the gas into the enclosure is illustrated in the diagram of connexions shown in Fig. 38. The compressing cylinder is first filled with glycerine or oil, which is then displaced by gas, from the storage cylinder or commercial bottle at a pressure of 100 to 200 atmospheres, the liquid passing into the funnel from which the pump is later supplied. The inlet valve at the top of the cylinder is closed, and, by means of the hydraulic pump, liquid is forced into the base of the cylinder, the gas being thus compressed to the required degree and then displaced into the furnace. When all is transferred and the cylinder is filled with liquid, the valve leading to the furnace is closed, the compressing cylinder again filled from the bottle and the compression and displacement repeated until the gas is brought to the required pressure in the furnace. The hydraulic pump is illustrated in Fig. 39. The two compression cylinders M and N are connected in series. By operating the handle A, glycerine

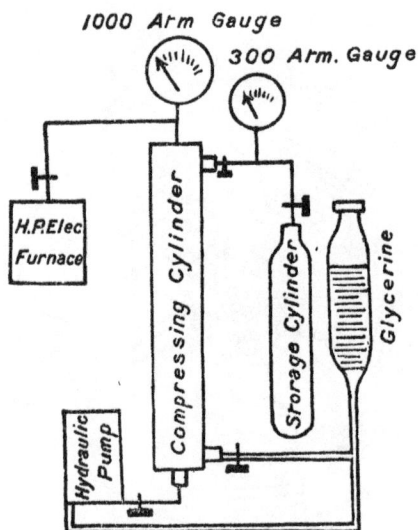

FIG. 38.

is drawn into the cylinder M through the inlet valve S, and, except for pressures above about 800 atmospheres, is forced through the cylinder N and into the reservoir through B. For the higher pressures, the cylinder N is filled after closing the valve D from the low-pressure side, the connecting valve C is closed, D opened, and by means of the screw plunger K, the liquid is forced from the high-pressure cylinder into the compressing cylinder when D is closed and the process repeated.

A view of the complete apparatus comprising pump, compressing cylinder and furnace is given in Plate III.

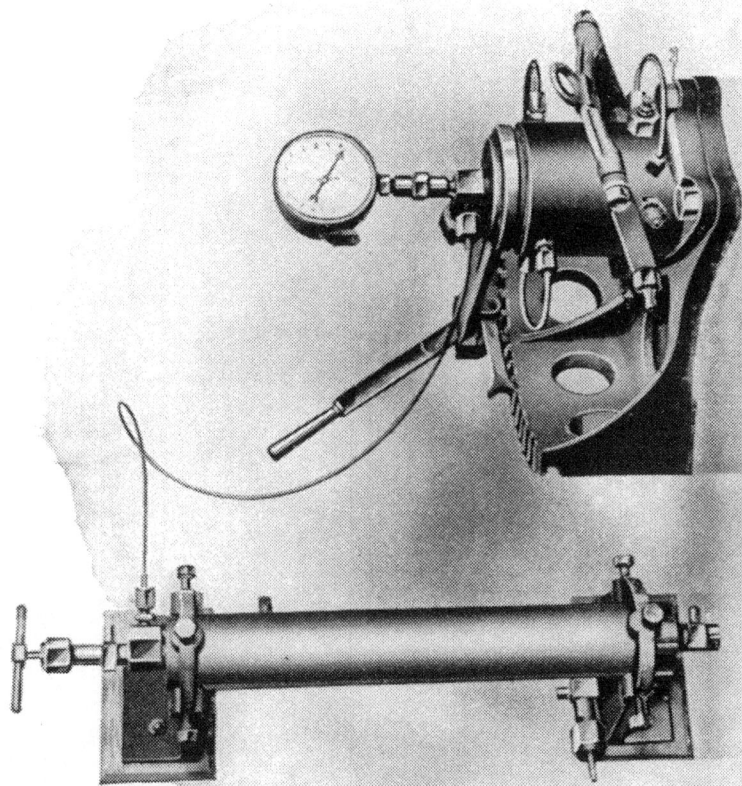

PLATE III.—HIGH-PRESSURE FURNACE AND COMPRESSION CYLINDER,

MODIFIED ELECTRIC FURNACE FOR PRESSURES UP TO 1000
ATMOSPHERES.[1]

The design of this furnace for the application of pressures up to
1000 atmospheres is a modification of the type introduced by Hutton
and Petavel, and was originally designed by the writer for the investigation of the synthesis of methane and other hydrocarbons at high
temperatures and pressures.   In the vertical cross section shown in
Fig. 40 the electrode leads consist of steel tubes E E, which are
cooled by water circulation.   Contact with the carbon is made by

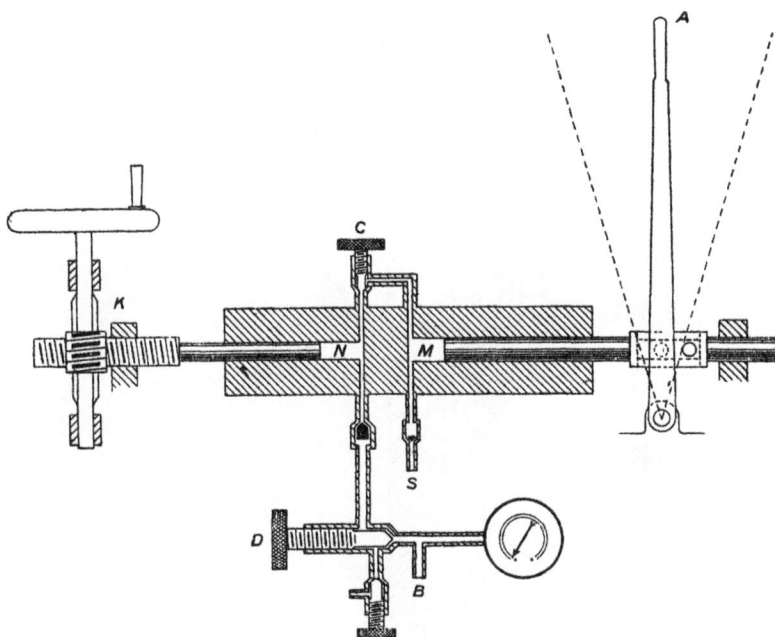

FIG. 39.

nickel clamps N N brazed to the ends of these tubes, which, through
intermediate plugs of graphite, establish electrical contact with the
carbon resistance heater consisting either of a tube or rod.   The
capacity of the furnace is about 750 c.c., the internal diameter being
3 inches and the length between the ends of the clamps, when withdrawn to their fullest extent, 6½ inches.

The electrode holders emerge from the furnace at the stuffing
boxes P P.   Small tubes of ebonite or fibre F F insulate the
electrodes from the walls of the latter.   Asbestos cord mixed with a

[1] Pring and Fairlie, "Trans. Chem. Soc.," 1912, **101**, 93.

little tallow or wax is used as the packing material for the boxes P P

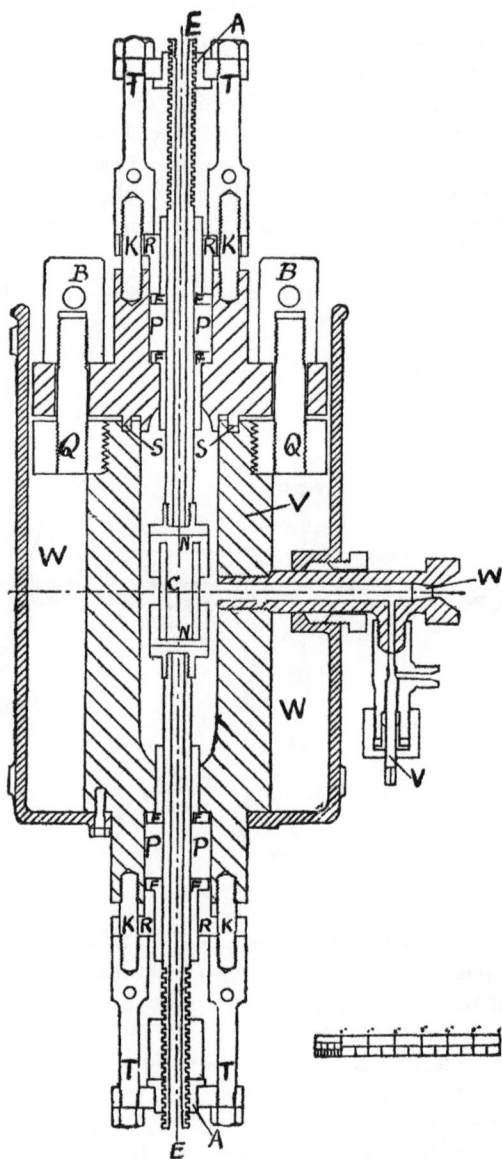

FIG. 40.

and is compressed by the brass rings R R, which are forced down on

FIG. 42.

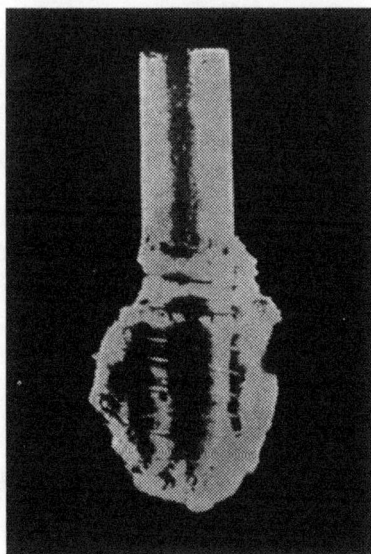

FIG. 44.

to the ebonite rings by six bolts, as at K K. The outward thrust of the electrodes is secured by the bolts T T, of which there are two at each end of the furnace. The top part of the furnace is fastened down by six bolts as at B B and a gas-tight joint is made by means of the lead spigot S S. By opening the bottom stuffing box and removing the packing, the bottom electrode can be lifted while clamped to the carbon rod and top electrode and removed with the furnace top. Water circulation is provided in the annular jacket W.

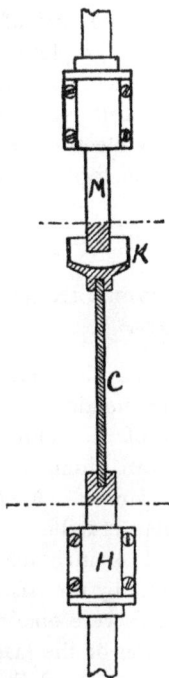

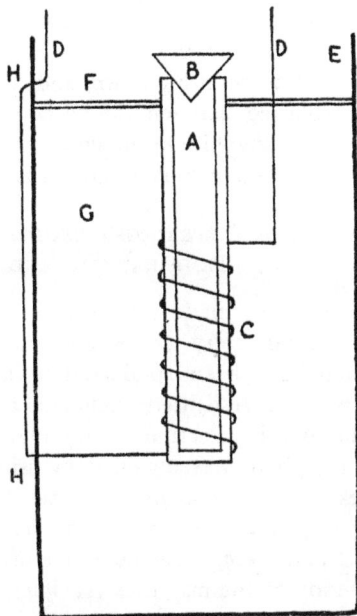

Fig. 41.   Fig. 43.

The glass window W is placed at a sufficient distance to be protected from the heat of the carbon.

Temperature readings are made by means of the light radiated from the carbon, after passing through the window and being re-flected from a mirror placed in front of the window at an inclination of 45°. The window could, as a rule, without fracture, be used with pressures up to 200 atmospheres and the remainder of the furnace is capable of withstanding a pressure of 1000 atmospheres.

The carbon used at C consists of amorphous retort carbon in

the form of tubes about 12 cm. long, and either 20 mm. external and 15 mm. internal diameter or 15 mm. external and 9 mm. internal diameter.

In some experiments rods of (Acheson) graphite were used. The connexions adopted in this case, by which definitely controlled and measured temperatures between 1100° and 2100° C. were obtained, are shown in Fig. 41. The carbon (C) of 6 to 8 mm. diameter and 8 to 12 cm. long, is inserted in graphite end-pieces, the bottom one of which is clamped by one of the nickel electrode holders H. The top graphite piece is turned in the form of a cup K, which is filled with pure graphite powder. A graphite rod M is fastened in the top clamp, and after the furnace top has been fastened down and the pressure of gas applied, this electrode can be lowered by screwing down, and a satisfactory electrical connection established between the carbon rod and the top electrode by means of the graphite in the cup.

A view of the complete furnace is shown in Fig. 42.

REDUCTION OF REFRACTORY OXIDES WITH HYDROGEN AT HIGH TEMPERATURES AND PRESSURES.[1]

Experiments made with a view of determining the reducibility of certain refractory oxides and obtaining the metals in a pure condition have been conducted in a furnace of the above type. The oxides investigated included those of vanadium, niobium, uranium, titanium, cerium, chromium and manganese. A carbon-free atmosphere was obtained by using a tubular crucible of fused magnesia or alumina, of $2\frac{1}{2}$ inches length and $\frac{1}{2}$ inch diameter, wound with tungsten wire, which could be heated by the passage of an electric current. Temperatures of over 2000° were obtained in the interior of the magnesia crucibles, while fusion of the magnesia on the outside in the neighbourhood of the wire denoted the local production of above 2500° C. In this way chromium and manganese were obtained in a metallic condition of a high degree of purity.

A diagram of the heating arrangement is seen in Fig. 43. A is the magnesia or alumina crucible containing the metallic oxide; B is a small cone made of fused magnesia, serving as a crucible cover but having small V slots filed in it to allow more free diffusion of gas; C, the tungsten wire, 0·5 mm. thick and 50 cm. in length, which was first wound round a glass tube of slightly smaller diameter than the crucible and then "sprung" on to the crucible, fitting

[1] Newbery and Pring, " Proc. Roy. Soc.," A., 1916, **92**, 276.

into a spiral groove which had previously been filed into the walls of the crucible. The nickel wire leads D D were twisted round the ends of the tungsten wire, and a small extra piece of tungsten wire twisted around the junction and continued for about 1 inch along the main wire, as it was found that fusion of the tungsten was specially liable to take place near the junction with the nickel. The outer vessel E was a porous earthenware battery pot, through the sides of which small holes H H were drilled. This was heated to redness, cooled, filled with freshly ignited magnesia powder, and covered with an asbestos disc F. The whole of this apparatus was then placed in the pressure cylinder and after admitting hydrogen at the desired high pressure, the temperature could be raised to the required degree. Up to moderate temperatures, readings were made with a suitably mounted thermo-element and higher temperatures were estimated by extrapolation from the current passed through the tungsten wire.

Fig. 44 gives a view of a crucible after use in this manner and shows the fusion of the magnesia in the neighbourhood of the tungsten spiral.

Furnace for Determining the Boiling-points of Metals.[1]

A careful determination of the boiling-points of metals has been made in a furnace which, as illustrated in Fig. 45, consists of a vertical carbon tube, electro-coppered at the ends, and soldered into brass castings provided with water circulation at A and B. Temperature readings were taken optically through the side tube of carbon, attached to a brass tube with a window at the end, a current of hydrogen being admitted at C to clear the tube of vapour. The whole furnace was packed in crushed wood-charcoal, while a thin-walled graphite crucible contained the metal to be studied.

In these measurements the possibility was foreseen of falsely high values of the temperature being obtained through the outside walls of the crucible being at a higher temperature than the metal during boiling also the possibility of inaccurate temperature readings through the reflection of radiation from the hotter surface of the heating tube. The results obtained were accordingly compared with those given by a system in which an annular graphite crucible was heated internally, as seen in Fig. 46, by means of a central carbon rod, which was supplied with current from the two thick graphite rods. The crucible was arranged inside a long wide carbon

---

[1] H. C. Greenwood, "Proc. Roy. Soc.," A., 1909, **82**, 396.

tube surrounded by kieselguhr and temperature readings taken
through a side tube, as with the apparatus shown in Fig. 45. The
two methods of heating gave, however, closely agreeing results.

Boiling-points of Metals under High Gaseous Pressures.[1]—The
influence of high gaseous pressures on the boiling-points of metals
was examined by means of an adaptation of the high-pressure furnace
of Hutton and Petavel (p. 30). This was arranged as shown in
Fig. 47. One graphite block D was soldered into a fitting screwed
on to the water-cooled insulated electrode F, passing through the
stuffing box at the bottom of the furnace and the other was held

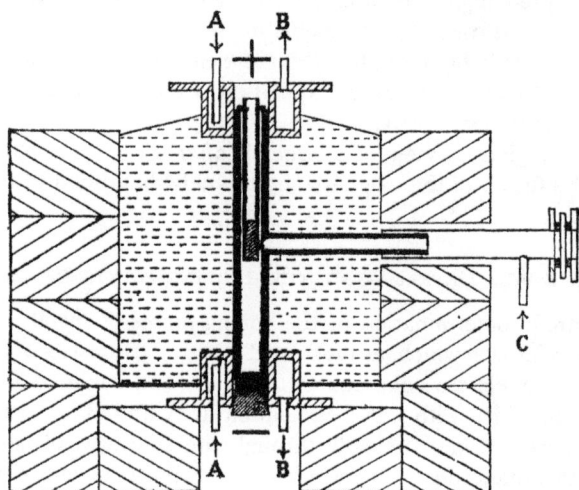

FIG. 45.

by stout steel strips attached to the cast-iron protection liner at G,
good electrical connexion to the main forging being ensured by
the use of brass wedges. The current was led in at the main forging
and out at the insulated electrode. Electrical insulation and approximate gas-tightness of the side tube were secured by means of
asbestos. The pressure was obtained with hydrogen compressed
into cylinders. For viewing the surface of the metal, a steel cover
was fitted to the top of the furnace and was provided with a thick
conical glass window A. A similar window B was employed for
taking temperature readings, being fixed in a gun-metal fitting

[1] H. C. Greenwood, " Proc. Roy. Soc.," A., 1910, **83**, 483.

which screwed into the side of the pressure enclosure. By means of a side opening C, connexion could be made to a cylinder of compressed hydrogen, from which a slow current of gas was admitted down the window fitting and the side tube of the furnace in order to keep the sighting tube clear from fumes. The range of temperatures employed in these measurements extended up to 2450° C., the boiling-point of iron at atmospheric pressure, 1510° C.,

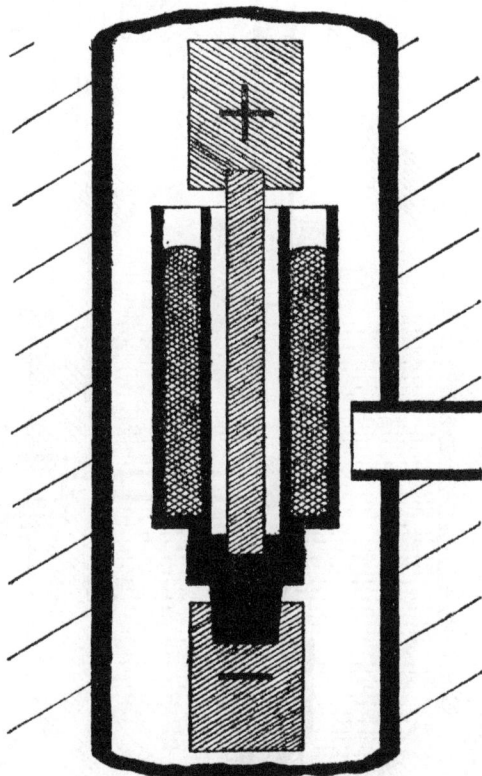

FIG. 46.

the boiling-point of zinc at 53 atmospheres and 2060° C., that of bismuth at 16·5 atmospheres.

USE OF SOLID ELECTROLYTIC CONDUCTORS IN FURNACE CONSTRUCTION.

At very high temperatures the refractory oxides such as magnesia and zirconia become more or less conducting by electrolysis while still in the solid condition. A mixture of zirconia together with a small percentage of yttria becomes appreciable conducting at

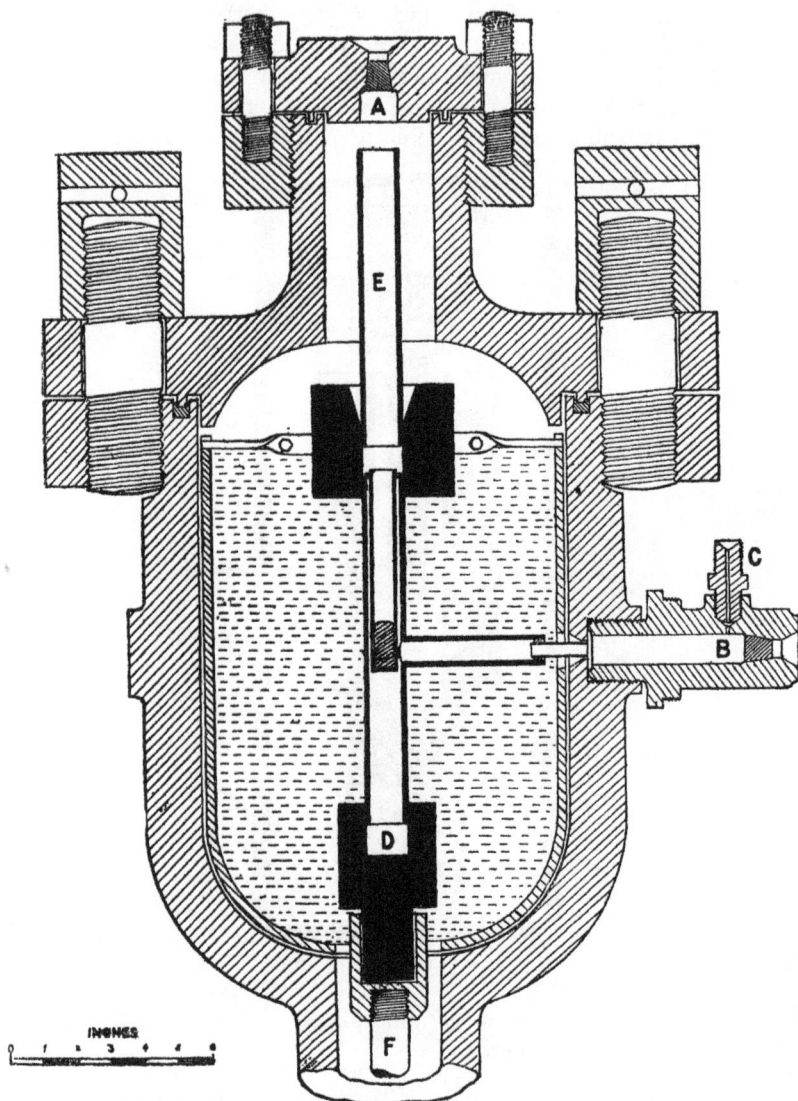

FIG. 47.

moderately high temperatures, while by the passage of the current it can be taken up to a high temperature. This mixture was applied by Nernst in the construction of a lamp containing a filament composed of these oxides which was first heated by a subsidiary circuit until it became sufficiently conducting to light up by the applied voltage.

An electric furnace operating on this principle, in which the heating medium is composed of a tube of conducting oxides (zirconia

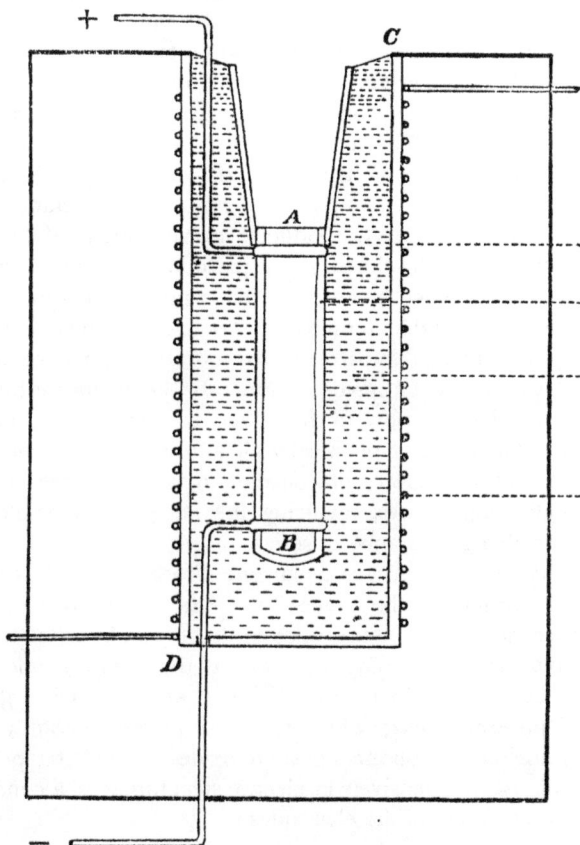

FIG. 48.

and yttria), was constructed for the purpose of pyrometer standardisation by J. A. Harker.[1]

On account of the difficulty to be anticipated through overheating at the point of contact of the tubes and the platinum

[1] " Proc. Roy. Soc.," A., 1905, **76**, 237.

connexions serving to lead in the current, the high temperatures were reached by means of a "cascade" arrangement. The furnace tube, together with its insulating packing, was first heated to a temperature of about 1000° C. by means of a surrounding nickel coil. At this temperature the central electrolytic heater tube became conducting and by the passage of the current enabled temperatures up to 2000° C. to be reached. Fig. 48 shows the arrangement adopted. A B is the central conducting tube, which was about 10 mm. internal diameter and 60 to 70 mm. long. C D is a tube of hard porcelain or other suitable refractory material, 30 to 40 mm. diameter, on which is wound the heating spiral of nickel wire, protected from oxidation by some suitable means. The space between is filled with zirconia which should not contain any appreciable admixture of any other substance, otherwise it gradually becomes a conductor at very high temperatures. The current is led into the tube by platinum flexibles enveloping it at A and B and joined by autogen soldering to nickel or platinum wires leading to suitable terminals. To light up such a furnace, a current sufficient to dissipate 160 to 300 watts is passed through the nickel spiral, the terminals of the tube being meanwhile connected to a voltage supply of from 200 to 500 volts. Since the temperature coefficient of practically all solid electrolytic conductors is negative and very large, the current in them becomes unstable at very high temperatures, involving the necessity of employing a sufficient steadying resistance in series.

With this type of furnace a current of 2 amps. suffices to take the central portion to about 2000° C., the voltage depending on the distance between the electrodes and the condition of the tube material, but rarely exceeding 100 volts at the higher temperatures, and often being only 60 to 80. With a well-designed regulating resistance, an exact control of the temperature can be readily made.

The principle of conducting solid oxides at high temperatures is now made use of extensively in electric steel furnaces, a conducting hearth serving as one of the electrodes.

SECTION IV.

CURRENT SUPPLY IN ELECTRIC FURNACE OPERATION.

In electric furnace operation on an industrial scale, use is invariably made of alternating current. In many cases direct current has been found to exert a deleterious action on the material under treatment on account of chemical changes brought about by electrolysis. An important advantage with alternating current is, moreover, obtained

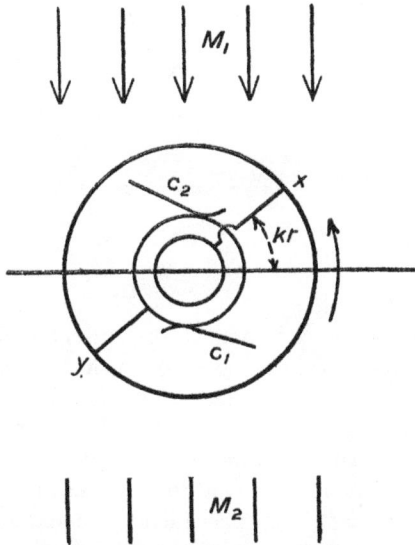

Fig. 49.

through the facility of transformation whereby the potential of the power supply can be readily transformed to any desired voltage. The current can thus be generated on a large scale at a high potential so as to obtain the maximum economy with the generator, transformed to a still higher potential for transmission to a distance, whereby a saving in the metal conductors and in the energy loss through resistance is obtained and finally stepped down to suit the individual furnace requirements. In the course of the furnace operation, an adjustment of the voltage can also be made by regulation of the transformer without loss of power, in accordance with the changing resistance or power requirement of the furnace.

The distinction between direct and alternating current is that with the former the current always flows in one direction, while with alternating current the intensity and direction vary as a periodic function of the time.

The principle of the generation of a single-phase alternating current is illustrated in the diagram in Fig. 49. In the path of a magnetic field $M_1 M_2$ a cylindrical drum or armature is represented as a circle and is arranged to rotate by means of an external force in a counter-clockwise direction. A single loop of wire xy is stretched diametrically behind the drum, and the ends extend over the front and connect respectively with two rings known as " slip rings " attached near the hub of the armature. Stationary brushes at c_1 and c_2 make

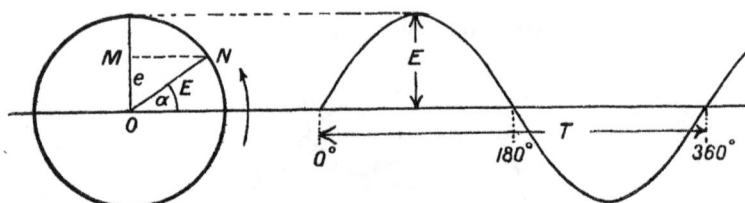

FIG. 50.

contact with these rings and conduct the current generated. When the circuit xy is in a horizontal position with reference to the lines of magnetic force, it follows that no lines are cut by a very small rotational displacement and the e.m.f. induced in the circuit is accordingly zero. After a rotation through 90° the line xy is parallel to the magnetic field and the number of lines of force cut by a given displacement and consequently the e.m.f., is at a maximum. This falls through the next 90° of rotation at the end of which it again becomes zero. Beyond this stage and through the following 180°, a similar change is followed but in this half cycle the polarity of any given slip-ring or the direction of the current, is reversed.

If, as in Fig. 50, a circle is described of radius ON corresponding to the maximum voltage E_m given during the cycle and the locus of N is the circumference of the circle to correspond with the rotation of the armature of the generator, then the potential of the circuit at any given time will be given by the projection of ON on the vertical axis, viz. OM. If a is the angle given by ON with the horizontal axis,

then $\qquad\qquad\qquad$ OM = ON sin a

or $\qquad\qquad\qquad$ $e = E_m$ sin a

where e denotes the instantaneous voltage at the time t. In place of the angle, however, the value of the angular velocity ω can be substituted, or

$$a = \omega t, \text{ whence } e = E_m \sin \omega t \quad . \quad . \quad . \quad (1)$$

In this theoretical case the relation between the voltage and the time is accordingly given by a sine curve as shown on the right in Fig. 50. The time taken to pass through one complete revolution or cycle is shown as T. The number of complete cycles passed through per second or $\frac{1}{T}$, is known as the periodicity or frequency. Since the angle of rotation corresponding to one cycle is 2π radians, it follows that

$$\omega = \frac{2\pi}{T} \text{ or } \omega = 2\pi n \quad . \quad . \quad . \quad (2)$$

where n is the frequency or periodicity.

It should be noted that in actual generators as used in practice, on account of the arrangement of the windings, the voltage curves, in many cases, depart considerably from the sine relationship, but large modern power-station generators give e.m.f. curves which approximate closely to the sinusoidal form.

DIFFERENCE BETWEEN MEAN AND EFFECTIVE VALUES OF ALTERNATING CURRENTS.

(a) *Mean Values.*—The mean value of a current or voltage can be determined from the area of the graph given during one half period and dividing this area by the length of base line, i.e. $\frac{T}{2}$. In the case of a sine relationship, the equation of this graph in terms of the maximum or peak potential is seen in equation (1) to be

$$e = E_m \sin \omega t.$$

The area bounded by the curve and the zero axis is therefore given by the expression

$$\int_0^{\frac{T}{2}} E_m \sin \omega t \cdot dt$$

$$= \frac{-E_m}{\omega} (\cos 180° - \cos 0°)$$

$$= \frac{2E_m}{\omega}.$$

4 *

Dividing the area by $\dfrac{T}{2}$, we have for the mean voltage E

$$E = \frac{4E_m}{\omega T} = \frac{2E_m}{\pi} = 0\cdot637E_m \qquad . \qquad . \qquad . \quad (3)$$

(*b*) *" Effective" or " Virtual" Values.*—In measuring alternating currents by means of electro-dynamometers, hot-wire voltmeters, and electrostatic voltmeters, the effect of the current varies as the square of its value. The instrument readings are consequently proportional to the mean square of the current, and the scale is graduated to give the root of the mean square. These so-called R.M.S. values are generally used in alternating current measurements since the heating value of a current is often required. An alternating current of given R.M.S. value produces the same heat as direct current of the same value.

The R.M.S., effective or virtual value of the voltage expressed by E_f is given by the square root of the mean value of

$$e^2 = E_m{}^2 \sin^2 \omega t$$

over one half period. E_m is a constant and to determine the mean value of the variable $\sin^2 \omega t$, it may be noted that the graphs of $\sin \omega t$ and $\cos \omega t$ are represented by the same curves but with a displacement of 90°.

Further $\qquad\qquad \sin^2 \omega t + \cos^2 \omega t = 1$

hence, $\qquad$ average of $\sin^2 \omega t =$ average of $\cos^2 \omega t$

$\qquad \therefore$ average of $(\sin^2 \omega t + \cos^2 \omega t) = 1$

$\qquad \therefore$ average of $\sin^2 \omega t = \tfrac{1}{2}$

$$\therefore E_f{}^2 = \frac{E_m{}^2}{2} \text{ or } E_f = \frac{E_m}{\sqrt{2}} = 0\cdot707E_m$$

similarly the effective current $I_f = \dfrac{I_m}{\sqrt{2}} = 0\cdot707I_m$. $\qquad$. $\qquad$. $\quad (4)$

It follows from this that the true average power

$$P = E_f I_f = \frac{E_m I_m}{2}$$

or is one half of the maximum (see, however, p. 57).

The ratio of the effective e.m.f. or current to the mean value in the case of a sine function with time is thus equal to

$$\frac{0\cdot707}{0\cdot637} = 1\cdot11 = \frac{\pi}{2\sqrt{2}} . \qquad . \qquad . \qquad . \quad (5)$$

CURRENT LAG, PHASE DISPLACEMENT AND POWER FACTOR.

In practice in an alternating current circuit, the flow of current does not immediately take place to the amount required by Ohm's law from the applied voltage. A lag in the current behind the voltage thus takes place and may be regarded as being due to the influence of a counter e.m.f., which is proportional to the rate of change of the magnetic flux produced by the current. This effect follows from the principle of electro-magnetic action, whereby an electro-motive force is produced whenever a conductor is moved in a magnetic field so as to cut magnetic lines of force. It follows, conversely, that if the conducting circuit remains stationary but is subjected to a varying magnetic field, so that magnetic lines of force are introduced into or removed from the circuit, then an e.m.f. is generated and current circulated. The e.m.f. and current thus produced are proportional to the number of lines of force introduced or withdrawn from the circuit in a unit time. Similarly, if two conductors are placed side by side and a current is passed through one of them and at intervals interrupted, or changed in direction, then the magnetic field which accompanies the passage of the current introduces magnetic lines of force which cut the path of the second circuit and during their introduction and withdrawal, while a change in the number of lines of force is imposed, an e.m.f. will be generated in the second circuit and if the circuit is closed a current will flow. The direction of an induced e.m.f. is always such as to oppose the change which gives rise to it.

The current which is induced in a second circuit acts through its magnetic field reciprocally on the primary circuit, generating an e.m.f. which opposes the primary e.m.f., and thus diminishes the effective voltage and retards the flow of current. In an electric circuit through which current is flowing, apart from the interaction on the primary circuit of an induced neighbouring path, the magnetic field of the primary circuit itself forms lines of magnetic force which during their growth and decline generate a counter e.m.f. and this opposes the original voltage.

The self-induced e.m.f. of a current is proportional to the rate of change of the magnetic field set up by the current and is therefore proportional to the rate of change of current, or

$$e_l = L\frac{di}{dt} . \qquad . \qquad . \qquad . \qquad . \quad (6)$$

where e_l is the self-induced e.m.f. and L the self-inductance or co-efficient of self-induction. L is equal to $\dfrac{\text{magnetic flux}}{\text{current}} \times$ No. of turns in circuit linked with flux, and may be assumed constant. If e is the impressed e.m.f. at time t, then the resultant e.m.f. at this instant is equal to

$$e - L\frac{di}{dt}$$

and the current, which in accordance with Ohm's law is determined by the resultant (or energy) e.m.f., is given by the expression

$$i = \frac{e - L\dfrac{di}{dt}}{r}$$

or
$$e = ri + L\frac{di}{dt} \qquad . \qquad . \qquad . \qquad . \quad (7)$$

According to this equation, the potential e which is applied to a circuit may be regarded as being resolved into two components:—

1. e_o, the value required to balance the drop of potential ri due to the resistance of the circuit.

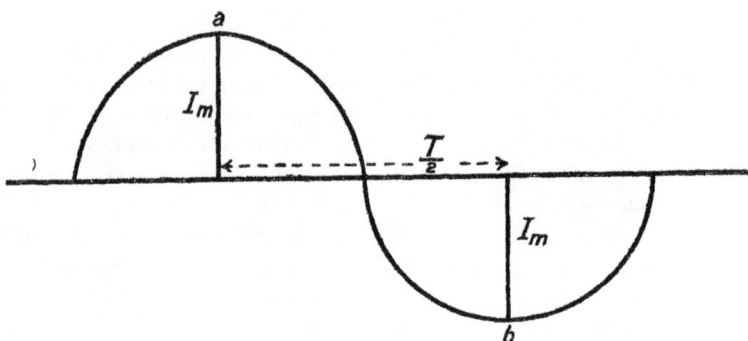

FIG. 51.

2. e_l, the value required to oppose the e.m.f. of self-inductance. Therefore, for instantaneous values,

$$e = e_o + e_l \qquad . \qquad . \qquad . \qquad . \quad (8)$$

The expression
$$e_l = L\frac{di}{dt}$$

can be evaluated by a consideration of the graph in Fig. 51. The *average* indicated e.m.f. of induction during the period a to b or $\dfrac{T}{2}$ is given by

$$L\frac{\text{change of current}}{\text{time}} = L\frac{2I_m}{\dfrac{T}{2}} = 4nLI_m.$$

This, from equation (4) $\qquad = 4\sqrt{2}nLI_f \qquad . \qquad . \qquad . \quad (9)$

The *effective* indicated e.m.f. of induction is, from equation (5), equal to average indicated e.m.f. $\times \dfrac{\pi}{2\sqrt{2}} = \dfrac{\pi}{2\sqrt{2}} \, 4\sqrt{2}\,n\mathrm{L}I_f$

or
$$E_l = 2\pi n\mathrm{L}I_f = \mathrm{L}\omega I_f \qquad . \qquad . \qquad . \quad (10)$$

If I_f is measured in amperes, n in cycles per second and E_l in volts, L is given in henries.

In the case of the instantaneous values, we have from equation (7) and (8) the following relation :—

$$e = e_0 + e_l = ri + \mathrm{L}\frac{di}{dt}.$$

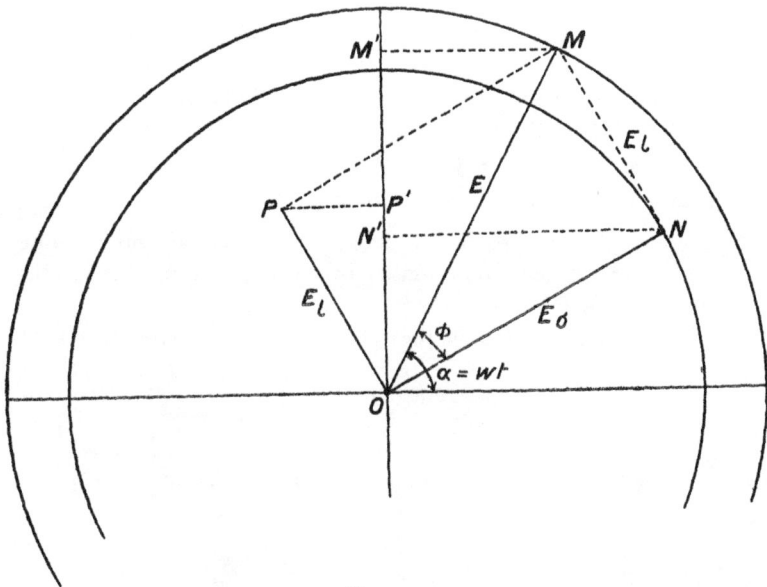

FIG. 52.

This, from equation (1)

$$= r\mathrm{I}_m \sin \omega t + \mathrm{L}\frac{d(\mathrm{I}_m \sin \omega t)}{dt}$$
$$= r\mathrm{I}_m \sin \omega t + \mathrm{LI}_m\omega \cos \omega t$$
$$= r\mathrm{I}_m \sin \omega t - \mathrm{LI}_m\omega \sin (\omega t - 90°) \qquad . \qquad . \quad (11)$$

It is seen from this that the component of impressed e.m.f. required to overcome the self-induced e.m.f. is 90° in advance of the component required to overcome the ohmic resistance.

The components and resultants of the e.m.f. and currents can be represented on a vector diagram as shown in Fig. 52. As in Fig. 51, the rotating vector OM represents the applied maximum potential E_m and may also be taken to refer to the effective value, while ON and OP represent the maximum (or effective) energy potential E_o and induction e.m.f. E_l respectively.

The lengths of OM, ON, and OP are drawn to represent the value of these respective components and in accordance with equation (11) OP is drawn at right angles to ON. By completing the square, MN is obtained equal to OP and OM results as the hypotenuse of the triangle. The wider circle represents the path of the point M and the inner circle that of the point N during the rotation of the vectors.

The instantaneous values of impressed e.m.f., energy e.m.f., and induction e.m.f. are given respectively by the projection of OM, ON, and OP on the vertical axis, viz. OM', ON', OP'. It follows from the construction of the figure that

$$OM' = ON' + OP'$$
or, as in equation 8, $e = e_o + e_l.$

The current is in phase with the energy potential and its value may be taken as being represented by the line ON and the instantaneous value by ON'.

From a consideration of the relations in the triangle, it is seen that

$$E^2 = E_o{}^2 + E_l{}^2$$
or $$E = \sqrt{E_o{}^2 + E_l{}^2} \quad . \qquad . \qquad . \qquad . \quad (12)$$

where E, E_o, and E_l represent either the maximum or effective (R.M.S.) values.

Since $$E_o = Ir$$
we have from (10) and (12) $E = \sqrt{I^2r^2 + I^2\omega^2L^2}$
$$= I\sqrt{r^2 + \omega^2L^2}$$

or $$I = \frac{E}{\sqrt{r^2 + \omega^2L^2}} \quad . \qquad . \qquad . \quad (13)$$

The expression $\sqrt{r^2 + \omega^2L^2} = z$

represents the apparent resistance of an alternating circuit, and is known as the *impedance*.

In the rotating vectors OM and ON in Fig. 52, the angle ϕ measures the angle of lag of the energy potential or current behind the impressed potential, and the relation is such that

$$\tan \phi = \frac{\omega L}{r} \quad . \qquad . \qquad . \qquad . \qquad . \quad (14)$$

Further $\qquad\qquad$ ON $=$ OM cos ϕ

or $\qquad\qquad\qquad$ $E_o = E_m \cos \phi$. $\quad$. $\quad$. $\quad$. (15)

The influence of the lag produced by phase displacement is also shown in Fig. 53 in the form of curves representing the function between voltage and time. The ordinate of the dotted line measures the impressed voltage and that of the continuous line the energy voltage or current I, while the horizontal axis denotes the time. The term cos ϕ is called the power factor of a circuit.

INFLUENCE OF POWER FACTOR ON POWER CONSUMPTION.

In the case of direct current, the power is measured by the product of the current or amperes and potential or volts, i.e. P $=$ EI.

However, with alternating current, the influence of lag or phase-

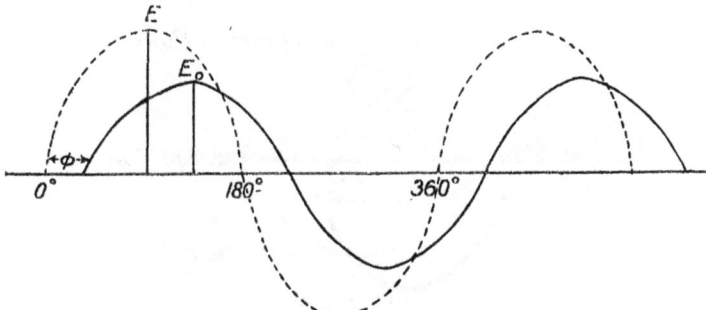

FIG. 53.

displacement leads to a value for the power consumed or effective power which is lower than the value of this product.

As ordinarily measured, the product of the virtual amperes and virtual volts, i.e. the volt-amperes or kilovolt-amperes, is thus different from the watts or kilowatts by the magnitude of the power factor.

It is seen from the diagram in Fig. 52 that

$$\text{ON}' = i = I_m \sin (\omega t - \phi)$$

in accordance with the lag of current behind e.m.f.

and $\qquad\qquad$ OM$' = e = E_m \sin \omega t$

$$\therefore \; ei = E_m I_m [\tfrac{1}{2} \cos \phi - \tfrac{1}{2} \cos (2\omega t - \phi)].$$

The power is given by the average value of ei over a complete period.

Since the term cos $(2\omega t - \phi)$ has double the frequency of the term $\tfrac{1}{2}$ cos ϕ, it follows that during any whole number of periods the value of the total area of the curve enclosed by cos $(2\omega t - \phi)$ is zero when the area below the horizontal axis is taken as negative

$$\therefore \text{ average value of } ei = \frac{E_m I_m}{2} \cos \phi$$

$$= \frac{E_m}{\sqrt{2}} \cdot \frac{I_m}{\sqrt{2}} \cos \phi = E_f I_f \cos \phi$$

where $E_f\, I_f$ are the effective volt-amperes and cos ϕ the power factor.

In the above derivation of the power factor as the cosine of the angle of phase-displacement, a sine relationship has been assumed as the function of e.m.f. and current with time. Where the waves are no longer of the sine form, the power factor is still used to express the relation

$$\text{Power factor} = \frac{\text{true power}}{\text{volt-amperes}}.$$

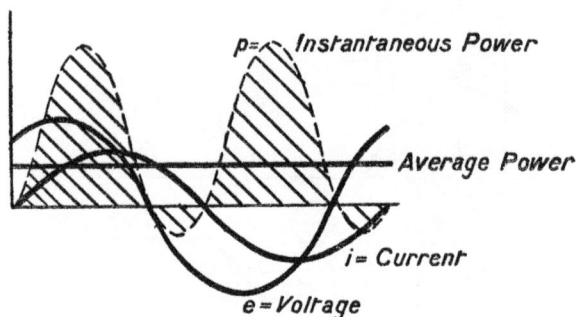

Fig. 54.

The value $I_f \cos \phi$, which corresponds to ON in Fig. 52, is known as the power component or load component of the current and the term $I_f \sin \phi$, which corresponds to NM or E_l, is known as the " idle " or " wattless " component of the current, as this is without any influence on the actual power and does not enter into the watts actually utilised. The net result of this wattless component is to diminish the amount of power which can be absorbed by the circuit. To consider further the nature of the " wattless " component of alternating current, it follows that the instantaneous value of the power is given by the product of the instantaneous values of voltage and current. If the current and voltage are not in phase, there will recur certain intervals during which the voltage and current are of opposite signs. At these stages the instantaneous power is a negative quantity, or power is being returned to the generator by the diminishing magnetic field. This condition is illustrated in Fig. 54, in which *e* represents the impressed voltage curve, *i* the instantaneous current, while the product of these two values is given by the shaded area under the curve *p* which represents the instantaneous power. The average power is represented by a horizontal axis and the negative power by the shaded area below the zero axis. This negative power and the corresponding amount of positive power which it neutralises in the total output represents the " wattless component ".

The net result of this " wattless " component is to diminish the amount of power which can be absorbed by the circuit and may be sufficiently great to counterbalance the full capacity of the generators and conductors although very little energy is being generated or transmitted.

From equation (10) we have

$$E_l = L\omega I_f.$$

The magnitude of this reactance is thus seen to depend on the frequency of the current, the amount of current and the inductance. The inductance is increased by the vicinity of magnetic circuits and in the case of electric furnaces the iron casing of the furnace and water jackets of the electrode holders are liable to have a large influence on this factor.

In furnace construction the following points are of importance in determining the power factor :—

1. Alternating current of as low a frequency as possible should be used.

2. Material of high permeability to be avoided for the furnace construction.

3. Iron masses to be kept as far as possible out of the electric circuit.

4. Conductors of different polarities should be interleaved and brought into close proximity to each other. In any portion of the circuit where the conductors are not so interleaved, any conductors parallel to the main current should not be short-circuited.

5. Metal parts such as the water jackets of electrodes, which can carry the magnetic flux, should be constructed of some non-magnetic metal such as copper or brass.

Influence of Low-power Factor.—The effect of inductance or phase displacement in choking down a portion of the power supply does not lead to an actual loss of energy, but for a given power increases the total current and thereby the heating of the conductors and involves an increased outlay of electrical plant. Generators which are constructed of a certain capacity are not capable of yielding their rated output at lower power factors than that for which they are constructed since the output of an alternator is determined by its temperature rise, and a wattless current increases the temperature without supplying a corresponding amount of power.

CAPACITY.

The influence of capacity in a circuit is due to the property possessed by conductors of retaining or storing a quantity of electricity. The capacity of conductors is largely determined by the presence of neighbouring circuits.

The charge Q of a condenser at any instant is given by

$$Q = Ce = CE_m \sin\omega t$$

where e is the e.m.f. applied and C the capacity of the condenser.

Since $\frac{dQ}{dt}$ represents the current, we have

$$i = \frac{dQ}{dt} = \omega CE_m \cos \omega t$$
$$= \omega CE_m \sin (90° - \omega t).$$

In a vector diagram, therefore, the current would be represented by a vector 90° ahead of the energy e.m.f. This component of capacity reactance accordingly opposes that of induction reactance and may cause the energy e.m.f. and current to lead in phase over the impressed e.m.f.

The introduction of capacity by means of condensers is in some cases, as with the high-tension arc, made use of to improve the power factor.

EDDY CURRENTS.

Eddy or parasitic currents are generated by alternating current circuits in adjacent masses of metal. The currents become dissipated in the metal with the production of heat.

Eddy currents circulate wholly within the metal and are not confined to particular paths along insulated wires. In the construction of transformers and dynamo armatures, the formation of eddy currents in the iron cores, which leads to a loss in efficiency, is reduced by making the cores of laminated plates insulated from each other. Eddy currents are produced in any conducting medium but are more marked when the metal is magnetic.

An application of the heating effects of eddy currents has recently been made in the construction of an electric furnace in which a current of very high frequency is passed around a circuit surrounding the conducting material to be heated (p. 350).

SKIN EFFECT.

When a current passes along a conductor of large cross section, the central portion of the conductor is surrounded by a larger total magnetic field than the outside layers, since in addition to the field produced in the space surrounding the metal, there is a certain magnetic flux in the metal itself. With an alternating current the opposing induced e.m.f. which is generated is therefore higher in the centre of the conductor and the flow of current is accordingly lower than on the outside layers. This uneven distribution of current is equivalent to a reduction in the cross section or an increase in the resistance.

The phenomenon is known as "skin effect," and is an important factor to be considered in the proportioning of conductors for carrying large currents. The effect is much greater with a magnetic metal such as iron than with copper.

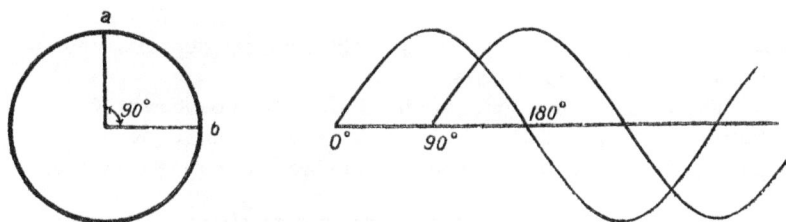

FIG. 55.

POLYPHASE CURRENTS.

The systems of electric power supply which are in most common use in power generation and distribution and in electric furnace operation are two-phase, three-phase, and, to a minor extent, four-phase currents.

Four-phase System.—Four-phase current is provided by a generator in which the armature is wound with two independent windings which are displaced relatively to each other so that there is a difference of phase in the e.m.f. generated in each by an amount equal to a quarter of one period, i.e. $\frac{\pi}{2}$ radians or 90°. The change of e.m.f. and current of the two phases with time is illustrated in the diagram in Fig. 55. As in the case derived in Fig. 52, the maximum or crest values of the potential and current are given by O*a* and O*b* for the two phases, while the instantaneous values are given by the projections of *a* and *b* on the vertical axis. As the angle of separation of the phases is 90°, it follows that the maximum of one curve occurs when the other is zero and if the potential O*a* or O*b* is 1, the value of *ab*, or the maximum phase voltage between

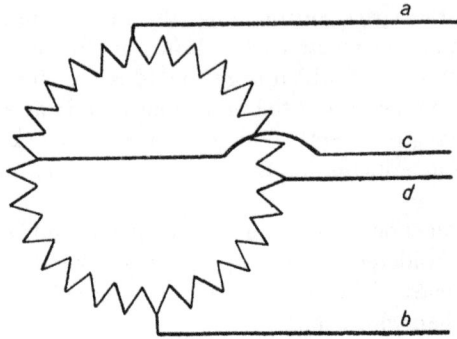

FIG. 56.

a and *b*, is equal to $\sqrt{2}$ or 1·414. The same relation also applies between the effective voltages.

The generator windings may be arranged in two different ways as shown in Figs. 56 and 57. In the method of ring connexions with two-pole machines the coils are connected in series to form a continuous

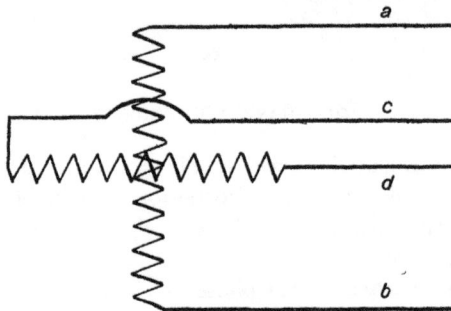

FIG. 57.

winding which is tapped at four points. The leads *a* and *b* constitute the circuit of one phase and *c* and *d* that of the second. The potential is E volts between *a* and *d*, *d* and *b*, *b* and *c*, or *c* and *a*, and $\sqrt{2}$E volts between *a* and *b* or *c* and *d*. The current in each line is $\sqrt{2}$I.

In the method of star connexions two separate windings are provided, as shown in Fig. 57, and connected in the centre to form a common point. In this system the potential between *a* and *b* or *c* and *d* is 2E volts, and that between *c* or *d* and *a* or *b*, while the current in each line is I amps.

Two-phase System.—With the star system of connexions, in place of employing two separate circuits with four wires, *b* and *d* (Fig. 57) may be connected together so as to form a common return to *a* and *c*, provided the two windings are independent and not joined in the centre. In this method, as represented in Fig. 58, the e.m.f. between the two outgoing wires is $\sqrt{2}$ or 1·414 times that between each outgoing wire and the common return. Similarly, the resultant current carried by the common return is equal to $\sqrt{2}$I, where I is the current carried by the outside wires. A two-phase

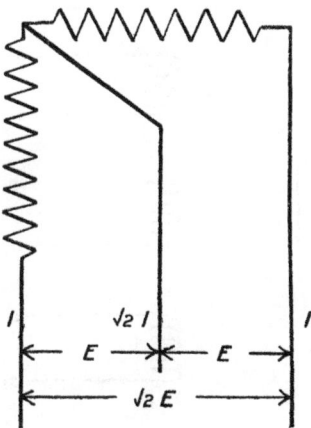

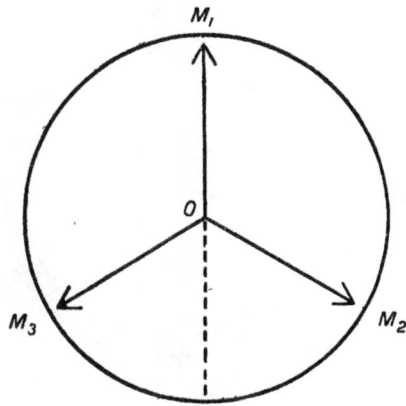

FIG. 58.   FIG. 59.

electric furnace will generally have three electrodes, one serving as a common return for the other two poles.

In addition to electric generators, two-phase current can, by a special type of transformer, be transformed from three-phase, or conversely three-phase from two-phase (cf. p. 68).

Three-phase current can be generated from alternators having three sets of similar windings spaced at a distance of 120° apart as seen in Fig. 60, when between consecutive windings is a phase difference equal to one-third of a complete period. With a balanced circuit, i.e. one in which the current in each phase is of equal value, the variation of the potential of each phase with time is shown by the vector diagram in Fig. 59, in which the axes OM_1, OM_2, and OM_3 represent the maximum or crest potentials and the vertical projections of these on the horizontal axis the instantaneous potentials. Fig. 76 shows at A, B, and C the instantaneous phase potentials (as ordinates) with a horizontal time axis.

In Fig. 60 the terminals of the coils of the generator are seen at aa_1, bb_1, and cc_1. The three separate circuits could be employed each with its two wires. However, the total number of wires necessary can be reduced to three by taking advantage of the fact that any single wire can act as a return for the other two. There are two distinct methods by which these connexions can be made so as to form a three-phase circuit:—

(a) By the so-called method of Star or Y-connexions. In this case a_1, b_1, and c_1 are joined together to form an electrical centre as shown in the upper diagram of Fig. 61. In some cases a wire is

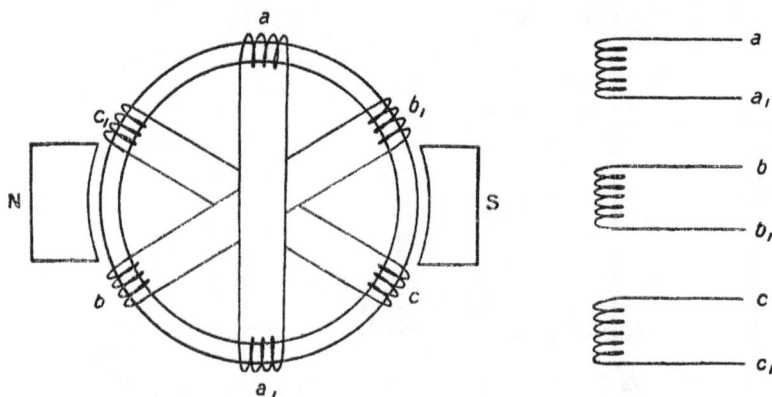

FIG. 60.

taken from the common or neutral point so as to form a fourth connexion.

(b) By the so-called Delta or Mesh connexions. In this system, as shown in the lower diagram of Fig. 61, a is joined to b_1, b to c_1, and c to a_1, and the three lines are taken from the common points. In the case of both of these systems the sum of the e.m.f.'s expressed with correct signs, on all three lines, at any given instant is equal to zero. Similarly, the sum of the instantaneous values of the currents on the three lines is equal to zero. This relationship applies both in cases where the load is equally distributed between the three phases, i.e. a balanced circuit, or when more power is taken from between two of the wires, used as a single-phase circuit, than between the second and third wire, i.e. an unbalanced circuit.

DISTINCTION BETWEEN PHASE AND LINE VALUES FOR VOLTAGE
AND CURRENT.

(*a*) *With Star Connexions.*—In Fig. 62, OM_1, OM_2, and OM_3 represent the phase voltage E_p or the difference between the potential of the neutral point and the line when the potential is at its maximum or crest value. The instantaneous values are given by the projections on the vertical axis and the maximum potential E_1

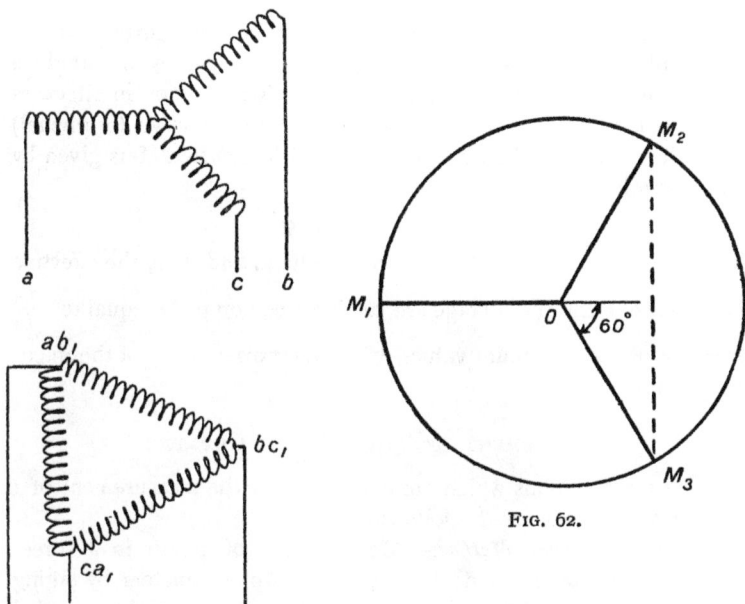

FIG. 61.

FIG. 62.

between any two lines is given by the line M_2M_3. The relation between the two values can be derived as follows :—

$$\frac{E_l}{2E_p} = \sin 60°$$

$$\therefore E_l = 2E_p \sin 60° = \sqrt{3}E_p = 1\cdot73E_p.$$

The above derivation is also applicable to effective in place of maximum values.

Accordingly, the line voltage is $1\cdot73$ times that of the phase voltage.

With regard to the current, it follows from the nature of these star connexions that the line current is equal to the phase

5

current, the same current flowing through the line as through the generator coil.

(*b*) *With Delta Connexions.*—The single coils are with this system joined in series as shown in the lower part of Fig. 61, and the points of junction of the coils led to the transmission lines. In this case the phase voltage is equal to the line voltage, while if I_p represents the current along each generator coil and I_l that along each phase on the line, then

$$I_l = \sqrt{3}I_p.$$

In the case of both systems of connexions the current is not necessarily in phase with the voltage but may lag by the angle ϕ which applies equally to all three phases. The relation in all cases with a balanced load between the power P, the effective (or virtual) line voltage E_l, and the effective or virtual line current I_l is given by the expression

$$P = \sqrt{3}E_l I \cos \phi.$$

In the case of a sine function between voltage and time, the effective values for current and voltage are, as described on p. 52, equal to $\dfrac{1}{\sqrt{2}}$ or 0·707 of the maximum values, while the power is 0·5 of the maximum value.

MEASUREMENT OF THREE-PHASE CURRENT.

The main systems which are employed for the measurement of a three-phase current are the following :—

Two Wattmeter Method.—Measurement of power in a three-phase circuit can be made by means of two wattmeters by taking advantage of the principle that any one of the lines may be regarded as a common return for the other two. From this standpoint there are in effect two independent circuits and the algebraic sum of the two readings gives a measure of the total power. This method gives a true result in cases where the load is not evenly balanced between the two phases. For this measurement the wattmeters are arranged as shown in Fig. 63, so that the current coil of one instrument is inserted in the line A and its voltage coil between the lines A and C; similarly, the second wattmeter is placed with its current circuit in line B, and its voltage coil between B and C.

Method 2.—With a star-connected system, if the neutral point is accessible and a balanced load is obtained on the three circuits, the measurement of power can be made by one wattmeter only by placing its current coil in one line and its voltage coil between

that line and the neutral point. The total power will then amount
to three times that indicated on the instrument. If the neutral point
is not accessible, one may be derived by connecting the three wires

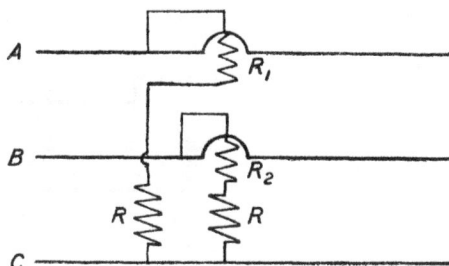

FIG. 63.

to the ends of three resistances or inductances joined in star form,
and taking the neutral point of this system for the voltage connexions of the wattmeter.

ADVANTAGES OF THREE-PHASE CURRENT SUPPLY.

In modern practice large-scale current generation and transmission are almost invariably conducted by the three-phase system.
The advantages gained are a saving of material in the generator
outlay for a given output and more particularly in the saving in
the transmission line. For the transmission of the same amount of
power, the total weight of cable in the case of the three-phase system
is only three-quarters of that required for single-phase current under
the same conditions of power factor and total loss.

FIG. 64.

THE APPLICATION OF TRANSFORMERS FOR USE WITH TWO-PHASE
CURRENT.

The principle of the main types of transformers used in connexion
with two-phase current is illustrated in Figs. 64 to 66. In Fig. 64 the
arrangement consists of two single-phase units, the phases being

5 *

separated in both primary and secondary circuits. In Fig. 65 two of
the secondary leads are joined making a common return for the other
wires. The two circuits being 90° apart, the voltage between a' and

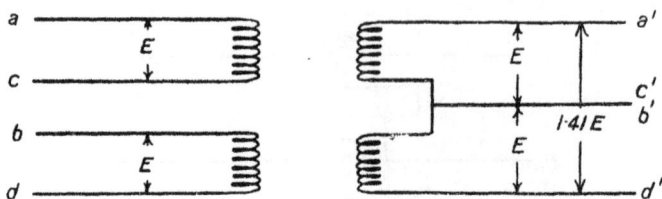

FIG. 65.

d' is $\sqrt{2}$ or 1·414 times that between the outside wires and the
common return, and the current in the common return is $\sqrt{2}$ or
1·414 times that in each of the outer wires.

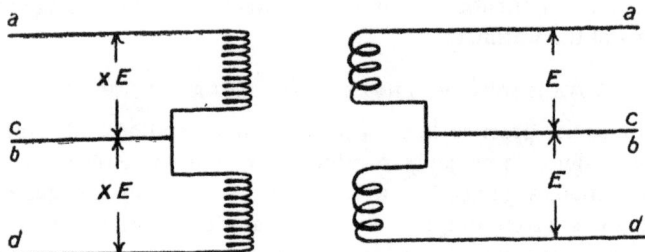

FIG. 66.

As shown in Fig. 66 common returns can be used on both the
primary and secondary of the transformers in cases where both sides
of the system are balanced.

CONVERSION OF THREE-PHASE TO TWO-PHASE CURRENT.[1]

The transformation from two-phase to three-phase or *vice versa*
is effected by arranging the windings as shown in Fig. 67 by the so-
called method of Scott connexions. The two windings on the two-
phase side are of equal voltage, while on the three-phase side AD
and BC are wound in the ratio of 86·7 to 100 volts. AD is shown
as the vertical axis in the triangle to the right of Fig. 67 and the end
point D is connected to the middle of BC. The transformation
can be performed by means of two single-phase transformers suitably

[1] From "Scott-connected Transformers," by J. L. Thompson, Special Publi-
cation No. 7370/2 (The Metropolitan-Vickers Electrical Co.).

wound and connected together. However, in practice, in order that only one spare transformer need be kept as a stand-by for failures, the two transformers are generally made duplicates of each other and provided with an 86·7 per cent and a 50 per cent tapping as shown in Fig. 68 where N is inserted to indicate the neutral point.

The principle of the derivation of a two-phase voltage from a three-phase voltage is illustrated in Fig. 69, in which A, B, C, N represents a three-phase voltage vector diagram for three star-connected transformers. A, B, and C are the angular points of an equilateral triangle, N is the neutral point and centre of the circumcircle. Producing AN to cut BC at D, it follows from the above conditions that D is the mid-point of BC, and that AD is at right angles or in quadrature to BC. This construction, therefore, has resolved the three vectors AN, BN, CN (or in the case

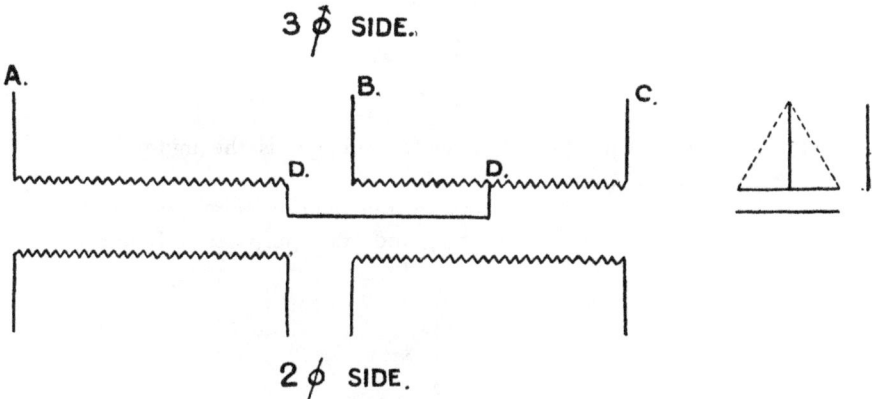

FIG. 67.

of the line voltage the three vectors AB, BC, CA), into two vectors AD and BC in quadrature.

Since the definition of a two-phase voltage is that the angular displacement between two generated voltages shall be 90°, the vectors AD and BC are equivalent to a two-phase voltage and the values of the two voltages are equal to their linear length.

The value of BC is equal to the line voltage and that of AD is

$$AD = AB \sin ABD$$
$$= AB \sin 60°$$
$$= AB \times \frac{\sqrt{3}}{2}$$
$$= 0·867 \times AB.$$

Hence AD is 86·7 per cent of the line voltage.

From this it is seen that a three-phase voltage can be obtained from a two-phase voltage, i.e. two voltages in quadrature, provided that on the three-phase side one voltage is equal to the line voltage and that the other voltage is equal to 86·7 of the line voltage and

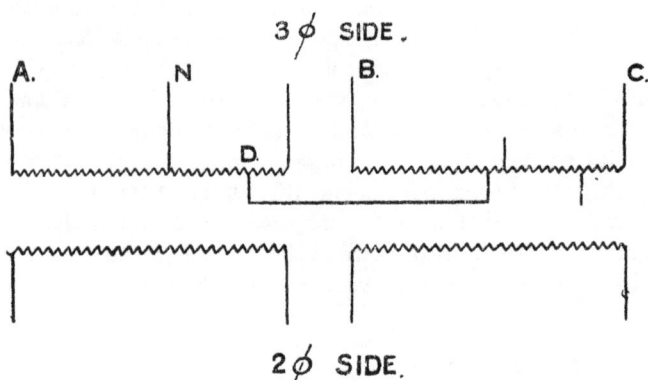

Fɪɢ. 68.

further, that the junction of these two voltages is the mid-point of the line voltage vector.

In Scott-connected transformers the neutral point can be obtained and can be used for earthing and other purposes.   Referring

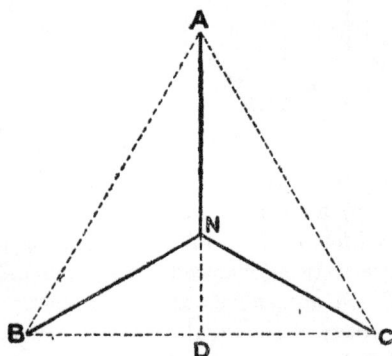

Fɪɢ. 69.

again to Fig. 69, since N is the centre of the circumcircle, then DN = 1/3 AD, and therefore, if a tapping is provided on the 86·7 per cent winding of the transformer at a third of the distance from the junction D, the neutral point is obtained as represented at N in Fig. 68.

For long distance transmission of power the generators are sometimes wound for two-phase current and the secondary distribution at
the receiving end is likewise by the two-phase system, while, on account of the saving in copper, the intermediate transmission is by
the three-phase system (cf. page 416).   Fig. 70 shows the transformer
connexions for changing two-phase to three-phase and back again.

The current and voltage relations in the system of Scott-connexions is further illustrated in the diagram in Fig. 71.

If two single-phase transformers are taken of the same k.v.a.
capacity but the voltage ratios of which are 100/86·7 and 100/100
respectively, then if the current carried by the 86·6 per cent winding,
which is represented by AD, is 100 per cent, the current carried
by the 100 per cent winding BC will be 86·6 per cent.   If a junction
of the two windings is made at D, so that the return from AD to

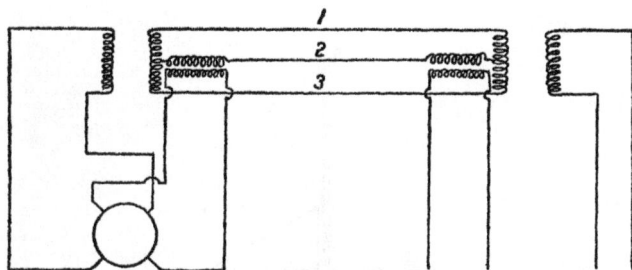

FIG. 70.

the generator or load point is through DB or DC, it follows that
since these two halves are symmetrical, half the current in AD flows
through DB and half through DC.   As a result two currents flow
in BC, so that in DB and DC the resultant current is the quadrature
sum of 86·7 per cent and 50 per cent, or $\sqrt{86\cdot6^2 + 50^2} = 100$
per cent.   It follows from this that 100 per cent current flows in
or out of the three-phase line at ABC, and gives a balanced circuit.

It will be seen from Fig. 71 that the current in BD is leading
its voltage by 30°, and that in DC is lagging its voltage 30°, while
at the same time they are in phase with the three-phase line voltage.

The lag and lead in the 100 per cent transformer involves an
increase in the k.v.a. capacity required, or lowers the power factor.
Thus if losses are neglected, the k.v.a. input and output on the
two-phase and three-phase side can be assumed equal, or if $E_{(s)}$ represents the two-phase voltage, $I_{(s)}$ the two-phase current, $E_{(p)}$ the

three-phase voltage and $I_{(p)}$ the three-phase current, the current flowing in or out of the two-phase side is given by

$$I_{(s)} = \frac{K.V.A.}{2} \times \frac{1}{E_{(s)}}$$

and that on the three-phase side

$$I_{(p)} = \frac{K.V.A.}{\sqrt{3}} \times \frac{1}{E_{(p)}}$$

With regard to the k.v.a. windings capacity, that on the two-phase side of both the 100 per cent and 86·7 per cent transformers is given by

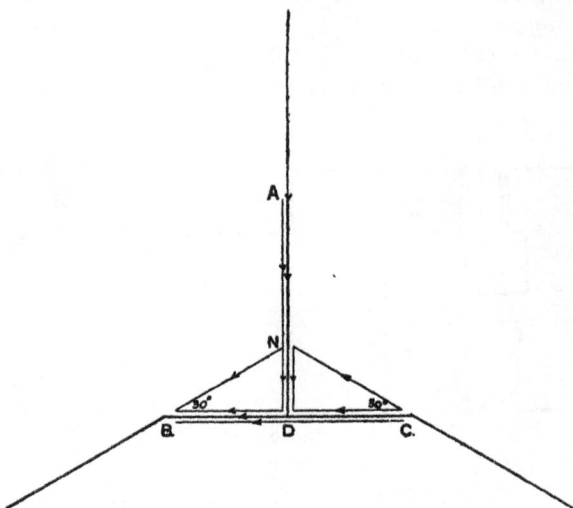

FIG. 171.

$$I_{(s)} \, E_{(s)} = \frac{K.V.A. \times E_{(s)}}{2 \times E_{(s)}} = 0\cdot5 \; K.V.A.$$

while on the three-phase side of the 100 per cent transformer we have

$$I_{(p)} \, E_{(p)} = \frac{K.V.A. \times E_{(p)}}{\sqrt{3} \times E_{(p)}} = 0\cdot578 \times K.V.A.$$

and on the three-phase of the 86·7 per cent transformer we have

$$I_{(p)} \times 0\cdot867 \, E_{(p)} = \frac{K.V.A. \times 0\cdot867 \, E_{(p)}}{\sqrt{3} \times E_{(p)}} = 0\cdot5 \times K.V.A.$$

The k.v.a. windings capacity of the 100 per cent transformer on the three-phase side is thus seen to be 15·5 per cent greater than in the 86·6 per cent transformer and consequently the 100 per cent transformer must be 7·75 per cent larger than a true transformer.

Reactance in Main Transformer.—In the main transformer (BC, Figs. 69 and 71), the current consists of two separate components, part in phase with the main volts and part in quadrature. As an example the following ratios of voltage and current in the two transformers may be taken :—

86·6 per cent transformer—100 amps., 86·6 volts.

Main transformer : $\sqrt{50^2 + 86\cdot6^2} = 100$ amps., 100 volts.

With regard to the influence of the 50 amps. component on the

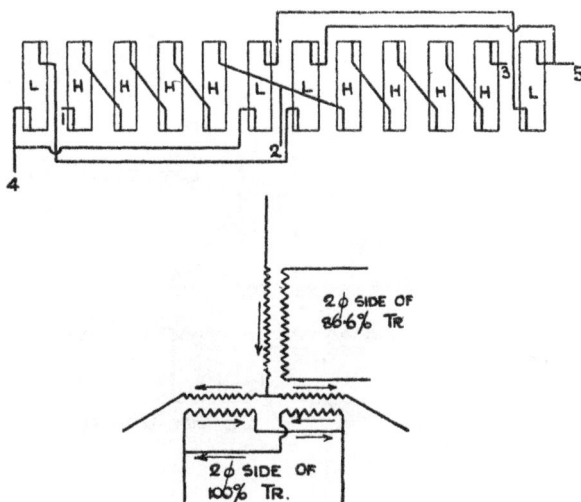

FIG. 72.

main transformer, the magnetic core remains unaffected if the three-phase coils are correctly designed and placed with reference to the two-phase windings ; however, in practice, there is a certain reactance between the two halves of the windings DB and DC due to the contra-currents passing along each from AD. Any reactance from this source will cause a voltage drop in quadrature with the current that produces it, and hence a voltage drop in phase with the main transformer voltage. The drop will affect the secondary voltage, so that the two-phase voltage from the main transformer will be lower than that from the 86·7 transformer and in value this reduction will be approximately equal to this reactance. The reduction and practical elimination of this reactance between the halves can be brought about by two methods :—

1. By interconnecting the three-phase coils, and so increasing the number of opposite groups. Coils of different line potential are brought adjacent to each other, the voltage between adjacent coils being half the line voltage. This method involves the use of extra insulation between the coils.

2. By connecting the two halves of the two-phase winding in parallel, as illustrated in Fig. 72, in which the arrows indicate the circulating current set up. This connexion provides a local circuit

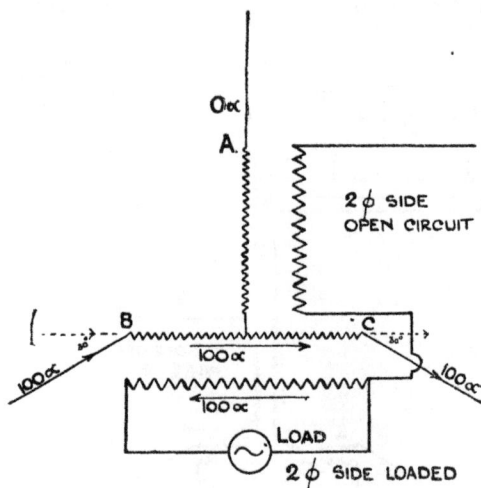

Fig. 73.

for circulating currents which eliminates the effect of the magnetic leakage.

This method obviously increases the current in the two-phase winding and hence, if the losses are to be kept to a practical minimum, extra copper section must be allowed.

Effect of Unbalanced Load.—In applying two-phase current for the operation of electric furnaces, there is a possibility of the load on one phase being greater than the other. To consider the two cases where one phase is fully loaded and the other idle.

1. With load wholly on the main transformer as in Fig. 73. In this case the main transformer only carries current, and the load on the three-phase side will be an ordinary single-phase load, while the current in the line at A is zero.

2. With load wholly on the 86·7 per cent transformer as in Fig. 74. In this case both transformers carry current, the 86·7 per cent transformer the full load circuit, and the main transformer the contra 50 per cent current.

If the windings of the three-phase side of the main transformer are correctly designed there will be no voltage drop. The contra-currents from the 86·7 per cent transformer winding encounter the resistance of the main transformer winding, but since this is positive and negative, and only of small value (approximately 14·4 per cent of the normal resistance drop) it is of no account. The current on the three-phase side here also is ordinary single-phase load in phase with the line voltage at A and leading and lagging the line voltage at C and B. In this case the lines are more evenly loaded

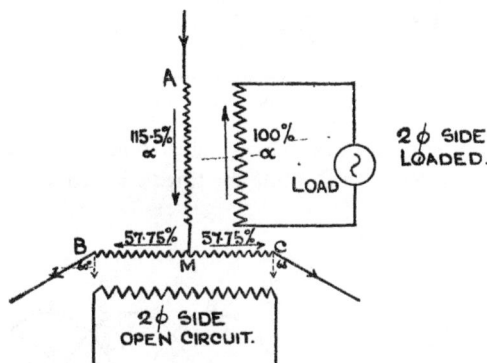

Fig. 74.

as far as the current is concerned, and the line loss will be lower than in case 1.

3. Loading across phases as in Fig. 75. In this condition the voltage across the load is EK, i.e. $\sqrt{2}$ times the two-phase voltage and the current, if load is at unity power factor, is in phase with EK. Taking the current on the two-phase side to be 100 per cent, then the current in AM is 115·5 per cent. The main transformer carries three currents; 57·75 per cent in direction MB; 57·75 per cent in direction MC; and 100 per cent in direction CB. Since all these currents are in phase, they result in 157·75 per cent flowing in MB, and 42·25 per cent in CM. This system of operation can be applied satisfactorily if the source of power can deal with the unbalanced load and the transformer can stand the heavy overload current of approximately 50 per cent in half of its winding.

Balanced Load.—The most usual arrangement of connexions for two-phase current when applied to electric furnace operation consists

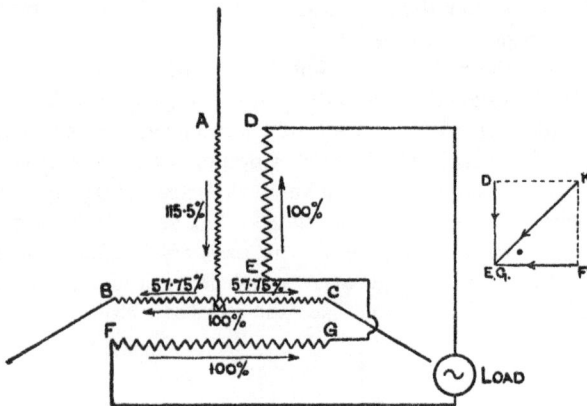

FIG. 75.

in the use of three electrodes. No. 1 electrode is connected to the point D (Fig. 75), No. 2 to F, while the third e lectrode is connected to

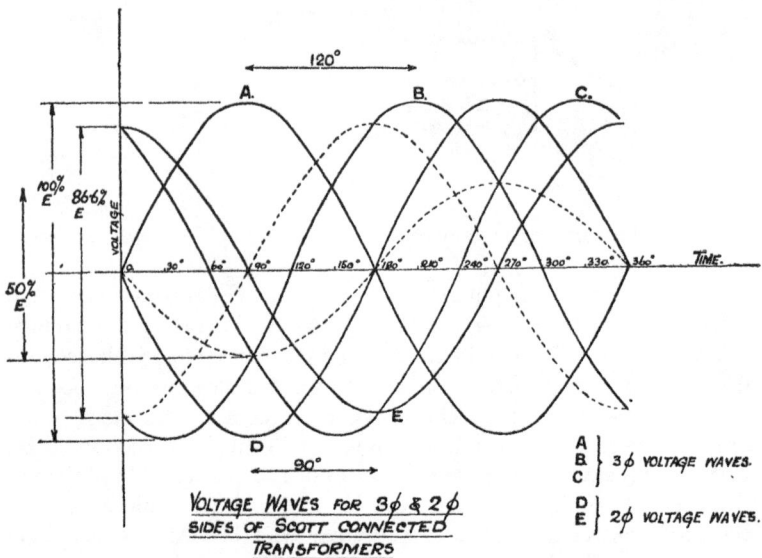

VOLTAGE WAVES FOR 3φ & 2φ SIDES OF SCOTT CONNECTED TRANSFORMERS

FIG. 76.

the common point of E and G. To obtain a balanced load on the three-phase supply, it is only necessary that the resistance between the electrodes D and EG should be equal to that between F and EG.

With this arrangement, as illustrated in Fig. 77, the two vertical electrodes, when in a state of balance, will each carry 100 amps.,

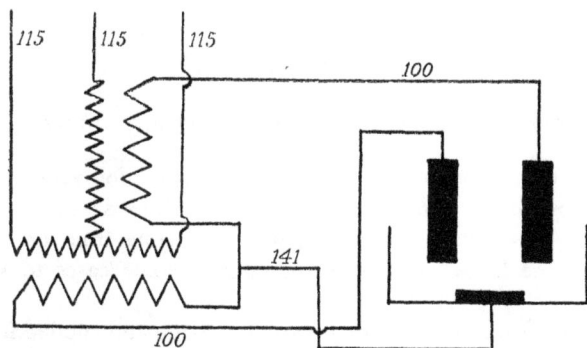

FIG. 77.

when the lower electrode carries 141 amps., while each circuit of the three-phase supply side will take 100 amps.

In some types, as in the Stobie furnace (p. 247), four electrodes are used, G and E being connected to separate electrodes. The condition for a balanced load in this case is that the resistance between D and E should be equal to that between F and G.

Fig. 76 is a cyclic diagram showing the formation of three-phase balanced voltage waves obtained from two-phase voltage waves.

SECTION V.

TRANSFORMERS FOR USE WITH ELECTRIC FURNACES.[1]

THE main characteristics of electric furnace operation which require special provision in transformer design are as follows :—

1. The low voltage required, generally ranging from 60 to 120 volts, and the large currents, amounting in some cases to 50,000 amps.

2. In the case of arc type of furnaces, large and sudden fluctuations of the load. These cause heavy mechanical and electrical strains to bear on the windings and necessitate a strong construction of windings, core, and leads and the presence of a certain amount of reactance as a safeguard during short circuits. The voltage surges on the high-tension side of transformers caused by the rapid fluctuations of load necessitate end-turn insulation and make it desirable that tappings for voltage regulation should be internal to the windings.

3. Heavy and frequent overloads must be allowed for and out-of-balance loads provided against in such a manner that the voltage does not fall in consequence.

For three-phase working a transformer is needed having three primary and three secondary coils wound on three connected cores or three separate transformers may be used.

Regulation of Voltage.—With arc furnaces a regulation of voltage is necessary according to the nature of the slag and furnace charge in the case of the direct arc and according to the degree of heat and amount of power to be applied. In the case of a steel furnace, the power factor varies through the change of the magnetic properties of the bath at different temperatures, whereby a lower power factor is obtained at lower temperatures leading to a voltage drop. During the refining stage the power required is less and hence the voltage of the supply can be reduced.

With a resistance furnace the increased conductivity of the charge with rising temperature has to be provided for.

[1] From article by J. L. Thompson, " The Metropolitan Vickers Gazette," 1919, 4, 237, 252.

A regulation of voltage is generally made in accordance with one of the two following principles :—

(a) *System of Tappings.*—In this method the voltage can be varied within certain limits by connecting the terminals to different connexions which lead to various parts of the windings and enable the voltage yielded to be varied in steps. For low voltages the tappings cannot be provided on the low-tension side, due to the few low-tension turns available and the very large currents involved. A typical design for this system is shown in Fig. 78. The high potential supply is applied at the ends of the windings 1 and 10, while at the centre a larger or smaller number of turns can be eliminated from the circuit by connecting 5 with 6, 4 with 7, 3 with 8, or 2 with 9, etc. On the low-tension side, a number of windings will, as a rule, be joined in parallel.

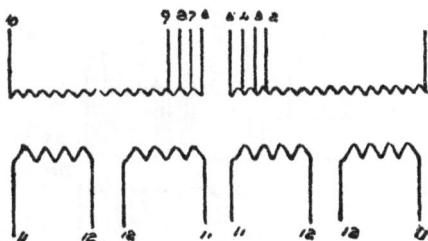

FIG. 78.

To change the voltage by this method a selector switch is provided, the arrangement of which is shown in Fig. 79. This switch is interlocked with the main oil switch so that it can only be operated when no current is circulating. A temporary suspension of the operation of the furnace is consequently necessary with this system.

In the connexions illustrated in Fig. 79 the transformer is supplied with three-phase current and yields two-phase by means of the system of Scott connexions described above (p. 68). The secondary windings on the low-tension side are wound in the form of two parallels for each of the two phases, the middle connexion forming the common return.

In the case of resistance furnaces such as those for carborundum and graphite, it is required to change the voltage during operation. For this purpose a system of connexions can be used as shown in Fig. 80. By the rotation of a cylindrical drum, a larger or smaller number of turns on the high-tension side can be short-circuited. In the case illustrated, the high-tension side is wound for delta-connexions and the low-tension for star-connexions.

(*b*) *Booster Regulator System.*—In this method the main transformer is wound for the lowest required working potential and in addition it is provided with an auxiliary winding connected in

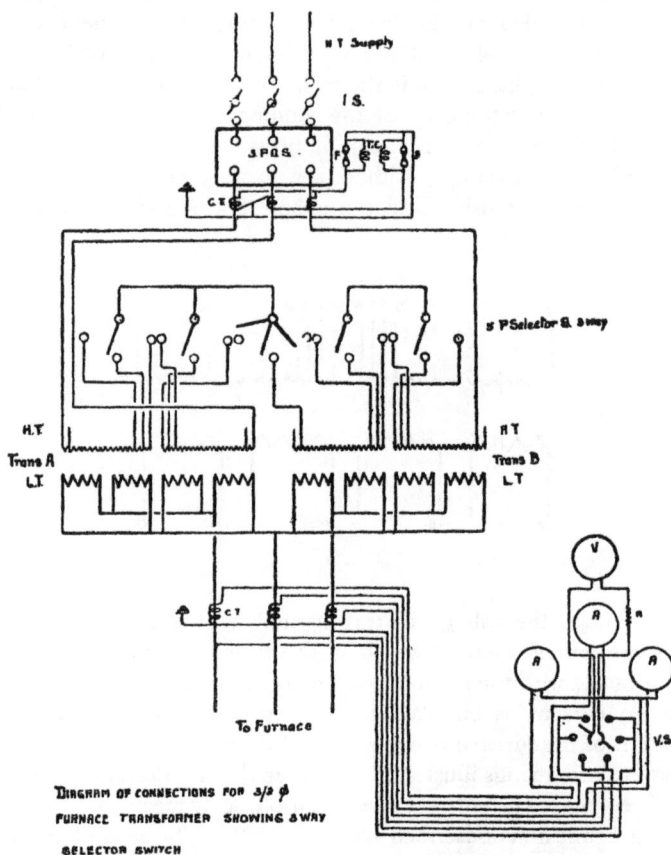

FIG. 79.

sections to a selector switch. This auxiliary winding feeds when required a booster inserted in the low-tension leads and thus increases the voltage of the low-tension winding according as the potential is applied to primary winding of the booster.

The auxiliary circuit may be arranged either to diminish or increase the potential and so give double the number of voltage variations. In this case the main transformers must be wound for the mean low-tension potential.

Fig. 81 shows a system of connexions used for transforming

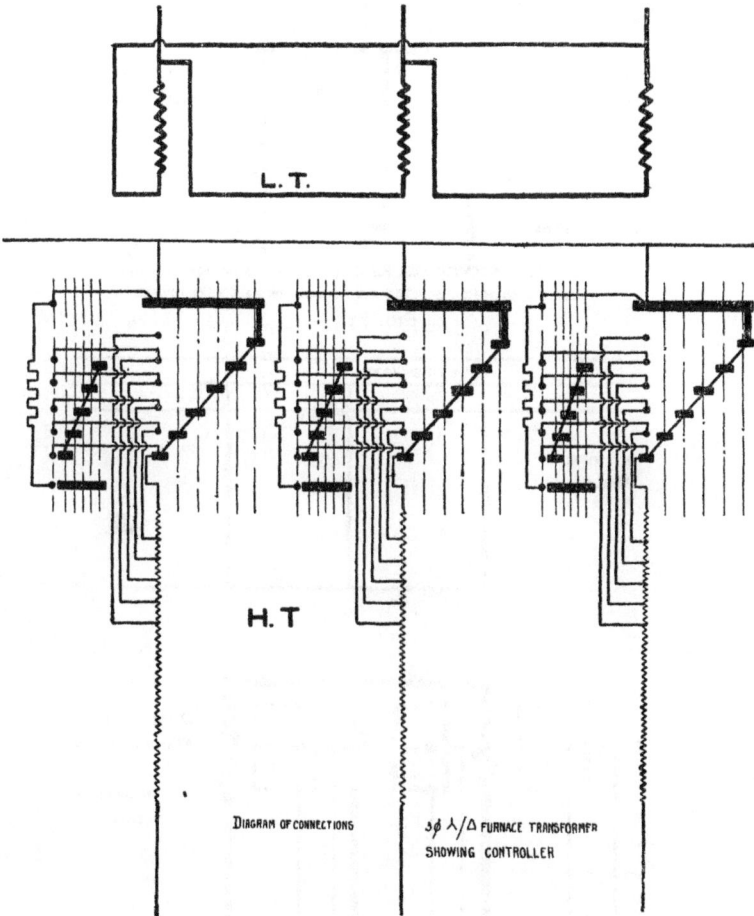

L. T.

H. T

DIAGRAM OF CONNECTIONS

3φ λ/Δ FURNACE TRANSFORMER
SHOWING CONTROLLER

FIG. 80.

from three-phase high-tension to two-phase low-tension, with a positive booster and auxiliary windings on the main transformer.

Fig. 82 shows a diagram of connexions for transforming high-tension three-phase star-connected current to star-connected low-tension by a transformer with a positive booster and auxiliary windings on the main transformer.

6

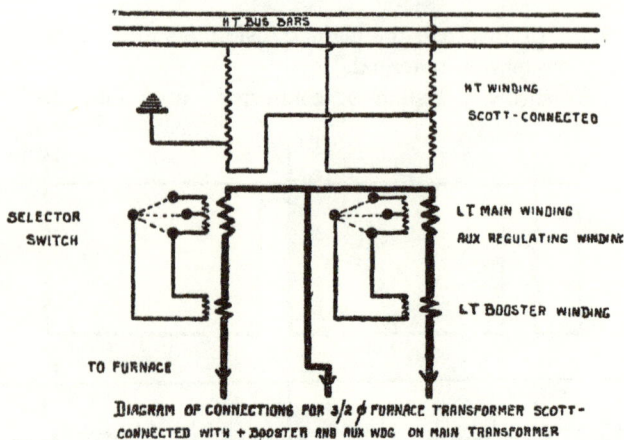

DIAGRAM OF CONNECTIONS FOR 3/2 φ FURNACE TRANSFORMER SCOTT-CONNECTED WITH + BOOSTER AND AUX WDG ON MAIN TRANSFORMER

FIG. 81.

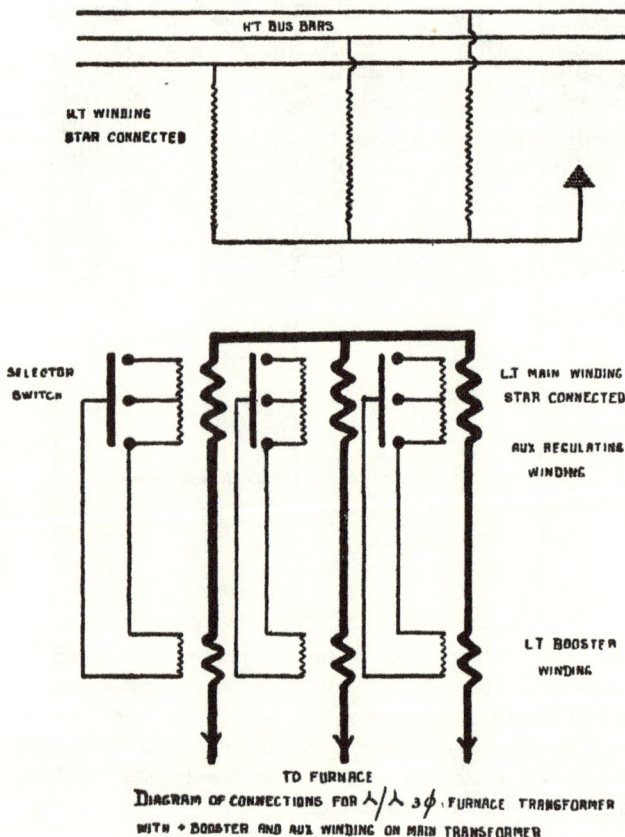

TO FURNACE

DIAGRAM OF CONNECTIONS FOR λ/λ 3φ FURNACE TRANSFORMER WITH + BOOSTER AND AUX WINDING ON MAIN TRANSFORMER

FIG. 82.

The advantages of this system employing auxiliary windings are that high potentials are not interrupted during operation, and connexions can be made more mechanically sound than where tappings are provided; the auxiliary switchgear need only be made for normal voltage; the balancing of load in the multiple low-tension windings is more easily maintained; if necessary, voltage change can be made without cutting off the load, the auxiliary winding acting as a choke coil during the change over.

Reactance.—With arc furnaces, a reactance within the range of 8 to 12 per cent is necessary for the purpose of steadying the arc and protecting the supply system. The reactance can be obtained by the grouping and arrangement of the high-tension and low-tension coils.

Heavy Current Leads.—Currents up to 6000 amps. can be led up to the furnace walls, along two heavy bars for single-phase, and three for two or three-phase and connected by flexible cables to the electrodes. A symmetrical arrangement is necessary in order to make the resistance and reactance of the leads equal on all phases, and thus ensure balanced voltages. For larger currents, however, in order to minimise the reactance of the circuit, multiple bus bars or groups of 4, 8, or 16 conductors should be used and interleaved so that different polarities are adjacent to each other.

SECTION VI.

THE MEASUREMENT OF HIGH TEMPERATURES.

In most chemical processes involving the use of high temperatures, it is generally of first importance to control accurately the temperature, a change of 25° to 50° C. having frequently a determining influence on the reactions produced. With fuel-heated furnaces, empirical methods of gauging temperatures are still mainly used, such as visual estimation from the colour of the light emitted or, in the method of Seger, by observing the softening points of cones of material of graded fusibility or by the old method, not yet altogether obsolete, of Wedgewood, consisting in measuring the permanent contraction which prepared clay cylinders undergo.[1] For more precise estimations, methods applied which are capable of the greatest accuracy consist in measuring either the change of resistance of a platinum wire or, by the thermo-electric method, the e.m.f. generated at the junction of two different metals such as platinum and an alloy of platinum with rhodium or ruthenium. For temperatures higher than about 1200° C., when, through the failure of gas-tight containing vessels, the gas thermometer cannot be applied, the temperature scale is itself defined in terms of the change of resistance of platinum with temperature, and is in consequence known as the "platinum scale". The assumption is made that the change of resistance with temperature follows the same relation as that which is derived at lower temperatures where a direct comparison can be made with a thermometer based on the change of volume of a gas. The thermo-electric values are derived by a comparison with the platinum resistance scale. For use with industrial furnaces, the above methods are attended by a number of disadvantages, in spite of the great precision of which they are capable. The indicating or recording instrument can be placed at any convenient place away from the furnace but the connecting wires must be efficiently insulated from each other and bad contacts avoided in the wires connecting the potentiometer with the heated wires. Serious inaccuracies may arise in practice through failure of these precautions. Unless efficiently

[1] "Trans. Faraday Soc.," 1917, **13**, 330.

protected, this type of pyrometer is liable to contamination and damage from the furnace materials. At temperatures above about 1200° C., since all refractories become porous, it is not possible in the presence of a reducing atmosphere to protect the wires against the deleterious action of certain vapours and gases. On account of the non-uniformity of temperature in different parts of the furnace which generally occurs, it is frequently necessary to determine the temperature in different parts of the furnace. With pyrometers in a fixed position, it is therefore necessary to instal a number thus introducing complications. At temperatures above 1500° C., such as are given in steel furnaces, this deterioration prevents the platinum type of pyrometer being used for long periods, while the melting-point of platinum limits its application under any circumstances to a temperature of 1750° C.

In laboratory work, use has been made of thermo-couples con-structed of iridium and iridium-ruthenium, but, on account of their comparative brittleness and high cost, these are hardly practicable in industrial furnaces.

For higher temperatures attempts have been made with some promise to apply combinations of graphite with amorphous carbon.[1]

RADIATION AND OPTICAL PYROMETERS.[2]

For the high temperatures utilised in many furnaces, methods for the estimation of the temperature based on changes in the properties of materials placed in contact with the charge are precluded. Systems so far developed in which no part of the measuring instrument is placed in the immediate vicinity of the furnace zone are based on the measure-ment of the energy radiated, either the total radiation, which is mostly of wave-lengths in the infra-red portion of the spectrum, or by determining photometrically the intensity of light of some par-ticular wave-length. A scale of temperatures has been derived for both of these cases, which is based on certain theoretical deductions in reference to the relation between radiant energy and temperature. Although this scale has only a theoretical basis and cannot at very high temperatures be directly compared with any other standard, yet, for practical purposes, it gives a basis which even if only arbitrary, can at all times be conveniently reproduced. By means of pyrometers which have now been constructed, the system

---

[1] Cf. A. Stansfield, "The Electric Furnace," 1914, p. 145.

[2] Cf. Waidner and Burgess, Bulletin No. 2, Bureau of Standards, Washing-ton, 1905; Burgess and Le Chatelier, "The Measurement of High Tempera-tures," J. Wiley & Sons, New York, 1912; "Trans. Faraday Soc.," 1917, **13**, 234.

enables temperature determinations to be made with a facility in
many respects superior to that with the thermo-electric method and
while for very high temperatures no alternative method is available,
for temperatures in some cases as low as 500° C. an accuracy is
obtainable, which, in general practice, is equal to that given by
any of the older methods.

The relation which is of particular advantage in this system of
pyrometry is the very large increase in intensity of radiation or
illumination which results from a small increase in temperature.

"BLACK-BODY" RADIATION.

An important property which it is necessary to take into
account as being liable to lead to large errors in temperature
estimations is that the radiation from an incandescent object
depends not only on its temperature but on the nature of the
substance and the condition of its surface. Different substances
at the same temperature may emit largely different quantities of
energy. In the case of polished platinum, for instance, the radiation of red light from this metal at its melting-point (1750°) is equal
to that from a substance which gives the full possible radiation,
known as "black-body radiation" at 1540° to 1545°, thus showing
a departure of 205° to 210°. The deviation from black-body radiation varies in different parts of the spectrum and thus gives rise to
what is known as *selective radiation*. The phenomenon of partial
radiation is related to the fact that when light falls on an object a
certain amount is absorbed or transmitted and the remainder reflected, and conversely when the body is rendered luminous, the
emitted light corresponds to the proportion formerly absorbed,
while the remainder is diffused, i.e. absorbed or retained. A case
where the emission and reflection factors are at a minimum is given
in the colourless gas flame which give no indication of temperature
on an optical pyrometer.

"*Black-body*" *Temperature.*—The expression *black-body temperature* is used to denote the apparent radiation or optical temperature
or the temperature an object would have when emitting the same
radiation if it behaved as a *black body* and emitted the full radiation
without retaining any. Thus in the case given above of polished
platinum, the *black-body* temperature for red light at the melting-
point of the metal is equivalent to 1540° to 1545°.

Kirchoff's Law.—A law of great importance in practical pyrometry, which was derived by Kirchoff, states that the conditions of

black-body radiation exist in the interior of any enclosure all the walls of which are at a uniform temperature. The light emitted from the interior of a furnace through a small opening is accordingly independent of the nature of the furnace interior and can be used as the basis of an accurate temperature estimation.

In accordance with Kirchoff's law, different materials such as magnesia, porcelain, platinum, and iron if heated to the same temperature in the open will emit very different amounts of light but if heated inside an enclosure of uniform temperature will all emit equal radiation by a process of continued reflection and absorption, and it follows that these materials will lose their contours when viewed against the bright background of the furnace walls.

In cases of hermetically sealed furnaces or where the interior is not at a uniform temperature, accurate estimations may be made locally by inserting a tube closed at its inner end and sighting the pyrometer on a beam of light emitted from the base and passing along the inner axis of the tube.

Radiation Law of Stefan and Boltzmann.—The relation between the total radiation (of all wave-lengths) and temperature has been derived by Stefan and Boltzmann, who formulated the law that the *total radiation is proportional to the fourth power of the absolute temperature.* Thus, if E denotes the radiation from a body at temperature T_1 to one at T_0, then

$$E = k \left(T_1{}^4 - T_0{}^4 \right).$$

Luminous Radiation.—The increase in luminous radiation from a heated body with rise of temperature was measured photometrically by Lummer and Kurlbaum, and the intensity of light emitted was found to satisfy the following formula :—

$$\frac{I_1}{I_2} = \left(\frac{T_1}{T_2} \right)$$

where I_1 and I_2 are the intensities of the light radiated at the temperatures T_1 and T_2 respectively, where the two temperatures are taken close together.

At 900° (abs.) x was found to have the value 30, and at 1900° the value 14. These measurements were further extended by Rasch, who found that the product Tx gives an approximately constant value, 25,000. The increase of radiation with temperature is also a function of the wave-length ; for shorter waves it is greater, and for longer waves less.

It is the very great increase in luminosity of a body on rise of temperature that makes it possible for experienced persons to gauge the temperature by mere observation of the colour.

Wien's Law.—The relation between the temperature and the intensity of the light of any particular wave-length can be expressed by a formula which was deduced by Wien.[1] According to this relation,

$$I = c_1 \lambda^{-5} \, e^{-\frac{c_2}{\lambda T}},$$

where I is the energy corresponding to wave length λ (region of spectrum λ to $\lambda + d\lambda$), T is the absolute temperature of the radiating black body, e is the base of the natural system of logarithms, and c and c_2 are constants.

This formula has been confirmed by experimental work conducted by Paschen[2] and by further deductions by Planck.[3]

The formula can be expressed in a simpler form as

$$\log I = K_1 - K_2 \frac{I}{T},$$

$$\text{where } K_1 = \log c_1 - 5 \log \lambda,$$

$$\text{and } K_2 = c_2 \frac{\log e}{\lambda}.$$

This is a linear relation between $\log I$ and $\frac{I}{T}$, so that it requires measurements of the intensity at two known temperatures only to calibrate an instrument. If the radiation is not black-body radiation, the same formula holds for calibration and use, but with different values of the constants K_1 and K_2.

CALIBRATION OF A RADIATION OR OPTICAL PYROMETER.

The readings of an optical pyrometer can be conveniently calibrated by sighting on to a heated substance which radiates as a *black body* or for which the amount of departure from this radiation is known and the temperature of which is simultaneously measured by some other standard method. For temperatures up to 1200° to 1300° C. the radiation from a heated enclosure may be made use of. The most convenient form of apparatus is an electric furnace, as shown in Fig. 83. The central porcelain tube is wound with platinum wire W (or wire consisting of some alloy of a high melting-point, such as " nichrome "), and the space between this and the outside case

---

[1] " Ber. K. Akad. Wiss." (Berlin), 1893, p. 55 ; " Ann. d. Phys.," 1896, **58**, 662.

[2] *Ibid.*, 1896, **58**, 455.   [3] *Ibid.*, 1900, **1**, 69, 719.

is filled with magnesia or kieselguhr. A plug of magnesia or other
refractory material is placed at P and is provided with two holes
to allow the wire of a thermo-junction to pass through, the junction
being made at K. The interior of the furnace can be slowly raised

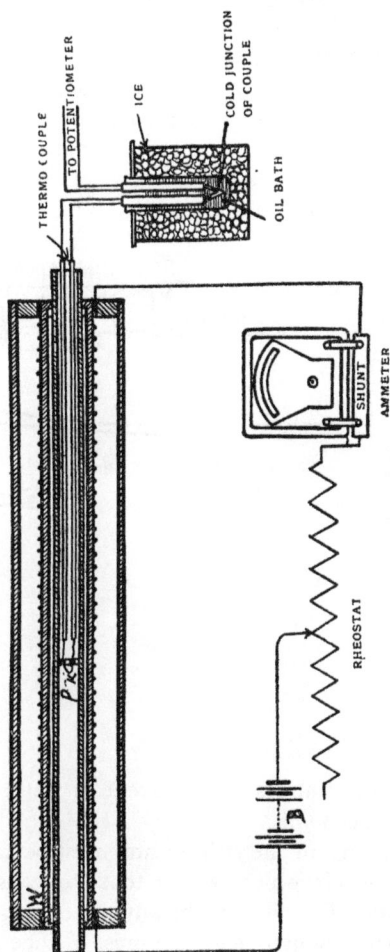

FIG. 83.

to incandescence by the passage of an electric current through the
wire coil. It is necessary that the temperature should be uniform
for some distance in the neighbourhood of the diaphragm and when
the thermo-junction wire is at the same temperature it becomes
invisible against the bright background.

The pyrometer is sighted into the porcelain tube and the temperature is given by the thermo-junction which is connected with a potentiometer or high-resistance galvanometer.

At higher temperatures a calibration can be made at certain fixed points by sighting the pyrometer on to a strip of pure platinum, rhodium or iridium, which is gradually heated to its melting-point by means of a carefully regulated current. The apparatus for this calibration is shown in Fig. 84. The brass tubes A and B, which can be cooled by water circulation, have small clamps attached at C and D. In this way a strip of metal S of 4 to 6 cm. long and about 4 mm. diameter, can be connected between the tubes, so as to form electrical contact. The current leads are

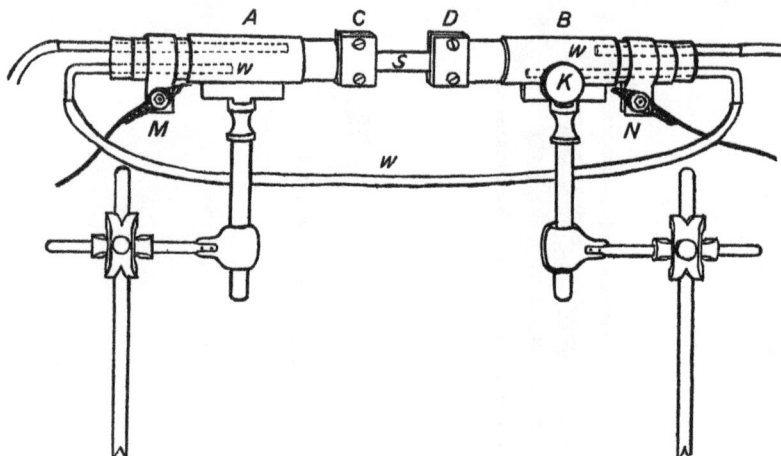

FIG. 84.

connected at M and N and a slow current of water is allowed to pass through the tubes w, w.

During the heating of the strip means must be taken to shield it from draughts of air which would tend to cause fluctuations of the temperature. The vicinity of any other bright light which would cause an error though reflection must also be avoided. The current through the strip is carefully regulated by a suitable rheostat and measured by an ammeter. Sometimes the metal does not heat uniformly and this is more apt to occur with a long than with a short strip. In this case, however, the place of highest temperature can easily be found by moving the pyrometer from side to side until sighted on the zone of highest temperature. The temperature is gradually and continuously raised and the pyrometer continually readjusted until the metal fuses.

With a pyrometer containing a polarising device it is necessary that the surface of the strip should always lie in a plane at right angles to the axis of the pyrometer. For this reason as the metal expands with the rising temperature, the extension should be taken up by rotating the screw K which works a rack and pinion.

For red light the apparent *black-body* temperature of platinum at its melting-point has been found by Waidner and Burgess [1] to be 1541°, while Holborn and Henning [2] found the *black-body* melting-point of platinum to be 1545°, that of rhodium, 1650°, and iridium 2000°.

Departure from Black-body Radiation.

Waidner and Burgess [3] measured the radiation from various substances at different temperatures and found the following values in the case of red light ($\lambda = 0.65 \mu$) :—

For platinum :

At 1750°(C.) departure from B.B. radiation corresponds to 209°(C.)				
,, 1500°	,,	,,	,,	126°
,, 1215°	,,	,,	,,	96°
,, 1064°	,,	,,	,,	91°
,, 782°	,,	,,	,,	65°
,, 723°	,,	,,	,,	57°

For iron oxide :

At 980° departure corresponds to	46°
,, 770° ,, ,,	23°

For fine grained Battersea crucible :

At 1050° departure corresponds to	54°
,, 770° ,, ,,	14°

For copper oxide :

At 1065° departure corresponds to	49°
,, 750° ,, ,,	25°

For unglazed porcelain :

At 730° departure corresponds to	30°

It has been shown that the radiation from graphite at 1250° C. is equal to the radiation from a *black body* to within 10° C.[4]

Holborn and Henning [5] made a series of measurements on the

---

[1] Loc. cit., p. 244.
[2] "Sitzungsber. K. Akad. Wiss.," Berlin, 1905, 12, 311.
[3] Loc. cit., p. 250.
[4] Greenwood, "Trans. Chem. Soc.," 1908, xciii., 1486.   [5] Loc. cit.

radiation from the noble metals at different temperatures, and conclude from these results that, for each method, $\dfrac{E_M}{E_T}$, the ratio of the light emitted by the metal to that which is emitted by a *black body* at the same temperature, is a constant for all temperatures.

In terms of this expression, it was found that platinum emits about $\frac{1}{3}$, gold $\frac{1}{8}$, and silver $\frac{1}{14}$ of the *black-body* radiation. The exact value of the coefficient varies with the wave-length, and becomes greater for light at the violet end of the spectrum.

According to Wien's law, the relation between the radiation and the wave-length is given by the expression

$$\log \frac{E_M}{E_T} = \log_{10} e \frac{k}{\lambda}\left(\frac{1}{T} - \frac{1}{M}\right)$$

where k is a constant which, for platinum, has the value 14,500.

Accordingly for a given wave-length

$$\frac{1}{M} - \frac{1}{T} = C$$

where T is the true temperature and M the apparent or optical temperature of the metal and C a constant for any particular metal.

The following values were found for $\dfrac{E_M}{E_T}$ and for C, with the different metals for red light :—

	$\dfrac{E_M}{E_T}$.	C.
Platinum	0·319	0·0000507
Gold	0·127	0·0000916
Silver	0·080	0·0001119

Total Radiation Pyrometers.

The Féry Pyrometer.[1]— The Féry is the only type of pyrometer based on the energy of total radiation which has been brought into industrial use. The construction of the original type of pyrometer is shown in Fig. 85, and contains a lens L whereby radiation from an incandescent body is focussed, by means of the rack and pinion adjustment P, on to a minute thermo-junction T, the temperature of which is thereby raised. The position of focus is judged by sighting through the eyepiece O of the telescope. A diaphragm DD gives a cone of rays of constant angular aperture independent of

[1] "Comptes rendus," 1902, **134**, 977.

the focussing and, for higher temperatures, the aperture can be reduced in order to keep the readings on the galvanometer scale.

The leads from the thermo-junction are connected with the terminals *aa'*, and the circuit is completed through a potentiometer or sensitive galvanometer or milli-ammeter, the readings of which bear a definite relation to the temperature of the source of radiation.

In applying the fourth-power law to the calibration of this pyro-meter, it is to be remembered that by far the largest part of the

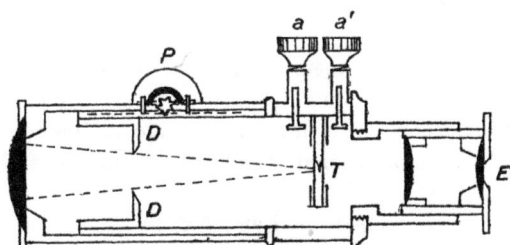

FIG. 85.

energy radiated from a heated body is in the infra-red part of the spectrum and that these rays are not transmitted through glass. By using a fluorite lens in the pyrometer, the radiations transmitted at temperatures above 900° C. are not appreciably diminished but, at lower temperatures, through the existence of an absorption band in the fluorite near 6 μ, the fourth-power law can no longer be assumed. For industrial use, this type of instrument is usually constructed with a glass lens, and can be calibrated by direct comparison with a thermo-couple or with a Féry pyrometer with a fluorite lens.

In a more recent type of pyrometer the use of a lens is obviated, and the radiations from the heated body are focussed on the junction by means of a concave mirror M placed behind the junction as shown in Fig. 86 and the position of which can be adjusted by the focussing screw P. The pyrometer is combined with a telescope with an opti-cal system, so arranged that on sighting through O, the field of view appears to be divided into halves. If the instrument is in focus, the two halves will form a continuous field, while, if not in focus, the two halves will be displaced relatively to each other. The

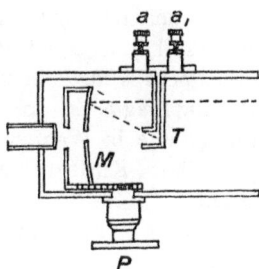

FIG. 86.

thermo-junction appears as a dark circle in the centre of the field and the image of the incandescent body must be sufficiently large to cover the whole of this circle. Temperatures as low as 500° C. can be measured by this form of pyrometer.

The relation between the galvanometer readings $R_1 R_2$ and temperatures $T_1 T_2$ with this instrument are given by the expression—

$$T_2 = T_1 \sqrt[4]{\frac{R_2}{R_1}}$$

so that errors in the galvanometer readings are divided by 4 when reduced to temperatures.

In the case of a pyrometer calibrated in accordance with the above equation, it was found by Féry that in a direct comparison with a thermo-electric pyrometer over a range of temperatures from 850° to 1450° C., the agreement at all temperatures was within 0·1 to 1·0 per cent.

OPTICAL PYROMETRY.[1]

Optical pyrometers are generally based on the photometric measurement of the intensity of the light emitted by incandescent bodies. In most pyrometers, by the use of a prism or red glass, only the red radiation is selected for the following reasons : Firstly, in order to eliminate inequalities in the colour of the incandescent source at different temperatures, and enable comparison to be made between the field under measurement and the standard source of the same colour, and secondly, to enable the measurement to be made at the lower temperatures, when the red light is the first to become visible.

In accordance with the principle at first suggested by Becquerel, optical pyrometers are constructed on the principle of taking a standard of light and varying the brightness of the photometric field illuminated by the red light emitted by the body under observation until it is equal in intensity to that of the standard. This adjustment is made in a pyrometer of Le Chatelier by means of an iris diaphragm, in one of Féry by means of an absorbing wedge, and in one by Wanner by a polarising device, while in pyrometers of Holborn-Kurlbaum and Morse, the standard light itself may be varied.

LE CHATELIER OPTICAL PYROMETER.[2]

The construction of this type of pyrometer is shown in Figs. 87 and 88. Light from the central portion of the flame of a small

---

[1] Cf. Waidner and Burgess, loc. cit.
[2] " Comptes Rendus," 1892, **114**, 214, 470.

standard lamp L, passes through the lens L_1 and the beam is reflected from the edge of a mirror M inclined at 45° and brought to a focus in the eyepiece, in front of which a red glass is inserted. At the same time, light from the source under measurement is admitted to the objective after passing through an iris diaphragm, *a*, or a Féry absorbing prism, *b*. The beam is focussed by the objective and, in the eyepiece, forms a red field immediately beside and touching the first. An adjustment of the diaphragm in front of the objective is made until the two red fields are of uniform brightness. For very high temperatures, one or more absorbing

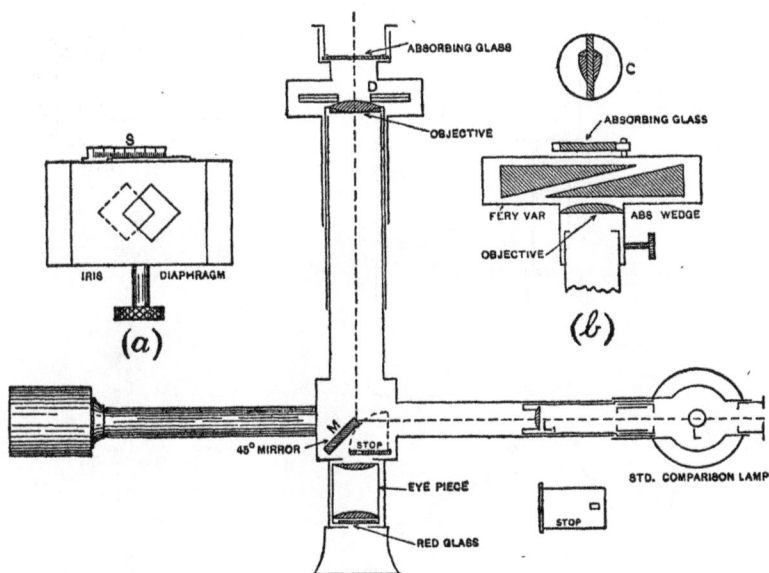

Fig. 87.

glasses, whose coefficient of absorption is known, are placed before the diaphragm and for relatively low temperatures before the comparison lamp. The opening of the diaphragm is read on an attached scale, the square of whose readings is a measure of the intensity of the light from the incandescent body.

The calibration of this pyrometer may be made in terms of Wien's equation (p. 88) according to which

$$\log I = K_1 - K_2 \frac{I}{T}.$$

A knowledge of the readings of the instrument for two temperatures

is thus sufficient for its calibration for all temperatures. The relation between log I and $\frac{I}{T}$ which is linear can be plotted graphically,
and from this a table constructed giving T in terms of or scale
readings.

The precision obtainable in the readings with this instrument
has been found to be within 5° at 1000° C.

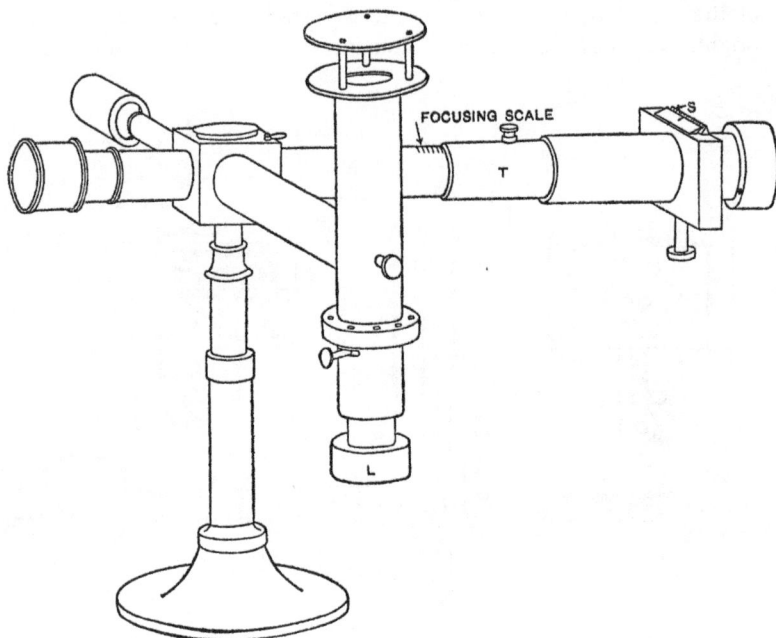

FIG. 88.

THE WANNER PYROMETER.[1]

In the Wanner pyrometer, as usually constructed, red light of
wave length equal to 0·65 μ is selected from the source under
measurement and compared photometrically with similar light
from a standard lamp of constant intensity. Rays of light from
the two sources enter the pyrometer side by side and after passing through a direct-vision spectroscope the red rays from each
traverse a Rochon prism where they are polarised in directions at
right angles to each other. A Nicol analyser is placed before the

[1] " Phys. Zeitsch.," 1902, 3, 112; " Ber. K. Akad. Wiss.," Berlin, 1893, 55·

eyepiece and by rotating the prism, the intensity of the light from one source is increased and that from the other diminished. On sighting through the pyrometer the light from the two sources is found to illuminate the two respective halves of the field and the Nicol analyser is rotated until the two halves are equally bright, when the field appears to be uniform. The degree of rotation necessary to produce this uniformity gives a measure of the temperature of the incandescent object.

As shown in Fig. 89, light from the standard lamp enters the slit S_1, after diffuse reflection from a right-angled prism placed before S_1, while light from the object whose temperature is to be measured enters the slit S_2. These two beams of light are rendered parallel by the lens O_1 and, by means of the direct vision spectroscope, the rays are resolved into a continuous spectrum. The Rochon prism R then separates each beam into two beams polarised at right angles to one another. The light now traverses the biprism B, which is constructed with such an angle that the red light of two images only of opposite polarity falls on the slit D. This biprism increases the number of images to 8 but the remaining 6 are cut off from the slit D. If, now, the analyser A is inclined at an angle of 45° to each polarised beam and if the intensities of the light entering S_1 and S_2 are equal, then the field seen through the eyepiece will appear of uniform brightness. If, however, instead of 45°, the Nicol is placed at an angle ϕ the following relation holds :—

$$\frac{I}{I_0} = \tan^2\phi \quad . \quad . \quad . \quad (a)$$

where I equals the intensity of light through S_2 from the body whose temperature is to be measured and I_0 equals that through S_1 from the standard source.

Since monochromatic light is used, the basis of the calibration of the Wanner pyrometer is given by Wien's equation

$$\log \frac{I}{I_0} = \frac{c_2}{\lambda}\log_{10}e\left(\frac{I}{T_0} - \frac{I}{T}\right) \quad . \quad . \quad . \quad (b)$$

FIG. 89.

7

where T_0 and T are the absolute *black-body* temperatures of the standard source and the incandescent body respectively.

For a *black-body* the constant c_2 has been evaluated at 14,500, and as the instrument is usually constructed $\lambda = 0.656\ \mu$.

According to this formula, all the data necessary for the calibration of the instrument is a knowledge of the apparent *black-body* temperature of the standard source, and the reading of the analyser at the normal point when $I = I_0$. Any other temperature can then be calculated by use of the above two formulæ (*a* and *b*).

It is very important to be able always to reproduce the comparison light at S_2 of constant intensity. A small filament lamp is used for this purpose and the current passing through is controlled by a regulating resistance and is measured by an ammeter. It is necessary to check frequently and readjust the intensity of this comparison source on account of the effect of "ageing" with the filament lamp. For this standardisation, use is made of an amyl acetate lamp, burning under carefully regulated conditions, on the flame of which the pyrometer is sighted. The Nicol prism is set at an angle of 45° to the two beams and the electric comparison lamp is adjusted in brightness until the light from both sources gives a uniform field. The apparent or *black-body* temperature of the flame is evaluated by sighting the pyrometer with the Nicol in the same position on to a substance which gives *black-body* radiation and whose temperature can be regulated and measured by some standard means. In Fig. 90, the incandescent lamp is seen at C, the leads to the battery at L connecting through the ammeter A and the regulating resistance R. A ground glass diffusing screen is interposed at D between the flame and the slit of the pyrometer, while a gauge F enables the height of the flame to be adjusted to a definite height. It is important that the ground glass diffusing screen should always be placed in the same position relative to the flame and the slit S. The amyl acetate lamp should be allowed to burn for five minutes before the readings are taken and the height of the flame adjusted, so that the tip is about 1 mm. above the level of the gauge F. For subsequent measurements of temperature, the pyrometer is removed from the stand and sighted directly on to the incandescent body. By referring to tables, the temperature is ascertained directly from the reading of the pyrometer scale indicated by the pointer P.

The sensitiveness of the instrument is greatest when the analyser is at the normal position or 45°, which corresponds to the *black-body* temperature of the amyl acetate flame when viewed through the diffusing screen. This *apparent* temperature is 1150° C., as usually constructed. A measurement with the pyrometer can generally be made to within 0·1 scale division, which is equivalent to an accuracy of 1° at 1000°, 2° at 1500°, and 7° at 1800° C. For higher temperatures a smoked glass of known absorption is placed in front of the slit S, and another tabulation of the temperatures, corresponding to the different positions of the Nicol, is provided.

The coefficient of absorption of the smoked glass is given by the following formula :—

$$\log_{10}K = \log_{10}\frac{I_1}{I_2} = \frac{c_2}{\lambda} \log_{10}e\left(\frac{1}{T_2} - \frac{1}{T_1}\right)$$

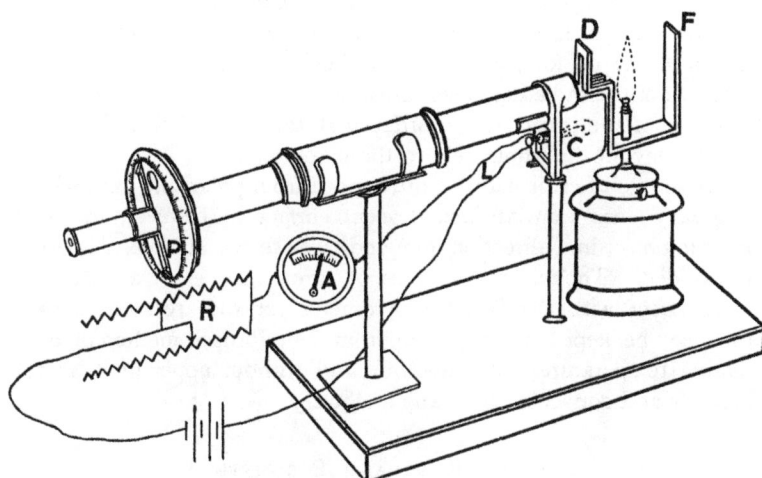

FIG. 90.

where K is the coefficient of absorption and T_1 and T_2 are the apparent *black-body* temperatures (abs.) sighted first without and then with the smoked glass.

Since $c_2 = 14,500$ and $\lambda = 0·65\ \mu$, then the above formula reduces to

$$\log_{10}K = 9983\left(\frac{1}{T_2} - \frac{1}{T_1}\right)$$

or

$$\frac{1}{T_2} - \frac{1}{T_1} = \frac{\log_{10}K}{9983} = C.$$

This constant C can thus be calculated by measuring the absorption at any one temperature.

In general, if a temperature reading is to be taken from light

7 *

which has passed through any medium which causes absorption, the correction to be applied can be determined by evaluating C in the above equation. A temperature reading is taken of an incandescent object at any constant temperature in the first place directly, and secondly, after the light has passed through the absorbing medium. The relation between real (T_1) and apparent (T_2) temperatures will then be given by the equation

$$\frac{1}{T_1} = \frac{1}{T_2} - C.$$

Range and Accuracy.—The optical system employed in the Wanner pyrometer leads to a great loss of light which prevents the measurement of temperatures below about 900° C.

No image of the object is formed in the eyepiece which obviates the necessity of focussing with varying distance from the object. The size of the radiating object must bear a minimum relation to the distance of the pyrometer in order that the rays of light from the source may fill the whole field in the pyrometer.

A source of error may be introduced from polarised light which is usually given from an incandescent surface. However, by viewing the object in a direction perpendicular to the surface, this error is avoided. Though the sources of error may exert a relatively large effect with the Wanner pyrometer yet with reasonable care they may be kept low when the pyrometer affords a method of estimating temperatures from 900° C. and without upper limit, as accurately and conveniently as any method.

HOLBORN-KURLBAUM PYROMETER.[1]

The pyrometer of Holborn-Kurlbaum, as with a similar type by Morse, is operated on the principle of interposing an electrically heated lamp filament in the field of view of the radiating body. The current passing through the filament is adjusted until the brightness becomes equal to the source when the outline of the lamp disappears against the background. The eye is particularly sensitive in recognising equality of brightness of two surfaces one in front of the other, and the system thus provides a delicate means of estimating temperatures.

In the Holborn-Kurlbaum pyrometer, as seen in Fig. 91, a small 4-volt filament lamp L is mounted in the focal plane of the objective and of the eyepiece of a telescope provided with stops DDD and a

[1] " Ann. d. Phys.," 1903, **10**, 225.

focussing screw S. An image of the object to be measured is brought into the plane AC and the current through the filament from the cell B is adjusted by means of the rheostat until the lamp filament becomes invisible against the luminous background. The temperature is then obtained from the readings of the milli-ammeter which has been previously calibrated against *black-body* radiation. For temperatures above 800° C. and particularly higher temperatures, readings are more conveniently taken by the use of one or more monochromatic red glasses in front of the eye-piece. For very high temperatures, it

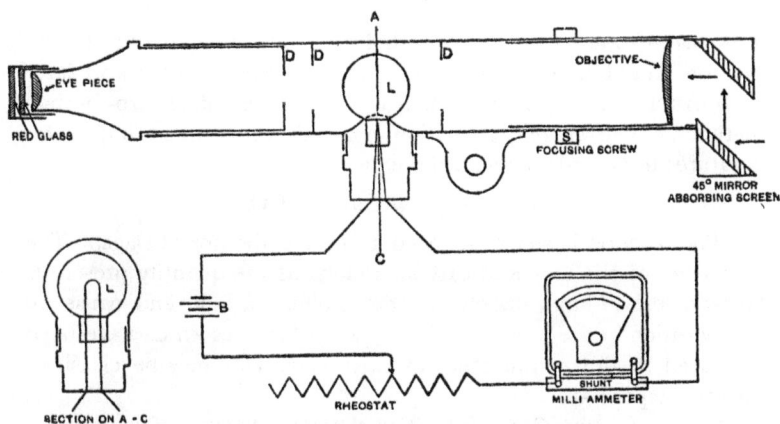

FIG. 91.

is necessary to interpose absorbing glasses or mirrors before the objective and calibrating to determine the absorption constant.

The relation between the temperature of the filament and the current circulating is expressed satisfactorily by the equation

$$C = a + bt + ct^2.$$

The lower limit at which readings can be taken with this instrument is about 600° C. and up to 1350° C., an agreement with temperatures indicated by a thermo-junction is obtained to within 5° and may be as good as 2° at 1500° C.

SECTION VII.

CALCIUM CARBIDE.

THE manufacture of calcium carbide was first carried out at the Willson Aluminium works at Spray, N.C., U.S.A., in 1891 and later at Merriton and may be regarded as the pioneer of electro-thermal methods. The chemical reaction proceeds smoothly at high temperatures in accordance with the equation

$$CaO + 3C = CaC_2 + CO.$$

The reaction is simple on account of the absence of slags. The impurities of the lime and carbon, so long as the quantity present is below a certain limit, dissolve in the molten carbide and cause no complication in the process. The types of furnaces which have been employed in the manufacture of calcium carbide may be classified into two types :—

1. *Intermittent Operation.*—For the manufacture of massive or block carbide, which is built up during the operation of the furnace, and removed after cooling, by dismantling the furnace.

2. *Continuous Operation.*—Apart from a few special designs, continuous operation is achieved by tapping the carbide from the furnace at intervals without interrupting the progress of the smelting and allowing the product to solidify and cool in a separate receptacle.

Intermittent furnaces, such as that of Willson (Fig. 92), are generally formed from an iron case with a conducting base of carbon. The Willson furnace was portable, being mounted on a truck on rails and wheeled into position in an enclosure in which the electrode is suspended vertically. The carbon monoxide generated in the reaction can be withdrawn. In this type of furnace, the charge is gradually admitted by shovelling or by hoppers, so as to keep pace with the production of carbide. The electrode is accordingly raised until the molten charge reaches the top of the furnace. In the case of the Willson process the mounted furnace is then wheeled away to be replaced by an empty one and, after cooling to some extent, the charge is removed from the case by tipping.

The main disadvantages of this intermittent method of working

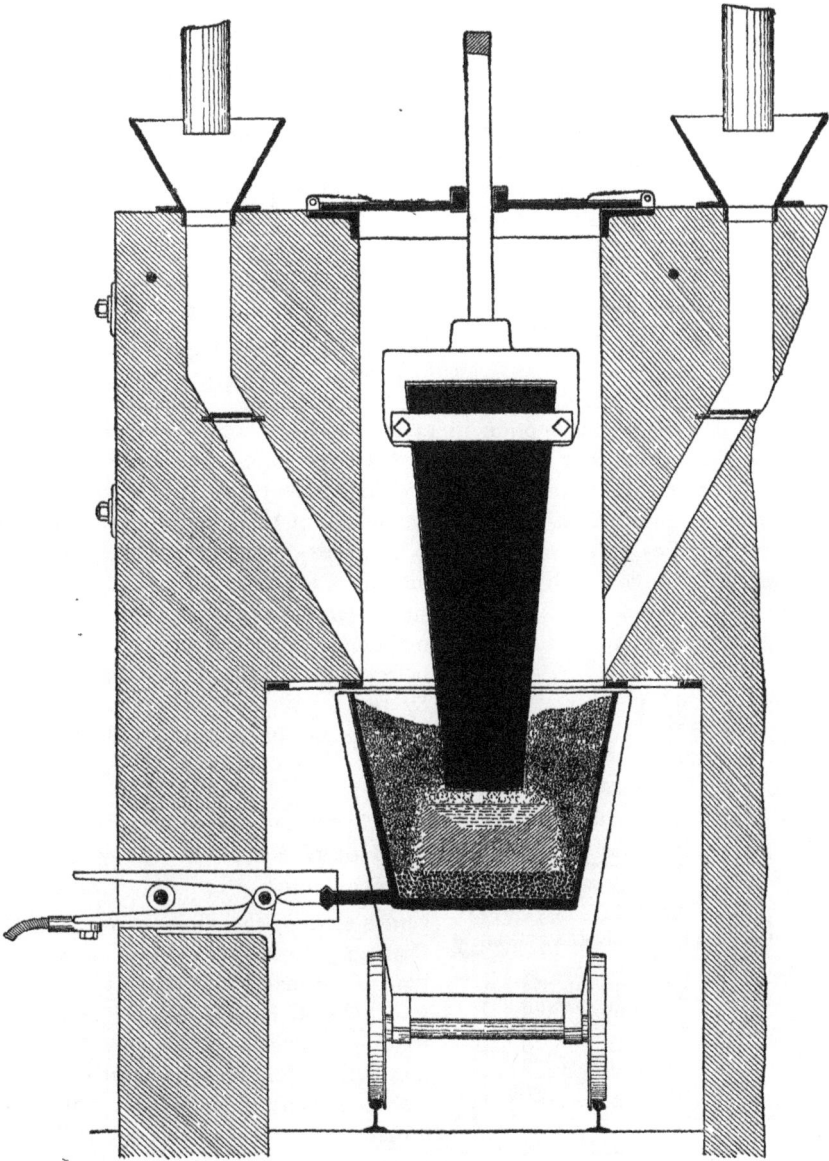

Fig. 92.

are the uneconomical utilisation of heat and loss of material by dispersal through evolution of vapour.   The conversion of the charge is not complete and in many instances the yield only amounts to half the charge added, while the rest consists of unchanged material. On removing the block of carbide the admixed carbon for the most part takes fire, while the lime is hydrated by weathering so that it cannot as efficiently be used again.   The consumption of materials for 1 ton of carbide was found to be as follows :—

Material.	Weight Consumed (Kilos).	Theoretical.
Coke	800 to 1050 ⎫	
Or Anthracite . .	800 to 1050 ⎬	562
Or Wood Charcoal .	1200 to 1500 ⎭	
Lime	1080 to 1200	875
Electrodes . . .	40 to 70	

The power consumption in the Willson furnaces was originally 100 h.p., which was later increased to 500 h.p.   The later type was provided with two electrodes, which at first gave two separate blocks of carbide, but in a later modification these were caused to fuse together.   The two electrodes were joined in series, and electrical contact with the base of the furnace thus avoided.   A voltage of from 40 to 80 was usually employed on each furnace. Advantages of the intermittent method are that the product is of better quality than that obtained by the tapping process, and the furnaces can be run intermittently to suit the power supply.

CONTINUOUS PROCESSES :
(*a*) HORRY ROTATING FURNACE.

The Willson Company was superseded by the Union Carbide Company at Niagara Falls.   The type of furnace at first adopted was that of Horry, which, by a rotating device, combined the production of massive carbide with continuity of operation.   This furnace, as shown in Figs. 93 and 94, is of iron and circular, with a recessed rim on the top of which are bolted segmental wings 24 inches deep.   The carbon electrodes are enclosed in a coni-

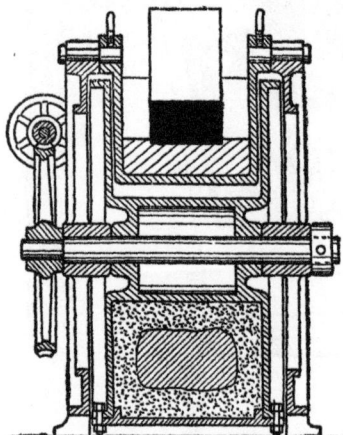

FIG. 93.

cal or cylindrical shaft which serves to lead off the evolved gas and impinge on the charge of lime and carbon contained in the recess. As molten carbide is produced more charge is admitted down the cylindrical shaft and the furnace is slowly rotated while plates are bolted on to the rim to keep pace, a complete rotation being effected in twenty-four hours. The carbide solidifies in cakes 6 to 9 inches thick and by removing the segmental plates is taken off the wheel comparatively cold at the opposite side. The electrodes used are 6 inches in diameter and placed 9 inches apart. The furnace is 8 feet diameter, 3 feet wide, and consumes 3500 amps. at 110

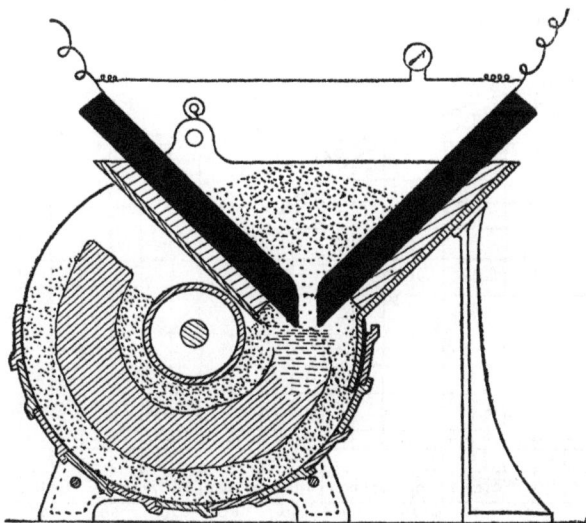

FIG. 94.

volts, or 500 h.p. The output is 2 tons per day. The materials used are burnt lime and ground coke in the proportion of 1 : 3.

This type of furnace has now been superseded by the tapping method.

(b) TAPPING PROCESSES.

All large furnaces at present in use operate continuously by provision for removing the carbide in the molten condition by periodical tapping and allowing the product to solidify and cool in receptacles which are withdrawn from the neighbourhood of the furnace. In this method, the furnaces are rectangular in form with open top for the admission of the charge. The electrodes

consist of two or three blocks of carbon, according to the nature
of current employed and suspended adjacent to each other above
the furnace charge.   The molten carbide as produced collects
in a common hearth underneath, or in three separate hearths under
each electrode and is removed through tapping holes at the base of
the furnace enclosure.   This opening is usually stopped by a cone
of clay, or solidified carbide.

The disadvantages of this system are that on account of the high
melting-point of the carbide, difficulty is often experienced in un-
stopping the tapping hole and, in any case, to increase the fusibility
of the product, it is necessary to employ an excess of lime in the
charge to lower the melting-point.   In this way a product is ob-

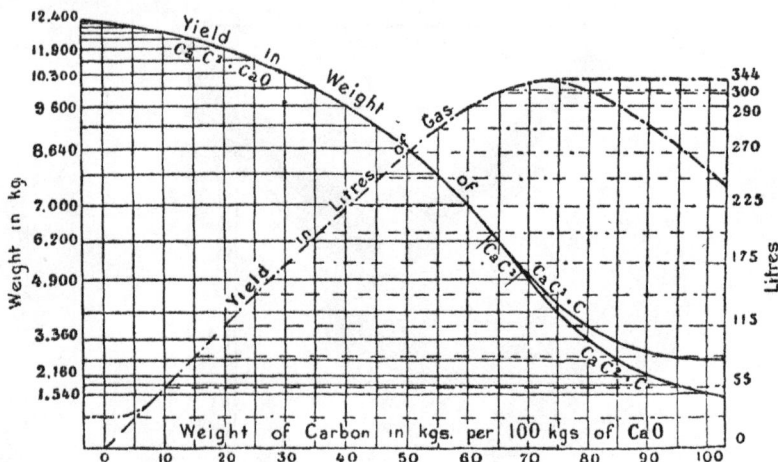

FIG. 95.

tained which yields a smaller volume of acetylene for unit weight
than with "block" carbide.   A further disadvantage sometimes occurring with this form of furnace is the accumulation of slag and
impurities which cling to the side and gradually fill up the furnace.

Fig. 95 reproduces a chart due to Keller showing the influence
of the composition of the charge on the power expenditure for a
given output of product and the amount of acetylene yielded by the
material.

The vertical ordinates on the left give the output per 24 kw.
hours, and on the right the yield in litres of acetylene of 1 kg. of
the product.

Output and Consumption of Materials.—Furnaces now constructed are from 3500 to 6000 kw. capacity, and generally work with three-phase alternating current at about twenty-five periods, and with a power factor of 0·9. The consumption of energy per metric ton of carbide produced amounts from 0·45 to 0·5 kw. year (8400 to 8500 hours) for a large scale plant.

Electrodes.—Electrodes of good quality carbon or of artificial graphite are used in the manufacture of carbide. The consumption of carbon electrodes when a high quality coke is used for the carbide melt is approximately one-third lower than that when anthracite is employed. In the latter case the consumption amounts to from 0·02 to 0·03 metric ton of electrodes per metric ton of carbide produced.

The materials needed for the production of 1 ton of carbide, in the best modern practice, are: 1·7 ton of limestone, 0·2 to 0·3 ton of coal for burning the limestone and 0·62 ton of anthracite, which can be replaced by coke.

Production Costs.—An estimate of the Nitrogen Products Committee[1] of the probable production cost of 1 ton of carbide on a pre-war basis in the case of a large scale factory is as follows:—

	£
Lime (burnt), 950 kg. at £0·75 per metric ton .	0·712
Anthracite 620 kg. at £0·8 ,, ,, .	0·496
Electrodes 25 kg. at £15 ,, ,, .	0·375
Repairs	0·226
Labour and expenses	1·087
Depreciation @ 8 per cent on £3 (Capital charges)	0·240
Interest @ 5 ,, ,, ,,	0·150
	3·286

This estimate is exclusive of cost of power and packing.

Power Cost.—The following table indicates in a comparative manner the influence of the cost of energy on the production costs of carbide:—

Production Costs.	£ Per Metric Ton of Carbide. Cost of Energy per kw. Year.		
	£2·0.	£3·0.	£3·75.
Chemical costs, excluding power	£ 3·286	£ 3·286	£ 3·286
Cost of energy (0·5 kw. yr.) .	1·000	1·500	1·875
Cost at factory per metric ton carbide, exclusive of packing .	4·286	4·786	5·161

[1] "Report of the Nitrogen Products Committee," p. 254. London, H.M. Stationery Office, 1920.

The figure of £2 and £3 per kw. year respectively represent the price of energy from average Norwegian water-power installations. The figure of £3·75 per kw. year of 8540 hours is an estimate on a pre-war basis of the cost at which it is estimated power could be obtained in this country from a steam power-station of 100,000 kw. maximum demand. The production costs with energy at higher figures can be readily determined from the above table, since each additional £1 per kw. year increases the manufacturing costs by £0·5 per metric ton of carbide. The packing cost, on a pre-war basis, is estimated to amount to £1·5 per metric ton. The charge, however, would not be incurred in the case of the utilisation of the carbide locally, as in the manufacture of cyanamide.

RAW MATERIALS IN CALCIUM CARBIDE MANUFACTURE.

In applying carbide for the generation of acetylene, it is very necessary that certain impurities should not be present. With regard to the lime, the important points are: (1) Suitability for burning. The lime must be such as to calcine without falling into powder. During the smelting process the gas evolution leads to a dispersal and waste of material present as powder, so that both the lime and carbon are used in the form of lumps 5 to 8 cm. diameter. The lime should be of a dense variety in preference to the crystalline form and should not crumble during any part of the process.

The natural impurities should not exceed 4 to 5 per cent of the burnt lime. These impurities will be reduced and for the most part transformed into carbides dissolving in the calcium carbide. The melting and reduction of these impurities causes a useless expenditure of current. Certain impurities such as magnesia and alumina make the carbide less fusible and so hinder the tapping and cause the formation of deleterious crusts and slags which hinder the raising of the electrodes. More than 2 per cent $MgO + Al_2O_3$ is not permissible for this reason.

Silica is not so harmful, as it forms calcium silicide, or in the presence of an excess of carbon, silicon. If magnesia and alumina are absent, the silica may amount to 5 to 6 per cent without injuring the product.

The iron content of lime is usually very low and in small quantities does not impair the product but only causes a small energy loss. The formation of iron silicide (ferro-silicon) is always to be guarded against, as being harmful to the condition of the furnace. When present in the carbide this compound is further liable, on account of its hardness, to damage the apparatus used for crushing. It has also been considered that the presence of ferro-silicon has been the cause of explosions obtained with calcium carbide.

Other impurities are detrimental to the quality of the acetylene when used for illuminating purposes, the most dangerous of these being phosphorus, which may be derived from phosphates present in the lime. The amount of phosphine present in the gas should not exceed 0·04 per cent. A larger quantity increases the risk of spontaneous ignition. The lime used in the process should not contain more than 0·006 per cent of this element. In the early days of the use of acetylene as an illuminant, the frequent explosions which occurred were probably due to the presence of this gas.

Arsenic is of less importance, as occurring more rarely, and in smaller quantities, than phosphorus in limestone.

Sulphur is a disagreeable but not a dangerous impurity in the lime used as raw material. The element occurs in the form of sulphate, which is reduced to sulphide in the furnace, and generates hydrogen sulphide together with the acetylene. On combustion this gives rise to sulphur dioxide. However, in the reaction between calcium carbide and water, the lime which is formed acts as a fairly efficient absorbent of hydrogen sulphide, so that this impurity is not encountered to any large extent in the gas. The presence of sulphur in the carbon used is of greater importance than that in the lime.

Carbon.—Carbon used in the charge is in one of the following forms :—

(*a*) *Anthracite.*—High quality anthracite containing not more than about 5 per cent ash is required for making carbide to be used for lighting or welding purposes. This provides the densest form of carbon and is very suitable for the large furnaces now used.

(*b*) *Coke.*—This form of carbon contains the largest amount of ash, which, if present above a certain amount, acts detrimentally on the carbide process. The quantity of ash present should not for this reason exceed 8 per cent and must be free from phosphorus. The amount of impurities permissible depends on those present in the lime. The content of moisture is also a drawback with coke. The material is, however, largely used for carbide manufacture and possesses some advantages, one being that the consumption of electrodes is in this case considerably lower than with anthracite.

(*c*) *Retort Carbon.*—This is the purest form of carbon which is occasionally applied in this manufacture.

(*d*) *Wood Charcoal.*—This has a low ash content and contains few impurities. Being voluminous it offers good contact with the lime, possesses a low electrical conductivity and gives a better product. On the other hand, its disadvantages are the high price, large consumption through access of air, the large space it occupies and danger of fire.

Manufacture of Calcium Carbide in Great Britain.

The manufacture of calcium carbide was commenced at Foyers in 1896 by the Acetylene Illuminating Company who acquired the Willson patent rights and leased power from the British Aluminium Company. During the boom which was caused by the interest at first attracted by acetylene as an illuminant, a number of other factories were erected in Great Britain, but on the decline of this industry in the years 1899 and 1900, these works all ceased operations with the exception of a small plant at Askeaton, near Limerick. Use was made here of a water-power generating 400 h.p. and the plant is still said to be in operation. The furnaces have a capacity of about 3000 amps. at 100 volts. A second similar plant is in operation in Ireland.

"British Carbide Furnaces."

In 1908 a carbide factory was erected at Thornhill, Yorkshire. The power was steam-generated and taken from the Yorkshire Electric Power Company. The works were designed for an output of about 2500 tons of calcium carbide per year. In 1916, in order to provide an increased output for war requirements, this plant was transferred and constructed on a considerably larger scale at Clayton, Manchester. With the assistance of a Government subsidy, a plant was constructed to give an output of 18,000 tons per annum. The Company is known as the British Carbide Furnaces. The production at these works in the early part of 1920 amounted to 12 tons per diem. Three single-phase furnaces were in operation with a total power consumption of 2000 kw. The furnaces are of rectangular form and the product is tapped in the molten condition.

British Cellulose Company, Spondon, near Derby.

A calcium carbide factory was, during the war, erected at Spondon, near Derby, at the works of the British Cellulose Company. The carbide produced was mainly applied to the synthetic preparation of acetic acid and cellulose acetate needed in aeroplane construction and other organic compounds.

The steam power-house which was specially constructed for this work is situated on the banks of the Derwent, from whence the cooling water for the condensers is supplied. Two turbo-generators of 6000 kw. each are now installed and the design of the building is arranged to allow of extension and the installation of future units of double the capacity. Each furnace is designed for a normal capacity of 6000 kw. and two are now actually in operation. The walls and base of the furnaces are constructed of firebrick covered with a layer of tamped-in carbon. Each of the three electrodes which are suspended above the open top of the furnace consists of six blocks of carbon 6 feet in length and, when new, covered for the greater part by a sheet-iron casing to protect from oxidation. The total cross section of each composite electrode is about 2 feet 6 inches by 3 feet 6 inches. The electrical connexions are made through copper clamps projecting above the head of the electrodes and cooled by water circulation. The charge consists of a mixture of burnt lime and coke and the fused carbide is tapped every hour through separate vents under each electrode. An average load factor is obtained of 90 per cent which is practically the same as the time factor.

The electrodes used are made in a special factory in these works; the material used consists of a mixture of anthracite, retort carbon, and the stumps of old electrodes. This is ground to a very fine powder, then placed in incorporating machines with screw agitators, in vessels which are steam-jacketed and heated. Molten pitch of about 1 or 2 per cent in weight is admitted and after thorough incorporation the product is withdrawn as a pasty mass, compressed in a mould in a hydraulic press and forced out by means of a plunger. The electrodes are then stoved by placing in brick enclosures, covering with small carbon to protect from oxidation and heating to over 1000° C., the total duration of each heat from beginning to end being about ten days. The main quantity of carbide produced is now being exported for use mainly in the generation of acetylene.

The production of synthetic products from acetylene at Spondon is, at the same time, being considerably extended.

MANUFACTURE OF CALCIUM CARBIDE IN NORWAY.

The manufacture of calcium carbide on a large scale was commenced at Odda, in Norway, in 1908, as a development of the Alby Carbide Company in Sweden, and constructed to produce 32,000 tons of calcium carbide per annum. A works was at the same time

installed on an adjacent site for the manufacture of calcium cyanamide by an associated company, which utilised the product from the
carbide factory as the starting-point for the cyanamide process (see
p. 137).

Odda is situated at the most southern point of the Hardanger
Fjord and is supplied with power generated at Tyssedal, 3 miles
north of Odda, on the Sorfjord, one of the branches of the Hardanger Fjord. The current is generated from water power by units
of 12,000 kilovolt-ampere generators, at 12,500 volts, 25 periods,
3-phase; and is transmitted at this voltage by means of aluminium
cables to the works.

A total power of 22,000 h.p. was originally employed at Odda
for the manufacture of carbide and cyanamide. In 1914 the power
applied was increased to 50,000 h.p. and the output of carbide to
85,000 tons per annum, 57,000 tons of which were delivered to the
cyanamide factory.

The raw materials used in the manufacture of the carbide are
imported, viz. in the installation of 1914, 50,000 tons of anthracite
per annum from Wales and 150,000 tons of limestone from the
company's quarries at various places in Norway and gas coal or coke
for burning the lime from England, though in the new furnaces the
waste gases produced in the reaction are utilised for this purpose.
The installation consisted originally of twelve furnaces, each of 1400
k.v.a. capacity and producing from 7 to 8 tons of carbide per day.
The installation of 1914 contains ten furnaces of 3000 k.v.a. capacity,
the production from each being 16 to 18 tons per day. The electrodes weigh 4 tons each. The furnaces are charged with lime and
anthracite through hoppers operated by hydraulic rams. An electrode
can, it is stated, be replaced by a new one in five to ten minutes.
The hot gases are led away and utilised in firing the lime-kilns.
The top of the furnaces is enclosed except for a small space to
admit the electrodes, and only allows a slight vapour to escape.

Current is supplied to each of the ten furnaces from a single-
phase core-type transformer which is air-cooled. The carbide is
tapped from the furnaces into iron crucibles standing on small
waggons and moved into the cooling house situated between the
two furnace houses. The carbide blocks are taken out of the
crucibles by overhead electric cranes and are then crushed, sorted,
and packed or conveyed to the cyanamide factory.

The Nitrogen Products and Carbide Company, which also includes the North-Western Cyanamide Company (*vide* p. 137), has now
amalgamated with the Alby United Factories, Ltd. In 1919 the

works at Odda were supplied with 65,000 h.p., while the earlier factory in Sweden has been extended to utilise 9000 h.p. for the purpose of carbide and cyanamide manufacture.

The Alby Company has acquired extensive rights amounting to a total of about one million horse-power in Norway and in Iceland for the generation of electrical power. A development for use in the manufacture of calcium carbide and cyanamide has been commenced at Sundalena, Norway. This water-power (Aura) will ultimately yield 250,000 h.p. The manufacture of cyanides is also being investigated by this company.

MANUFACTURE OF CALCIUM CARBIDE IN AMERICA FOR
MUNITIONS.

In 1918 a large plant was constructed at Muscle Shoals, Alabama, by the Air Nitrates Corporation, under the direction of the U.S. Government and as part of the state enterprise for the provision of nitrates required for munitions.

A steam-power plant was installed for the immediate operation of the process, but the situation was selected in order to utilise later the large amount of water-power available in the district. The erection of the works was begun in 1917 for the manufacture of calcium carbide and cyanamide (see also p. 138). In 1918 twelve furnaces were brought into operation for the preparation of carbide, with a daily capacity each of 50 tons or an aggregate output of 200,000 tons per annum.

Description of Furnaces.—The carbide furnaces are rectangular and each is 12 feet wide, 19¼ feet long and 29 feet interior depth. The electrodes consist of carbon blocks of 16 square inches section, and 6½ feet length. Three of these blocks are clamped together to form one electrode. Each electrode complete with clamps weighs 3171 kilos. The connexions are provided with water-cooling. Three of these assembled electrodes are suspended vertically in the furnace and take from a three-phase supply a current of 15,000 to 20,000 amps. at a voltage of 150. Each group of transformers supplying a furnace has a capacity of 8325 kw.

The electrode consumption is 31 kilos per ton of carbide. The bed of the furnace is provided with a lining of electrode graphite and above this, a layer of sand and tar. To commence the heating, a layer of coke is fed in, then the charge is emptied on to the coke and the current applied. The charge consists of a mixture of lime and coke in the proportion of 453 kilos of lime to 280 kilos of

8

coke. The limestone used is obtained from the district and contains 98·2 per cent $CaCO_3$, 0·97 per cent $MgCO_3$, 0·49 per cent SiO_2, 0·30 per cent AL_2O_3, and Fe_2O_3, and 0·07 per cent water. The first tapping of the furnace is made six hours after the start and subsequently every three-quarters of an hour, the furnace being kept filled to the top. The tapping is made by piercing the tap-hole by the formation of an arc from a portable electrode which is connected to one of the main electrodes and the molten carbide flows into containers of ½ ton capacity. After the tapping the hole is stopped with the help of pulverised carbide. When cold, the carbide is conveyed to the cyanamide building (*vide* p. 138).

SECTION VIII.

THE SYNTHESIS OF NITROGEN COMPOUNDS FROM THE ATMOSPHERE.

THE FIXATION OF ATMOSPHERIC NITROGEN.[1]

THE fundamental importance of combined nitrogen in agriculture has long been recognised, and the war has served to emphasise its vital bearing in relation to munitions. The main source of combined nitrogen in nature appears to be the fixation of atmospheric nitrogen direct through the aid of bacteria either acting alone or in symbiosis with leguminous plants. The amount of combined nitrogen thus rendered available is far from adequate for meeting the food requirements of the world, and as the nitrogen does not accumulate to any extent in the soil, the question of the world's supplies of nitrogenous fertilisers is seen to be of the first importance. Nitrogen is also an essential constituent of nearly all high explosives and propellants, and of many products which play an important part in industry under normal conditions. Before the war the industrial demand for combined nitrogen was quite small in comparison with the agricultural demand. To provide for national defence, very large reserves of nitrogen compounds are required which can be diverted from agriculture to the production of munitions. Until recent times the main sources of nitrogen compounds were nitrate of soda, vast natural deposits of which occur in Chile ; ammonium sulphate, obtained as a by-product in the destructive distillation of coal and shale ; and waste materials of organic origin.

Nitrogen gas constitutes about 75 per cent by weight of the atmosphere, and it is calculated that the air over a single square mile of land contains about 20,000,000 tons, equivalent to about thirty times the quantity of combined nitrogen contained in the world's production of Chile nitrate and ammonium sulphate in the year 1913.

[1] " Metall. and Chem. Engineering," 1911, **9,** 73, 99 ; " Final Report Nitrogen Products Committee," 1920.

The first modern nitrogen fixation process to be established on a commercial basis was the arc process, in which, at the high temperature of the high tension arc, nitrogen and oxygen are caused to combine.

In the early years of the nineteenth century, numerous attempts were made, both in this country and elsewhere, to manufacture cyanides and ferro-cyanides by fixing atmospheric nitrogen with the aid of mixtures of carbon and alkalies or alkaline earths heated to a high temperature.   Although such cyanide processes have hitherto failed to attain to commercial success, important results have accrued from the researches undertaken in connexion with them, the study of the behaviour of metallic carbides towards nitrogen gas being directly responsible for the technical development of the calcium cyanamide process.

The third of the established fixation processes resulted from the researches of Haber and others upon the synthesis of ammonia from hydrogen and nitrogen at high pressures and temperatures by the action of catalysts, and was brought into commercial operation in Germany in the year prior to the war.

In a fourth method the union of atmospheric nitrogen and oxygen is brought about by the high temperature transitorily produced in the explosion of a mixture of a hydrocarbon gas and air or oxygen. In this method, which has been developed industrially by Hausser, the mixture is fired electrically in a bomb, and the small amount of nitric oxide is removed and converted into nitric acid.

THE ARC PROCESS.

The possibility of causing the union of nitrogen and oxygen was discovered by Cavendish as early as 1785.   These gases were found to combine under the influence of electric sparks, yielding oxides of nitrogen.   This method was made use of by Lord Rayleigh[1] as a means of separating nitrogen from atmospheric air in the course of his work on the preparation of argon.

The union of nitrogen and oxygen takes place in accordance with the equation

$$N_2 + O_2 \rightleftarrows 2 NO$$

a reaction, which, at any given temperature, proceeds to an equilibrium stage, where there exists a fixed ratio between the quantity of nitric oxide and that of oxygen and nitrogen.   As the reaction leading to the formation of nitric oxide is accompanied by the absorption

[1] "Journ. Chem. Soc.," 1897, 71, 181.

of heat, it follows from the laws of thermodynamics that the higher the temperature the larger the proportion of nitric oxide at the equilibrium stage. The temperature of the high-tension arc is higher than out of the low-tension carbon arc, but has not been estimated with any accuracy.

Measurement of Equilibrium Values.—In attempting to determine the equilibrium values at definite high temperatures, the main difficulty encountered is that the whole of the nitric oxide present in the system at the high temperature cannot be cooled down to room temperature without undergoing some decomposition into its elements at intermediate temperatures. An approximation can only be reached by causing as rapid a cooling as possible of the products of reaction on withdrawing from the heated zone. In industrial practice, in order to obtain the highest yields, it is for this same reason necessary to arrange for rapid cooling to a temperature at which the *rate* of decomposition of the nitric oxide is inappreciable. The concentration of nitric oxide in equilibrium with air at high temperatures has been estimated by Nernst and others to have the values given in the following table. An interesting calculation has been made on thermo-dynamical grounds of the theoretical values by taking into account the heat of reaction and the change of the specific heats with temperature, the results of which are included in the table below :—

Temperature.	Per Cent of NO in Air. Observed.	Per Cent of NO in Air. Calculated.
1538° C. (1811° abs.) . .	0·37	0·35
1760° C. (2033° abs.) . .	0·64	0·67
1922° C. (2195° abs.) . .	0·97	0·98
2402° C. (2675° abs.) . .	2·23	2·35

Rapidity of Dissociation.—In addition to the equilibrium value, a factor of determining importance is the rapidity of dissociation. Below a temperature of 1500° this is very slight, but increases very rapidly with the temperature. In industrial plants, by arranging for a rapid circulation of the gases through the zone of reaction, yields can be obtained of 1·5 to 2 per cent NO, when working with air. These values correspond to equilibria at average working temperatures of 2200° to 2400° C.

Haber's Theory of Nitric Oxide Synthesis.—The view has been put forward by Haber that the formation of nitric oxide is not determined solely by the thermal equilibrium, but is also controlled by an ionising influence in the border of the arc flame. Experiments were made by the use of quartz tubes, which were cooled externally, and the discharge was passed through the tubes so as to fill the whole of the bore. By passing air through the tube under the reduced pressure of 100 mm. and exposing it to a discharge of 0·165 amps., a concentration of 10 per cent nitric oxide was obtained, while with a gas composed of equal volumes of nitrogen and oxygen, a product resulted containing 14·5 per cent of nitric oxide. The temperature is inferred to be not above the melting-point of iridium (2300° C.), and a concentration of nitric oxide is obtained several times higher than the value deduced from the Nernst equation.

With regard to the efficiency of processes working with gases at reduced pressure, the best yields obtained in the system of operation employed by Haber amounted to 57 grm. of HNO_3 per kw. hour, with a concentration of 3·5 per cent nitric oxide. By altering the conditions of the experiment, the yield has finally been increased to 90 grm.

INDUSTRIAL PROCESSES FOR NITROGEN FIXATION.

The industrial application of the electrical fixation of nitrogen was first carefully worked out by McDougal and Howles.[1] By means of a high-tension alternating arc a yield was obtained of 300 grm. of nitric acid per 12 h.p. hours, whilst with a mixture of two volumes of oxygen to one of nitrogen, the yield rose to 590 grm. per 12 h.p. hours.

This reaction was next applied with more favourable results by Bradley and Lovejoy[2] at Niagara Falls. An apparatus was designed by them to provide the greatest possible length of arc so as to bring it into contact with the maximum amount of air, and also to provide for sudden cooling once the products of the reaction had formed. In the apparatus used arcs were formed between projecting electrodes arranged on a rotating framework and stationary platinum electrodes in the walls of the case. A company known as the Atmospheric Products Company was formed, and a small trial factory built at Niagara Falls. The undertakings did not, however, meet with success, probably on account of the somewhat complicated nature of the apparatus, and its cost of erection and maintenance per kw. capacity being disproportionately large.

---

[1] " Proc. Manchester Lit. and Phil. Soc.," 1900, 44, part 4, pp. 1-19.
[2] " Electrochem. Ind.," 1903, 1, 20, 100.

BIRKELAND-EYDE PROCESS.

The Birkeland-Eyde method which was developed in Norway is the process which has been most extensively applied for the synthesis of nitric oxide.   An alternating arc is formed between metal electrodes arranged equatorially to the poles of an electro-magnet through which a constant direct current is passed.   As each arc forms it extends radially outwards under the influence of the magnetic field, until its resistance becomes sufficiently high to cause the formation of a second arc, when the first breaks.   A series of arcs thus follow each other in rapid succession, and as shown in Fig. 96 give two half discs of flame,

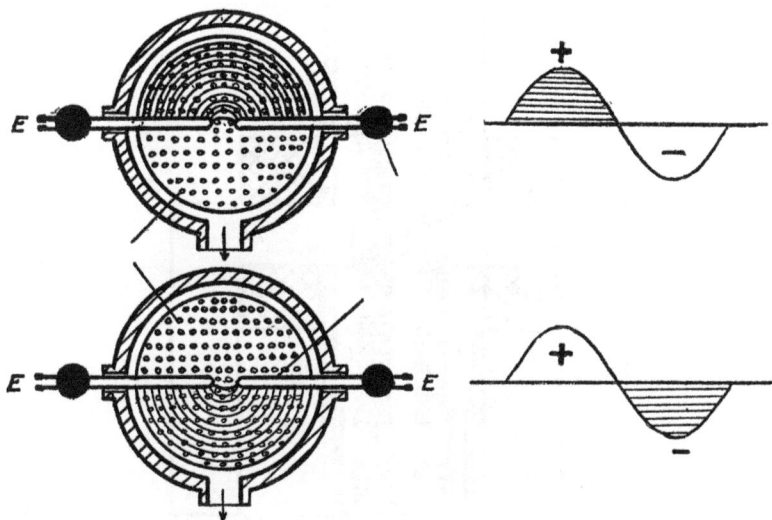

FIG. 96.

which alternately rise and break in the top half and bottom half of the reaction chamber according to the polarity of the phase.   The discharge thus has the appearance of a luminous wheel filling the space in a narrow circular chamber of refractory material.   The electrodes (EE, Fig. 96) are of copper tubes cooled by water circulation supported on ball insulators, and separated at a distance of about 1 cm. The walls of the furnace chamber are formed of firebrick, the temperature of which, on account of the rapid circulation of air, does not exceed 700° to 1200° C. in different types of furnaces.   Air is admitted through holes in the refractory lining and, after striking the arc flame at right angles, is rapidly withdrawn from the periphery of the reaction chamber.

In Fig. 97 the magnets are shown at M and the electrodes, which are aligned at right angles to the paper, at E.   Air enters through the pipes P and after passing through holes in the refractory lining L, traverses the arc zone and is withdrawn at B.   In the earliest furnaces, a power of 500 kw. was expended, a disc flame obtained measuring 6½ feet diameter, and a concentration of 1·5 per cent. NO obtained, with a yield of 500 kg. HNO_3 per kw. year. The furnaces are now constructed of 3000 and 5000 kw. capacity. A photograph of the interior of a Birkeland-Eyde furnace is shown in Plate IV.

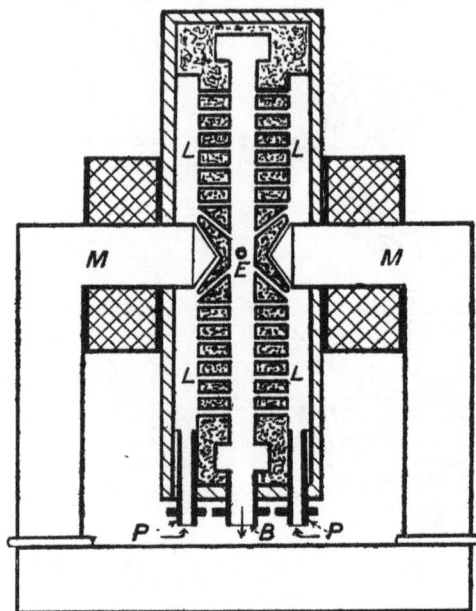

FIG. 97.

MANUFACTURE OF NITRIC ACID IN NORWAY.

Experiments with the Birkeland-Eyde furnace were made early in 1903 with a furnace using a current of 3 h.p.   These were followed later in the same year by a small factory with 150 h.p. at Ankerlokken, near Christiania.   The following year a plant using 1000 h.p. was opened near Arendal.   In 1905 the works at Notodden were commenced and 2500 h.p. utilised.   The works were enlarged in 1907 to utilise 40,000 h.p., while in 1911 a total of 50,000 h.p. was in use.   The enterprise now forms part of the *Norsk Hydro-Elektrisk Kvaelstofaktieselskab* of which company Dr.

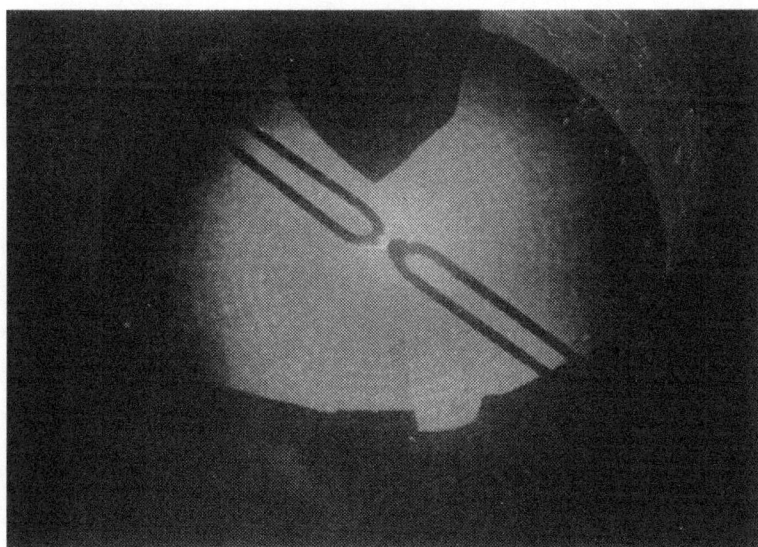

PLATE IV.—THE BIRKELAND-EYDE ARC.
(The Norwegian Hydro-Electric Nitrogen Company.)

PLATE V.—NITRIC ACID ABSORPTION TOWERS AT RJUKAN.

(The Norwegian Hydro-Electric Nitrogen Company.)

S. Eyde is the president. Other branches of the factory have been installed at Vemork and Saaheim, Rjukan. In all cases the plants are entirely operated from water powers.

At Notodden there are now in operation in furnace house 1, thirty-five furnaces, each of 900 kw. capacity, operated at a voltage of 4000 to 6000 volts. A direct current of 20 to 25 amps. is used for the magnets. Observation of the arc can be made through mica windows. The electrodes consist of tubes of pure electrolytic copper about 2 inches in diameter and $\frac{3}{16}$ inch thick, bent into the form of a U, each limb of which is about 8 feet long. The cooling of the electrodes is said to account for about 7·5 per cent of the applied energy. The gases from the furnaces, which are at a temperature of about 1000° C., pass through tubes, 3 to 4 feet diameter, to Babcock & Wilcox tubular boilers, where the heat energy of the gas during cooling is employed in raising steam. It is stated that the oxides of nitrogen do not exert a corrosive action on the boilers as long as the temperature of the gases is not allowed to fall below 200° C. It is estimated that more than half the total electric energy expended in the furnace is recovered as heat for the purpose of generating steam, or about 13 per cent of the applied energy can be recovered in the form of mechanical energy in a turbo-electric plant. After passing through the steam boilers, the gases are led by an iron pipe to the cooling house. Each cooler consists of a large number of aluminium tubes, over which cold water runs, and through which the gases are passed and then admitted to the oxidation tanks. The oxidation tanks are vertical iron cylinders lined with acid-proof stone, and their object is to give the cooled gases a sufficient interval of rest in which the nitric oxide can react with the excess of oxygen present to give peroxide. From the oxidation tanks the gases are passed by blast engines to a series of large absorption towers, each 110 feet high, 25 feet diameter, and built of thick red granite blocks mitred at the corners, luted with cement and bound with iron hoops (cf. Plate V). The interior of the towers is filled with quartz down which water is run, and acid of about 30 per cent concentration obtained. About 85 per cent of the combined nitrogen is absorbed in this way, and the remainder combined with alkali to give nitrate or a mixture of nitrate and nitrite. A second furnace house at Notodden contains four 3000 kw. Birkeland-Eyde furnaces worked from a separate power house. These newer furnaces each have their own transformer arranged immediately underneath, which step down the current from 10,000 to between 4000 and 6000 volts. The gases from the furnaces, at a temperature of about 1200°, pass directly into the tubes of boilers and from there through a cooling system, consisting of a series of aluminium tubes passing in parallel through a large tank in which water is circulated.   From here the gases are led to absorption towers as above.

Rjukan Works.—Two additional and newer factories of the Norwegian Hydro-Electric Nitrogen Company at Rjukan are situated at Vemork and Saaheim, and known respectively as Rjukan 1 and 2. In No. 1 there are seventy-two furnaces of the Schönherr type (p. 123), and eight of the Birkeland-Eyde type.   The former are each of 1000 kw. and the latter of 5000 kw., making a total of 112,000 kw.

The factory at Rjukan 2 was opened in 1915, and is operated from its own power house (cf. p. 425).   The furnace house, shown as the higher building in Plate VI, is situated on a terrace on the side of the hill above the power house.   As shown in Plate VII giving a view of the interior of the furnace house, there are about forty-five furnaces of the Birkeland-Eyde type installed, each of 3000 kw. capacity, of which about forty are in operation at a time.

The steam generated from the waste heat of the gases is·employed to drive all accessory plant and for evaporation and condensation, while still leaving a margin of 15,000 h.p. of steam, which, at present, passes to waste, but it is being arranged to utilise this in turbines for the generation of more electrical power.

The capacity of the three works in 1916, on completion of the Saaheim branch, amounted to 30,000 to 40,000 tons of concentrated pure nitric acid, 5000 to 10,000 tons of nitrate of ammonia, and 6000 tons of refined nitrate of soda per annum.

For the production of ammonium nitrate, ammonia is obtained partly from gas liquor and partly from cyanamide, which is manufactured on a small scale at Notodden and imported from Odda.

Since the termination of the demands for war requirements the production of calcium nitrate has been substituted for that of ammonium nitrate.

Yield of Furnaces.—The usual Norwegian practice is to pass about 2·1 cubic metres (74 cubic feet) of air through the furnaces per kilowatt per hour.   In the older types of furnaces the air is circulated by admitting at a slight pressure.   In the newer type the air is drawn through under a slight suction by means of fans placed beyond the cooling system.

The concentration of the gases from the furnaces is stated to amount to about 1·25 per cent, and an output based upon the combined nitrogen actually condensed and sold, equivalent to 515 kg.

PLATE VI.—The Saaheim (Rjukan 2) Power House and Furnace House,

(The Norwegian Hydro-Electric Nitrogen Company.)

PLATE VII.—BIRKELAND-EYDE FURNACES AT SAAHEIM (RJUKAN 2).

(The Norwegian Hydro-Electric Nitrogen Company.)

HNO_3 per kw. year of 8500 hours, measured at the switchboard, or 535 kg. per kw. year of 8760 hours. This estimate includes an allowance for losses during the absorption and concentration stages, and is equivalent to a consumption of energy amounting to 1·87 kw. year per metric ton of nitric acid (estimated as 100 per cent) produced, or 8·41 kw. years per metric ton of nitrogen fixed.

Concentration of Acid.—The dilute nitric acid from the main towers contains some 30 per cent by weight of acid and can be concentrated to a strength of about 50 per cent in evaporators heated by means of the furnace gases. The production of a concentrated nitric acid of 96 per cent strength is a more difficult operation and is commonly effected by distilling a mixture of nitric and strong sulphuric acids, the latter being recovered and re-concentrated by means of the heat of the furnace gases. Instead of concentrating the dilute acid, the larger part is neutralised directly with ammonia or lime and the corresponding solutions of the nitrates concentrated and crystallised by the heat of the furnace gases.

Power Factor.—The power factor obtained with the Birkeland-Eyde furnaces is 0·65, the current having a periodicity of 50 cycles.

SCHÖNHERR PROCESS.[1]

The main distinction of this from other types of furnaces is that in place of leaving a given volume of air momentarily in contact with the arc and then passing it into a zone of colder air or over a chilled surface, the entire volume is allowed to remain for an appreciable length of time within the range of activity of the arc. The reaction chamber, as illustrated in Fig. 98, consists of an iron pipe arranged vertically, about 30 inches diameter and 30 feet long. The lower electrode L is insulated from the furnace chamber, and consists of a strong copper body pierced through its axis and cooled with water. Through the axis passes an iron bar E, which constitutes the actual electrode from which the arc proceeds. Through the intense heat of the arc the electrode becomes coated with a layer of melted ferric oxide which slowly evaporates, and the position of the terminal is adjusted by feeding forward the rod. The iron bars used for electrodes will, in this way, last for about 2000 working hours. The second electrode is formed by the walls of the reaction chamber itself. A potential difference of 4500 to 5000 volts from a source of alternating current is maintained between the two, and an arc is struck in the first place by

[1] " Trans. Amer. Electrochem. Soc.," 1909, **16**, 131.

means of an igniting rod connected to the lever Z, which can be moved until the rod touches the walls of the tube. On forming the arc, a stream of air is blown tangentially into the tube through a number of holes at the base. The air passes upwards with a whirling motion and raises the point of contact of the discharge with the walls of the tube until finally a rod-like flame is formed and maintained in the centre of the tube for a distance of 23 to 25 feet, and finally impinges on the walls of the water-jacketed section of the chamber at the top or on to the top metal plug *g*. A continuous or "standing" arc is thus maintained along the axis of the tube, which makes intimate contact with the gases during their passage upwards. The introduction of the air tangentially through a number of small openings causes a very pronounced vortical motion which is imparted to the entire ascending column of air. As a result of this a reduced pressure exists at the centre of the vortex and there is evidence which indicates that a more favourable yield of nitric oxide is thereby obtained.

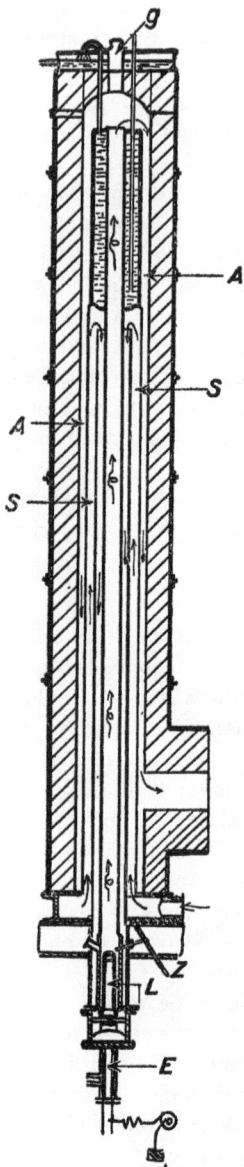

The gases are withdrawn from the top of the reaction chamber and led downwards along an annular space A, thus serving to pre-heat the entering air which is passed through an intervening annular space S, before admitting to the reaction chamber at a point opposite the lower electrode. The gases leave the actual flame-tube at a temperature of about 1200° C. and the exit flue at about 850° C. while the air entering the reaction tube is pre-heated to about 400° C. Furnaces of 1000 kw. capacity are stated to give a concentration of about 2·25 per cent nitric oxide and a yield of about 65 grm. of HNO_3 per kw. hour or 550 kg. per kw. year.

Fig. 98.

The Schönherr furnace was first developed at the Badische Anilin und Soda Fabrik at Ludwigshafen

PLATE VIII.—SCHÖNHERR FURNACES AT VEMORK (RJUKAN I).

(The Norwegian Hydro-Electric Nitrogen Company.)

in 1905, a single furnace of 300 kw. being installed. This was followed in 1907 by the erection of an experimental factory at Christiansand in the south of Norway, where, from a neighbouring water power, 1300 kw. was available in the form of three-phase current. By joining in star-connexion, this enabled the operation of three furnaces of 600 h.p. each.

At Rjukan (see p. 122) there are seventy-two of these furnaces in operation, each of 1000 kw., and taking a current of 300 amps. at 4000 to 4200 volts. A choking coil joined in series with each furnace absorbs about one-third of the applied voltage. The power factor of the furnaces is about 0·65. The lower part of the furnaces is about 2 feet diameter and the upper portion, for about 20 feet, jacketed with a cylinder about 3 feet 6 inches diameter to contain the annular passages for the pre-heating of the air. A view of this installation is seen in Plate VIII.

For some time the Badische Anilin und Soda Fabrik and allied works combined with the Norwegian company in the development of these processes, but prior to the war the German interest was disposed of.

THE PAULING FURNACE.[1]

The Pauling furnace employs a fan-shaped arc flame which forms between horn-shaped castings, inclined at an angle of 90° (cf. Figs. 99 and 100). The arc is carried upwards by a blast of air admitted through a pipe T at the base, until the resistance of the extended path becomes so high that a new arc forms at the apex, while the higher one breaks. In this manner a series of arcs are formed in rapid succession and travel upwards giving a discharge which has the appearance of a continuous flame about 30 inches high. The electrodes are 40 mm. (1·57 in.) apart at the narrowest section. Narrow metal plates known as "kindling blades" K are arranged to project in the horn gap and are adjusted to a distance of separation of 2 to 3 mm., and as they wear away are continually advanced by screw adjustments. The "kindling blades" are made narrow so as to offer no obstruction to the blast which fills the space between the main electrodes. A steam boiler is placed directly over the furnaces in order to utilise the waste heat of the

FIG. 99.

---

[1] "Zeit. fur Elektrochem.," 1907, 13, 225; 1909, 15, 544; 1911, 17, 431; " Electrochem. and Metall. Ind.," 1909, 7, 430.

gases. The main supply of air is blown in directly under each arc by means of tuyères T, and is pre-heated by first circulating through pipes exposed to the exit gases from the furnace. A rapid cooling of the gases, after leaving the arc zone, is brought about by passing

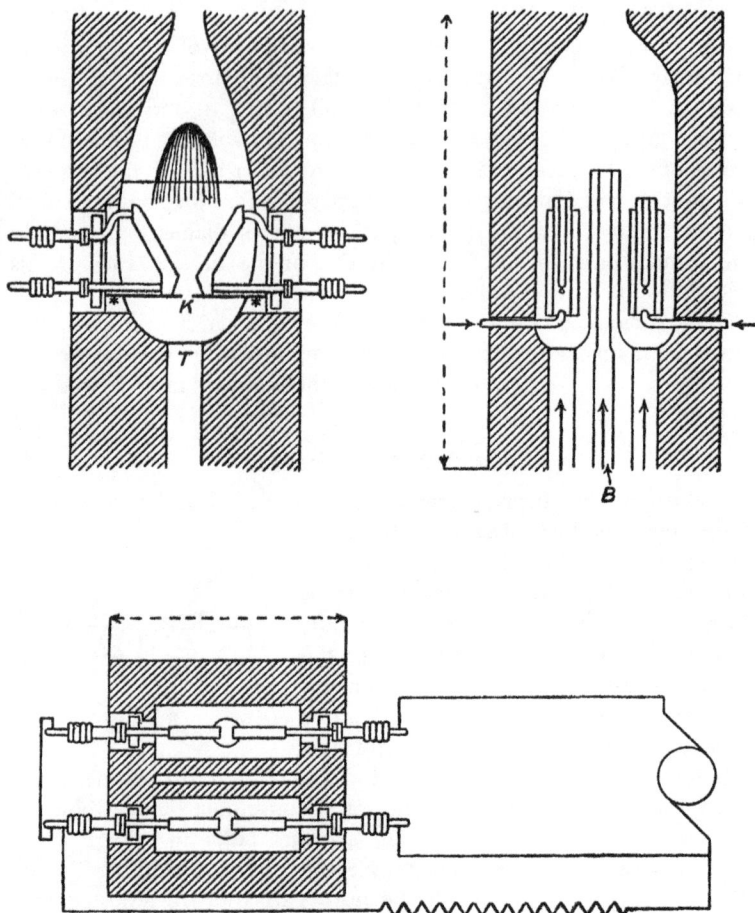

FIG. 100.

"cooling air" into the upper part of the arc flame from the side. The "cooling air" is taken from the cooled gas mixture, containing nitric oxide, before it enters the condenser, and is supplied through B into the flame at a lower speed than the main current. A suction effect is thus produced causing the arc to be lengthened and broadened, while the increase in the rate of cooling of the gases after leaving the arc zone which is thus produced improves the yield.

The furnaces are built of brick and each unit contains two arcs side by side and connected in series to the current supply, while a shunt circuit leads through a high resistance to the middle point of the two arcs. By this arrangement, on starting the arc the full potential is applied across each spark gap.

In the early stage of the development of this furnace, it is stated that unsatisfactory results were obtained, the yield not exceeding 25 grm. of HNO_3 per kw. hour. This was eventually considered to be due to iron oxide retained in suspension in the gases, and exerting a catalytic effect in accelerating the decomposition of the nitric oxide during the cooling. The defect was obviated by using as electrodes an alloy of aluminium (cf. p. 132).

The first installation of the Pauling process was made by the Saltpetersäure-Industrie-Gesellschaft, near Innsbruck, in the Tyrol. The factory was erected in 1904 and now contains twenty-four furnaces with a total capacity of 15,000 h.p. Current is obtained from the Sill water-power plant of the Municipal Electrical works of Innsbruck. The concentration of nitric oxide obtained is said to amount to about 1·5 per cent and the yield to 60 grm. HNO_3 per kw. hour, while the acid from the absorption towers has a concentration of 35 to 40 per cent.

In 1908 a French company, the "Société La Nitrogène," erected a works at Roche-de-Rame near Briançon (Hautes Alpes) to work the Pauling process.[1] Nine furnaces are in operation, each of 600 kw., utilising current at 4000 to 6000 volts, and connected in groups of three to a three-phase supply. The furnaces are of rectangular cross section 3 feet 3 inches × 3 feet 9 inches × 10 feet high. The arc flames formed are from 2 feet to 2 feet 6 inches high. The projecting iron plates at the point of ignition of the arc are from 2 to 3 mm. apart and have a life of 20 hours, while that of the main electrodes is 200 hours. The pre-heated air is supplied to the furnaces by means of high powered compressors and issues through the tuyères at a velocity of 400 metres per second.

At Legnano, Italy, nitric acid is made by a modified Pauling process which was introduced by C. Rossi. Since 1911 there have been in operation at these works six furnaces with a total capacity of 4000 h.p. The modification which has been introduced consists in employing an arc of 12,000 volts between special electrodes which

[1] "Metall. and Chem. Engineering," 1911, 9, 196.

are cooled by water circulation. In addition to the above, an installation of the Pauling furnace with an aggregate capacity of 9000
kw. is in operation in Italy.

A further installation has been erected in the United States by
the Southern Electrochemical Company at Nitrolee, S.C.[1] This
plant has a capacity of 4000 h.p. and the power is derived from the
Southern Power Company at Great Forks and Rocky Creek. Calcium nitrate is manufactured here for use as a fertiliser in the

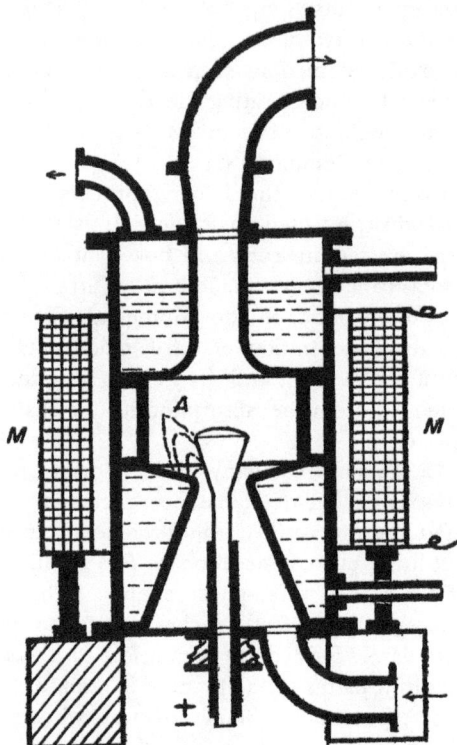

Fig. 101.

Southern States. The output of the factory is estimated at 8 tons
of calcium nitrate per twenty-four hours.

According to information collected by the Nitrogen Products
Committee of the Ministry of Munitions[2] the concentration of nitric
oxide in the furnace gases amounts to only 0·8 per cent, and if the
velocity of the air through the furnaces is diminished in order to

[1] " Metall. and Chem. Engineering," 1912, 10, 126.
[2] " Final Report Nitrogen Products Committee," p. 238.

obtain a higher concentration of nitric oxide, the net recovery decreases. Very large oxidising and absorbing towers are required, and the dilute nitric acid from the main towers has a strength of 23 to 25 per cent. Only 80 per cent of the combined nitrogen recovered is in the form of nitric acid, and the remainder in the form of sodium nitrite.

The products actually recovered from the towers are stated to be equivalent to a yield of 75 grm. of HNO_3 per kw. hour measured at the furnace terminals. This is equivalent to 657 kg. of HNO_3 per kw. year of 8760 hours.

The Moscicki Furnace.[1]

The Moscicki furnace (cf. Fig. 101) contains two concentric electrodes, one being a central insulated tube and the other the inner wall of the reaction chamber. The arc A is caused to rotate round the annular opening through the influence of a powerful magnetic field formed by the coils MM with lines of force parallel to the common axis of the electrodes. The reaction chamber and air chamber below are both surrounded by water. The arc is maintained by an alternating current of 8000 volts. The electrodes last on an average about four weeks and the furnace itself, if operated continuously, lasts about two years. The concentration of nitric oxide in the furnace gases is stated to vary from 1·5 to 2 per cent, and the yield of HNO_3 to amount to 50 grm. per kw. hour.

The Moscicki furnace is in operation at Chippis, in Switzerland, the installation comprising twelve furnaces each of 1000 kw. capacity.

The Guye Furnace.

This process combines the main features of the Pauling and Schönherr type. A number of V-shaped electrodes, from each of which a long slender arc is obtained, are joined in series. Five of these arcs are arranged in a furnace and form a single circuit. The metal electrodes are cooled by internal circulation of water. In a furnace of five arcs, each arc being a little over 1 metre in length, a voltage of 5000 is applied, 50 kw. consumed, and the total arc length amounts to about 6 metres (19·6 feet), which is equivalent in a Schönherr furnace with a single arc to the arc length obtained with a unit of 440 kw.

In 1909 an installation of this process was made by a Geneva company, La Société de Nitrogene. Furnaces were constructed of

[1] " Elektrotech. Zeitschrift," 1907, **28**, 1055,

200 and 400 kw., and arranged to work during times of off-peak load.   The yield obtained is given as 64 grm. of HNO_3 per kw. hour.

The process is now controlled by "La Société Electrometallurgique Francaise," with a view to its operation in the Pyrenees.

THE KILBURN-SCOTT FURNACE.[1]

The Kilburn-Scott furnace, as shown diagrammatically in Fig. 102, has three wedge-shaped electrodes arranged with intervening refractory material so as to enclose a six-sided conical space, having its apex at the bottom.   Three-phase current supplied to the electrodes produces a combined arc which is flared out by the air.   From the three sine curves drawn with a phase displacement

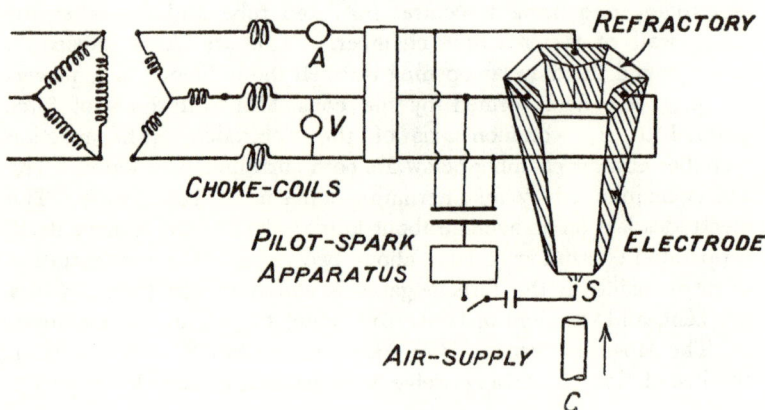

FIG. 102.

of 120° (Fig. 103), it is seen that current is always flowing in the reaction chamber, and it follows that the electric energy varies between 0·86 and 1·0 during a complete cycle.   On the other hand, with a single-phase current the electric energy varies from zero to maximum and twice in each cycle there is no current.

The three-phase currents interact in such a way as to produce an arc flame which rotates rapidly in accordance with the periodicity of the supply, viz. 50 or 60 cycles.   Compressed air entering the bottom from the pipe C expands and fills the cone before passing away at the top.   The arc flame appears to the eye like a double cone having one apex at the bottom, where the electrodes are nearest

[1] "Journ. Soc. Chem. Ind.," 1917, **36,** 771 : "Trans. Amer. Electrochem. Soc.," 1918, **34,** 221.

together, and the other at the top, where the flame tapers off. The flames move with great rapidity in different planes and so are constantly intercepting fresh particles of air. The employment of an auxiliary magnetic field for spreading the arc, as in the Birkeland-Eyde furnaces, is thus rendered unnecessary. It is claimed that a more thorough exposure of the whole volume of air to contact with the arc is obtained with this system than in any of the previously designed furnaces using single phase. The arcs are started by means of pilot or trigger sparks, passed between the electrodes and a wire S placed about midway between the points of the three electrodes. The

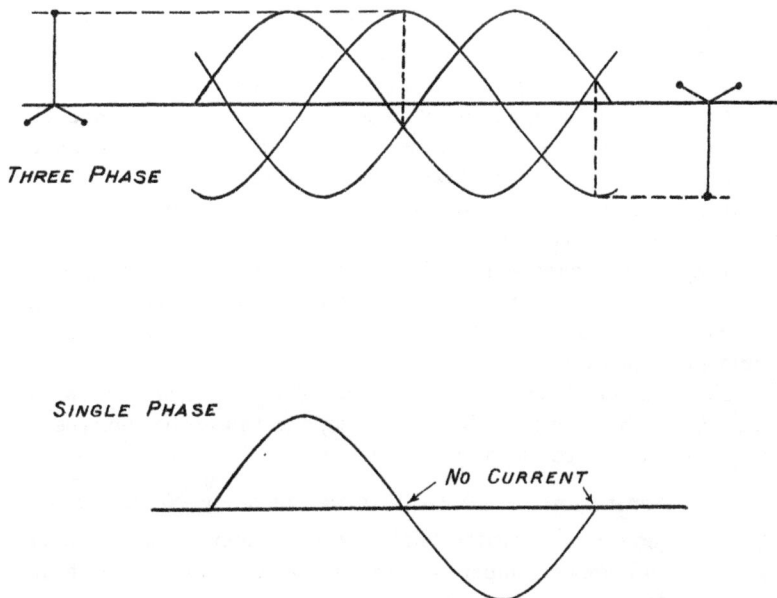

FIG. 103.

sparks are formed by means of an independent high-tension circuit and maintained during the operation of the furnace, the yield being thus improved and continuity of working maintained. It is found advantageous to pre-heat the air to a temperature of 250° C. or higher. The reaction chamber is surmounted by a steam boiler on the tubes of which the arc flame plays. The furnace gases are thus rapidly cooled and the dissociation of the nitric oxide minimised. The boiler is earthed to avoid electrical leakage and the metal is not attacked by the furnace gases at the temperature prevailing. The steam raised in cooling the gases will, it is estimated, when utilised in a steam turbine, represent a regenerative gain of over 10 per cent of the energy.

Electrodes.—The electrodes consist of water-cooled copper tubes of a special alloy.   Under the influence of the arc, particles of metal are severed from the electrode, and these, becoming incandescent and oxidised, are considered to accelerate the decomposition of nitric oxide during cooling.   Much less erosion takes place with copper electrodes than with steel, and less active vapours are released. Electrodes of copper alloyed with other metals have been used to advantage.   The process of "calorising" has also been applied in this connexion.   This process depends on the property possessed by powdered aluminium, when in a neutral atmosphere, of entering into combination with a metal and forming a homogeneous alloy.   The depth of immersion depends on the length of time of treatment, and by it the oxidising temperature of steel can be raised to over 1000° C.

Size of Furnace.—As the three-phase arc flame extends in three dimensions with increase in power, it follows that the kilowatt capacity increases very rapidly with increase of size, thus enabling the utilisation of high-powered units.

Output of Furnace.—It is claimed for the Kilburn-Scott furnace that a yield of 90 grm. of HNO_3 per kw. hour can be obtained, which is about 50 per cent higher than the usually assumed output from single-phase furnaces.

Installations of Kilburn-Scott Furnace.—One unit of 300 kw. of the Kilburn-Scott has now been installed, and two further units each of 500 kw. are in course of installation.

THE FORMATION OF METALLIC COMPOUNDS OF NITROGEN.

Nitrogen can be caused to combine with a number of bases, under suitable conditions of temperature and in presence of carbon, to form compounds from which the nitrogen can generally be recovered subsequently in the form of ammonia.   A means is thus provided for the fixation of atmospheric nitrogen in a utilisable form.   The products of technical importance which are formed in these processes vary with the nature of the base and the conditions of the experiment and are of the following three types :—

1. Cyanides, of the general formula XCN.
2. Cyanamides, of the general formula XCN_2.
3. Nitrides, of the general formula X_3N.

1. CYANIDE PROCESSES FOR FIXING ATMOSPHERIC NITROGEN.

Since the early part of the nineteenth century it has been known that when a mixture of alkali or alkaline earth and carbon is heated to a sufficiently high temperature in an atmosphere containing free nitrogen, the latter combines with the carbon and metallic base to form a ·cyanide and that this is decomposed when heated in the presence of steam, the nitrogen being evolved in the form of ammonia. It was shown in the course of early investigations that carbonates and hydroxides are equally suitable for the synthesis of metallic cyanides, but that the nature of the metallic base had an important influence upon the facility with which the nitrogen is fixed. It was later found that the products consist of mixtures of cyanide and cyanamide, the proportion of the former increasing with the temperature. In the case of the alkali carbides, the product was mainly cyanide, but with barium carbide the proportion of cyanamide in the product was larger. When calcium carbide was used practically the whole of the nitrogen fixed up to temperatures of about $1100°$ C. was in the form of cyanamide.

Reaction with Alkali Bases.—The complete reaction for the formation of sodium cyanide from sodium carbonate, carbon, and nitrogen is given by the equation

$$Na_2CO_3 + 4C + N_2 = 2\,NaCN + 3CO - 138,500 \text{ cals.}$$

The combustion of the carbon monoxide evolved would yield 61,500 cals. in excess of those needed for the main endothermic reaction.

Reaction with Barium Salts as Bases.—When baryta or barium carbonate is employed the absorption of the nitrogen results in the formation of cyanide and cyanamide, the latter reaction being accompanied by the liberation of free carbon. Under favourable conditions, however, the cyanamide combines with the liberated carbon to form barium cyanide in accordance with the equation

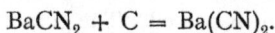

$$BaCN_2 + C = Ba(CN)_2.$$

Formation of Ammonia from Cyanides.—The combined nitrogen in the crude cyanide products can readily be obtained in the form of ammonia by treating with steam, or when aqueous solutions are boiled, in accordance with the following equations which have been attributed to the reactions :—

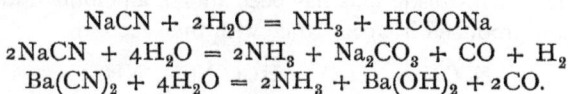

$$NaCN + 2H_2O = NH_3 + HCOONa$$
$$2NaCN + 4H_2O = 2NH_3 + Na_2CO_3 + CO + H_2$$
$$Ba(CN)_2 + 4H_2O = 2NH_3 + Ba(OH)_2 + 2CO.$$

The regeneration of the alkali or barium base of the cyanide reaction during the production of ammonia introduces the possibility of devising a continuous process for fixing atmospheric nitrogen in the form of ammonia through the intermediary of metallic cyanides.   Though much effort has been devoted to the development of such commercial processes, their establishment on a large scale still remains to be effected.

A process which has been established by Bucher [1] on a semi-technical scale in America consists in heating sodium carbonate with a mixture of carbon and powdered iron and treating with a gas containing free nitrogen at a temperature of 900° to 1000° C., when reaction takes place in accordance with the equation

$$Na_2CO_3 + 4C + N_2 = 2NaCN + 3CO.$$

The iron exerts a catalytic influence in lowering the temperature of the reaction.   To obviate the disadvantage of using powders, briquettes can be made by mixing the charge of coke, soda-ash, and iron with water at a temperature of 105° C., by means of a steam-jacketed kneading machine, and cutting the hot dough-like mass into suitable lengths and drying rapidly.   On heating the briquettes in tubes in an atmosphere of nitrogen, a product containing up to 28 per cent of sodium cyanide can be obtained.   In place of nitrogen, air, after passing through heated coke or producer gas, can be used in the cyanising tubes.

An electric furnace has been designed by Bucher for conducting this process which obviates the great disadvantage of using retorts at these high temperatures, and consists of a cylindrical furnace with a basic lining and a perforated bottom, like a Bessemer converter.   The furnace contains molten iron in which graphite fragments are deeply imbedded, while sodium vapour and nitrogen are blown into the bottom, and sodium cyanide is distilled at the top.   The heat is produced by the resistance of a graphite column through which a current is passed.   The graphite also serves to keep the iron saturated with dissolved carbon.   The conductivity increases with the temperature and by applying the heat internally in this manner, much greater efficiency and rapidity of operation are obtained than with a system of external fuel heating.

Conversion of Sodium Cyanide.—By boiling sodium cyanide solution to which caustic soda has been added, ammonia and sodium formate are produced in accordance with the reaction

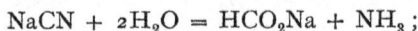

$$NaCN + 2H_2O = HCO_2Na + NH_3;$$

the caustic soda prevents hydrolytic dissociation to form hydrocyanic acid, and "salts out" the sodium formate, which may either be

[1] "Journ. Soc. Chem. Ind.," 1917, 36, 451; "Journ. Ind. and Eng. Chem.," 1917, 9, 233.

returned to the cyanising process or may be heated *in vacuo* to convert it into sodium oxalate. Other products which may be obtained from the sodium cyanide by suitable chemical reactions include urea, oxamide, oxalic acid, and formic acid.

Power Consumption.—An approximate calculation appears to indicate that the power requirements of the sodium process when using iron in the mixture would not exceed about 19 kw. per kg. of nitrogen fixed, or 1 kw. hour would produce 52 grm. of fixed nitrogen.

Heating by Decomposition of Carbon Monoxide.—An alternative to the electrical method of heating is proposed by the decomposition of carbon monoxide in producer gas in the presence of iron according to the equation

$$2CO = CO_2 + C + 38,080 \text{ cals.}$$

Finely divided carbon would thus be produced in the reaction mass itself and the preliminary grinding and mixing of this element obviated.

MANUFACTURE OF POTASSIUM CYANIDE AT NEWCASTLE.

A plant was erected at Newcastle in 1844 for the preparation of potassium cyanide as an intermediate product in obtaining potassium ferrocyanide. The method used consisted in heating together potash and charcoal in an atmosphere of nitrogen.

SCOTTISH CYANIDE COMPANY.

The preparation of barium cyanide, to be subsequently converted into the sodium salt, was for a time undertaken in Scotland by the Scottish Cyanide Company by a method devised by Readman. Internal electrical heating was employed by the use of carbon cylinders mounted upon water-cooled iron cylinders into which the furnace product was discharged after the reaction. The carbon cylinder constituted one electrode, the other being attached to the top of the furnace so that the current passed through the charge which acted as a resistance. The mixture employed consisted of barium carbonate and carbon, briquetted with pitch and was treated with nitrogen. The barium cyanide was converted into the sodium salt by double decomposition with sodium hydroxide and the baryta produced used again in the furnace.

2. Calcium Cyanamide.

Calcium cyanamide is most readily prepared by the reaction between calcium carbide and nitrogen, which takes place at temperatures between 900° and 1200° C. in accordance with the equation

$$CaC_2 + N_2 = CaCN_2 + C.$$

The discovery of the fertilising value of calcium cyanamide led to its commercial manufacture on a large scale, and in this connexion calcium carbide, which is used as the raw material in this process, has found its largest application.

The reaction is effected by passing nitrogen gas under a slight pressure into a furnace in which the finely ground carbide is maintained at the most favourable temperature for the reaction. The process is exothermic, and the heat evolved helps to maintain the temperature. The preliminary heating may be effected externally, or, as is more usual, by internal electrical heating, which is better adapted for controlling the reaction and preventing an excessive rise of temperature through the heat of reaction. To obtain a high efficiency it is necessary that the nitrogen should be pure, and the temperature maintained as constant as possible. The finely powdered carbide used in the manufacture is transformed during the reaction into a material of black-grey colour, either in powder form or agglomerated into a solid mass according to the type of process adopted. Use is made of the addition of certain reagents which facilitate the absorption of nitrogen at a reduced temperature. The most important of these addition agents are calcium chloride, as proposed by Polzenius,[1] and calcium fluoride, as proposed by Carlson.[2] Various forms of furnaces have been employed for the conversion process and two principal types involving different methods of working are in successful commercial operation.

The first or *discontinuous* method of working is carried out by the employment of batches of comparatively small furnaces, each having a capacity of 0·5 to 2 tons. When the reaction is complete, the furnaces, after being allowed to cool for some hours, are emptied and recharged with fresh carbide.

The second, or *continuous*, method of working is carried out by the employment of larger furnaces, each having a capacity of from 15 to 18 tons a day, the carbide being introduced and the cyanamide removed in a continuous manner.

The nitrogen employed in the process was formerly separated

---

[1] German Patent, 163,320.   [2] "Chem. Zeit.," 1906, 1261.

from the atmosphere by passing air over heated copper in order to fix the oxygen, the copper being afterwards regenerated by means of a reducing agent such as water-gas. The modern practice, however, is to employ the air liquefaction process.

The industrial manufacture of cyanamide was accomplished first in Italy at the Piano-d'Orte plant, near Pescara, where there is available 25,000 h.p. from water power, then in Switzerland, Norway, and in France.

PROCESS OF FRANK AND CARO.[1]

The commercial manufacture of calcium cyanamide was commenced at Odda in 1908 by a process of Frank and Caro, the original discoverers of this compound. A works was installed by the North-Western Cyanamide Company on a site adjacent to the carbide works (cf. p. 111), and was constructed for an annual production of 12,000 tons of calcium cyanamide, using the Linde system of fractional distillation of liquid air for the separation of the nitrogen required. In 1914 the plant was enlarged to give an annual output of 70,000 tons of cyanamide. The North-Western Cyanamide Company has since been reconstituted and allied with the carbide company to form the Nitrogen Products and Carbide Company Ltd.

Calcium carbide for this process at Odda is obtained from the adjacent carbide factory. After breaking into rough pieces, the material is loaded on to waggons on an inclined railway and conveyed to the crushing department of the cyanamide works. Here the carbide is reduced to pieces of approximately $\frac{1}{2}$ inch cube, and taken to the fine-grinding mills which are of the ball or roulette type. The finely ground carbide is sieved and collected in closed screw-conveyors under the mills and delivered to the retort-filling machines situated in the furnace house. The retorts are constructed in the form of cylindrical baskets from flat-iron strips and hold about 450 kg. of ground carbide. Before filling, they are lined with corrugated paper. The two filling machines consist each of a large rectangular steel hopper mounted upon a framework. To the mouth of each hopper, a feed screw of fine pitch is fitted, to transfer the ground carbide to a distributor which delivers it evenly through eight vertical tubes into the retorts placed below. In this manner, a uniform and level filling of the retorts is secured. The furnace house is divided

---

[1] " Engineering," 1914, **98**, 267 *et seq.; "* Met. and Chem. Engineering," 1909, **7**, 212, 309, 360.

into three sections, containing in all 604 furnaces.   The furnaces
are mounted upon a steel framework so as to provide head room
underneath for the furnace operators.   On the top of the framework
a continuous platform is provided a few inches above the furnace
tops.   The furnaces consist of cylindrical steel shells lined with firebrick and provided with external heat insulation.   The removable
covers, which are of sheet-iron lined with firebrick, rest in a sand
lute.   Each furnace is provided with an independent nitrogen
connexion, and an electric circuit communicating with a switchboard
placed under the platform above mentioned.   The retorts, after
being filled, are placed in the furnaces, and the covers secured.   A
carbon electrode is then fitted through a vertical hole left for the
purpose in the centre of the powdered carbide, and is secured in
terminal clamps on the furnace cover and bottom respectively.

The electrode, which is connected to an alternating current
circuit of 70 volts, quickly heats up the carbide in the immediate
neighbourhood of the carbon rod to the temperature necessary for
the chemical reaction to take place and nitrogen being at the same
time applied, the heat rapidly rises until the whole mass becomes incandescent.   An average temperature of 1100° to 1200° C. is reached,
and the conversion of 450 kg. of carbide requires some thirty hours.
After a short period of cooling the retorts are removed at a dull red
heat to the cooling house, and when cold the product is crushed
and then ground finely in roulette mills.   The small proportion of
carbide which remains untransformed is decomposed by adding
water in the form of a spray.   The hydrated product is then cooled
and mixed with a little mineral oil to render the product dustless.

THE MANUFACTURE OF CALCIUM CYANAMIDE IN AMERICA.[1]

The manufacture of calcium cyanamide has for some time been
carried out by the American Cyanamide Company at Niagara Falls,
the production during the war amounting to 64,000 tons annually.
As the starting-point in the preparation of ammonium nitrate for war
requirements a large extension of the process was, in 1918, installed
at Muscle Shoals in Alabama, by the Air Nitrates Corporation, working under the direction of, and with the assistance of a subsidy
from the United States Government.   The plant was worked in
association with the calcium carbide plant described above (p. 113),
and was of a capacity to utilise 200,000 tons of calcium carbide

[1] " Journ. du Four electrique," 1919, **28**, 61.

annually.   The outlay of the plant is similar to the one at Odda. The cyanamide building contains sixteen rows of ninety-six furnaces each, or 1536 furnaces in all.   The furnaces measure 4 feet 4 inches external and 2 feet 9 inches internal diameter, and are 5 feet 3 inches deep, with a refractory lining and steel case.   A paper cylinder 2 feet 6 inches diameter is placed in the furnace, and inside this a vertical paper tube 3 inches diameter.   The carbide is reduced to a fine powder in an atmosphere of nitrogen and a charge of 700 kilos admitted to the wider paper cylinder, which leaves a space of 1½ inches between the paper case and the furnace lining.

A carbon electrode, 15 mm. diameter and 1 metre long, is inserted in the paper tube.   The furnace is covered with two roofs, the outside one being luted with sand.   Nitrogen, which is produced from the distillation of liquid air, is led into the furnace under a pressure of 100 mm. of water.   When all is ready the carbon electrode is heated with a monophase current of 200 to 250 amps. at 100 volts for twenty minutes and then reduced to 100 to 115 amps. at 50 volts for twelve hours.   The current is then shut off and the reaction, being exothermic, proceeds for a further twenty-eight hours. The temperature attained is about 1100° C.   The product is fritted together to form a solid mass containing 63 per cent calcium cyanamide, 2 per cent calcium carbide, 13 per cent lime, 1 per cent calcium sulphate, 11 per cent carbon, 3 per cent silica, 2 per cent magnesia, 2 per cent oxides of iron and aluminium, and 3 per cent other elements.

The product, after cooling, is pulverised in an atmosphere of nitrogen to a fineness such that 95 per cent passes through a sieve of 200 mesh.   The material is then treated with water in amount sufficient to decompose the carbide and hydrate the lime, while the remaining cyanamide, in a dry condition, is conveyed by a transporter screw to the autoclave chamber.

Preparation of Ammonia.—A solution of 2 per cent sodium hydrate is first added to the autoclaves and the cyanamide admitted. Any residual carbide is thus decomposed, and the acetylene allowed to escape.   The autoclave is then closed and steam admitted at a pressure of 110 lb. for twenty minutes.   The ammonia is allowed to accumulate, and when the pressure reaches about 220 lb. the escape valve is carefully opened and the pressure held constant for three hours, when the liberation of ammonia abates.   The valve is then closed and more steam admitted, which causes the evolution of ammonia to continue for a further one and a half hours.   The ammonia can be recovered as sulphate by absorbing in sulphuric acid, or is passed directly to a plant for oxidation to nitric acid. This method of catalytic oxidation in presence of heated platinum is carried out as described on page 157.

CYANAMIDE MANUFACTURE IN FRANCE.

Calcium cyanamide is manufactured in France at the "Société des Produits Azotes," at Notre Dame de Briancon.[1] The conversion is made by charging carbide into crucibles holding 400 to 500 kg. These are of iron and heated internally electrically. Absorption of nitrogen continues during thirty to sixty hours, and a product is obtained of from 15 to 20 per cent nitrogen.

CYANAMIDE MANUFACTURE IN SWEDEN.

In Sweden cyanamide is made at the Alby Carbide works, situated at Alby, on the Ljunga river, about sixty miles west of Sundswall. About 12,000 electrical h.p. is generated from the river. The carbide works at Alby was installed in 1901, and in 1913 the production of carbide amounted to 14,000 tons. A cyanamide factory was established at Alby in 1912, and in 1913 had an output of 16,350 tons.

The Stockholms Superfosfatfabriks Aktiebolag established a small carbide factory at Mánsbo, near Avesta, at about the same time as the installation of the works at Alby. Experiments made at Mansbo on the formation of cyanamide led to the establishment of a works at the rapids of the Ljunga River known as Johannesbergfors and Hangstafors, which are about 15 kilometres below Alby. By leading the water for a distance of 2 or 3 miles by a pipe line, a head of water of 130 feet is obtained on the turbines and during the greater part of the year a power of 18,000 h.p. is generated. Calcium carbide and from it, by means of the Carlson continuous process, calcium cyanamide, are produced at these works.

CYANAMIDE MANUFACTURE IN GERMANY.[2]

During the war the manufacture of calcium cyanamide was carried out at the Bayer nitrogen works at Trostberg by the Frank-Caro process, and at Kaapsack and Gross-Kayna by a process of Polzenius. The output in Germany in 1913 amounted to from 30,000 to 40,000 tons. In 1915 three more works were erected in Germany, and in 1916

[1] Cf. "Met. and Chem. Engineering," 1911, **9**, 99.
[2] "Journ. Soc. Chem. Ind.," 1917, **36**, 1081; "Journ. du Four electrique," 1919, **28**, 15.

works were built at Izentmarton in Hungary, making use of the natural gas of Siebenburg. In 1917 the Austrian Nitrogen Works was founded at Vienna, for the manufacture of cyanamide.

CONSUMPTION OF POWER AND MATERIALS IN CYANAMIDE MANUFACTURE.[1]

In the *discontinuous* system the quantity of carbide required per metric ton of nitrogen fixed amounts to about 4 metric tons. In the case of cyanamide containing 20 per cent of combined nitrogen this is equivalent to 0·8 metric ton of carbide per ton of cyanamide. The power requirements in respect of the nitrogen plant, the crushing, the preliminary heating of the carbide and the crushing of the cyanamide amount, according to several independent authorities, to 3 h.p. years (2·24 kw. years) of 8400 to 8500 hours, per metric ton cyanamide (20 per cent nitrogen). According to these figures the power requirements for the *conversion* stage alone amount to 0·3 kw. year per metric of nitrogen fixed.

With the continuous cyanamide process, giving a product containing 19·5 per cent nitrogen, the quantity of carbide required per metric ton of nitrogen fixed is stated to amount to 3·62 metric tons, while the overall requirements of power are specified as 1·97 kw. years of 8400 hours per metric ton of nitrogen fixed.

STATISTICS ON OUTPUT OF CALCIUM CYANAMIDE.

The production of calcium cyanamide during the years 1913 and 1918 for different countries is estimated to be as follows :—[2]

Country.	Actual Production in 1918.	Production in 1913.
Germany	300,000	24,000
Norway . . . . ⎫	200,000	22,110
Sweden . . . . ⎭		18,352
France	100,000	7,500
Canada . . . ⎫	58,000	48,000
United States . . . ⎭		
Japan	50,800	7,000
Switzerland	40,000	7,500
Austria-Hungary	24,000	7,500
Italy	15,000	14,982
Total	787,800	156,944

---

[1] " Final Report of the Nitrogen Products Committee," p. 256.
[2] " J. du Four electrique," 1919, **28**, 103.

In the above returns, installations in the course of construction are not included, the most notable of these being the plant at Muscle Shoals, Alabama, U.S.A., for a designed output of 300,000 tons per annum.

Manufacture of Cyanides from Calcium Cyanamide.

Calcium cyanamide is now employed on a considerable scale for the manufacture of cyanides. The process generally employed consists in heating together a mixture of calcium cyanamide and salt to the fusing-point. In the early processes many difficulties were presented, the main one being due to the foaming of the charge,

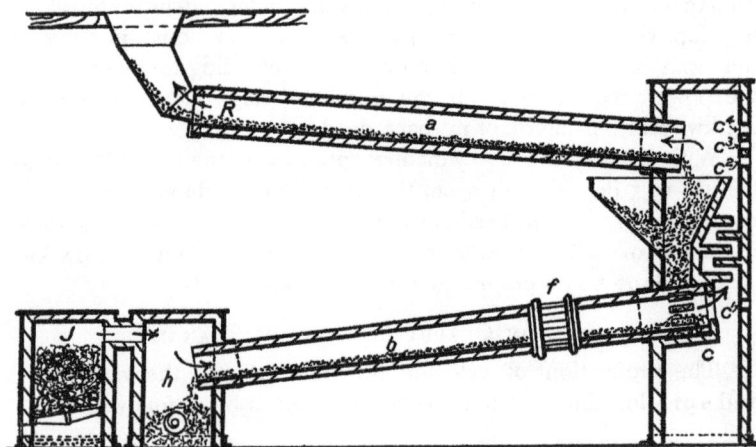

Fig. 104.

which for some years precluded the successful commercial development of this process.

A plant has, however, been in operation at the works of the Cyanamide Company at Niagara Falls since 1916,[1] and differs from earlier types in that a higher temperature is employed. The crucibles are heated electrically by single-phase current. One terminal consists of a suspended vertical electrode and the second of the conducting hearth of the crucible. A mixture of cyanamide, salt, and calcium carbide is fed continuously to the furnace which has a small cubical capacity compared with the power input. Melting takes place rapidly, calcium cyanamide reacts with its accompanying carbon to give

[1] W. S. Landis, " Trans. Amer. Electrochem. Soc.," 1920, 87.

cyanide, and the fused product is removed from the furnace almost continuously to a cooling device which instantly chills the product and thus prevents reversion of the initial reaction. The proportions of cyanamide and salt added vary with the product to be made.

Each furnace has an output of 30 tons a day, giving a product containing on an average the equivalent of 36·5 per cent sodium cyanide. The material produced appears to consist of a mixture of calcium cyanide, sodium chloride, and free lime with a fraction of 1 per cent of calcium carbide, cyanamide and other impurities, the cyanogen being in combination with the calcium.

The product is largely used in the mining industry, while a certain amount is utilised for the manufacture of ferrocyanide and anhydrous prussic acid.

3. THE APPLICATION OF NITRIDES TO THE FIXATION OF ATMOSPHERIC NITROGEN.—ALUMINIUM NITRIDE BY THE SERPEK PROCESS.[1]

The Serpek process depends upon the fact that a mixture of bauxite or alumina and carbon reacts with nitrogen at high temperatures in accordance with the equation

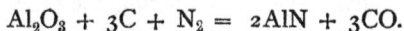

$$Al_2O_3 + 3C + N_2 = 2AlN + 3CO.$$

The most favourable reaction temperature is stated to be from 1800° to 1900° C. Alumina, in the form of bauxite, is fed into a revolving cylindrical furnace of the resistance type as shown at f in Fig. 104, the rotating resistance serving continually to agitate the material as it passes through. In the apparatus described in Serpek's United States patent (No. 996,032 of 1911) the plant, as shown in Fig. 104, consists of two superposed rotating cylindrical kilns, similar in construction to a cement kiln. Bauxite alone is passed through the upper kiln a, from left to right, and there calcined; then mixed with the necessary carbon in the hopper m, and the mixture treated by nitrogen at a high temperature in the lower kiln b, which is provided with a detachable electric resistance furnace at f. By this means the charge is subjected to the reacting temperature of 1800° to 1900° C., and the material is discharged from the lower end of the kiln into an air-tight receiver h. A large gas producer j furnishes producer gas (about 1/3 CO and 2/3 N_2) to the lower end of the lower kiln. The gas entering

[1] J. W. Richards, "Trans. Amer. Electrochem. Soc.," 1913, **23**, 352; L. L. Summers, *ibid.*, 1915, **27**, 368.

at a temperature of about 400° C. becomes highly heated as it passes through the kiln in a direction contrary to the descending charge, and at the electrically heated zone reacts upon the mixture forming nitride. After leaving the high-temperature zone, the gas enriched in carbon monoxide coming from the reaction pre-heats the descending charge, issues from the upper end of the kiln into a vertical closed chamber, and passes through it to the opening of the upper kiln, where it meets with a blast of air which burns it for the purpose of heating the contents of the upper or calcining kiln. The product is discharged in *h* as aluminium nitride, with a content of 26 to 34 per cent of nitrogen. Neglecting the specific heats of the solids the endothermic reaction requires 3 kw. hours of heat per kg. of aluminium nitride. It would therefore require 9 to 10 kw. hour per kg. of nitrogen under the best conditions if the coal and producer gas were capable of supplying all the heat energy necessary to produce the required temperature in the gaseous and solid products. In the case of cyanamide, it requires 16 kw. hours under favourable conditions for 1 kg. of nitrogen. If all the energy were supplied from an electric source, the Serpek process would require practically the same electric energy as the cyanamide process. There is a distinct advantage in being able to use producer gas in place of preparing purified nitrogen, and there is a further advantage in conducting the process in one operation. The aluminium nitride can be treated with steam and dilute alkali whereby decomposition takes place with great facility, yielding ammonia and a solution of aluminate of soda from which pure alumina can be recovered.

The alumina thus recovered is in a refined condition, and though much slower to react with carbon and nitrogen than impure bauxite, would be suitable for use in the preparation of aluminium, a manufacture which might be worked advantageously in conjunction with the Serpek process. Alternatively, the material which apparently acts as catalyst in the bauxite might be ascertained, and with its addition the recovered alumina could be used again in a cyclic process.

A large amount of experimental work upon a semi-technical scale was carried out by Serpek during the year 1905 and onwards with an installation near Mulhause, in Alsace. The process is said to have been working successfully during 1911-12, but the cost is stated to have been prohibitive.

The chief attraction of this process is the prospect of combining the manufacture of pure alumina for the aluminium industry with the production of synthetic ammonia as a by-product.

THE SYNTHESIS OF AMMONIA BY THE HABER PROCESS.[1]

At moderately high temperatures the union of nitrogen and hydrogen can be brought about as expressed by the equation—

$$3H_2 + N_2 \rightleftarrows 2NH_3$$

in accordance with which a definite equilibrium stage is reached, which depends only on the temperature and partial pressures of the different gases. According to the law of chemical mass action, the ratios of the substances obtained at this equilibrium stage, at any particular temperature, are determined by the following relationship :—

$$\frac{p \cdot NH_3}{p^{\frac{1}{2}} \cdot N_2 \cdot p^{\frac{3}{2}}H_2} = K$$

where p represents the partial pressure of the particular gas, and K is a constant at any given temperature. If c is the relative concentration of each gas, and P the total pressure of the system, then $p \cdot H_2 = P \cdot c\,H_2$; $p \cdot N_2 = P \cdot c\,N_2$; and $p \cdot NH_3 = P \cdot c\,NH_3$ and accordingly

$$\frac{c \cdot NH_3}{c^{\frac{1}{2}}N_2 \cdot c^{\frac{3}{2}}H_2} = KP.$$

From this equation it is seen that for low concentrations of ammonia, the percentage of ammonia in the resulting gas is directly proportional to the pressure.

By using high pressures, Haber determined the equilibrium values to be as follows :—

Temperature (Centigrade).	Pressure.	
	100 Atmospheres.	200 Atmospheres.
500°	10·7 per cent	18·2 per cent
600°	4·5 ,, ,,	8·3 ,, ,,
700°	2·1 ,, ,,	4·1 ,, ,,
950°	0·5 ,, ,,	1·0 ,, ,,

The formation of ammonia is exothermic as shown in the equation :—

[1] Haber, " Zeit. für Elektrochem.," 1910, **16**, 244; Haber and Le Rossignol, *ibid.*, 1913, **19**, 53; Haber, " Journ. Soc. Chem. Ind.," 1914, **33**, 49; Crossley, *ibid.*, 1914, **33**, 1140; E. B. Maxted, *ibid.*, 1917, **36**, 777; 1918, **37**, 233 ; Knox, " Fixation of Atmospheric Nitrogen," London, 1914, Gurney & Jackson; E. K. Rideal, " Electrometallurgy," London, 1918, Balliere, Tindall & Cox; " Report of the Nitrogen Products Committee of the Ministry of Munitions," 1920 ; Norton, " Utilisation of Atmospheric Nitrogen," 1912, Washington.

$$N_2 + 3H_2 = 2NH_3 + 2 \times 11,950 \text{ cals.}$$

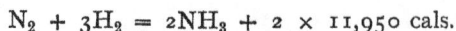

At low temperatures the reaction proceeds very slowly, but it was found by Haber that a marked acceleration was produced by the presence of a number of solids which act as catalysts. An increase of temperature raises the reaction velocity, but lowers the final equilibrium value.

The apparatus used by Haber in his original experiments consisted of two types :—

1. For lower temperatures (500° to 600° C.) the furnace consisted of a strong steel cylinder containing the catalyser, tightly closed with a cap, provided with a steel capillary inlet tube for the gases, and heated externally.

2. For higher temperatures the reaction chamber consisted of a thin iron tube A (Fig. 105), wrapped in asbestos and wound with a nickelin wire C as heating resistance. Into the iron tube passes a quartz tube, which is narrowed towards the lower end and contains the catalyst B between asbestos plugs at this point. The gases enter the quartz tube and leave at the bottom of the iron tube by the steel capillary which is cemented to the quartz reaction tube. The outer steel tube is capable of withstanding pressures up to 200 atmospheres. The space between inner and outer tube is filled with heat-insulating materials. The arrows show the direction in which the gases circulate. Measurements of temperature were made by the thermocouple connected to the leads EE. The pressure in the inner and outer tube is maintained equal by the auxiliary tube D. The ammonia formed in the gases can be conveniently separated by passing through a cooling tube at a temperature of − 75° C. An optical method was applied by Haber for making immediate determinations of the concentration of ammonia in the issuing gas. The method consisted in a measurement of the refrac-

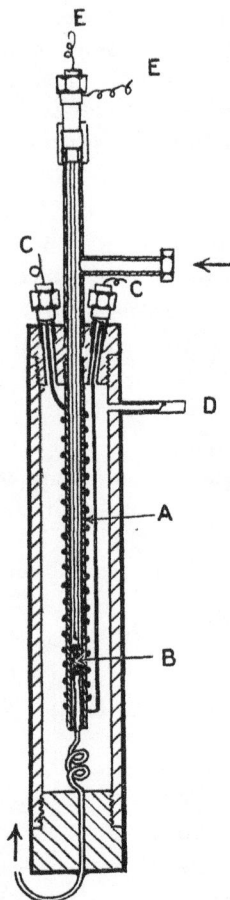

Fig. 105.

tive index of the gas by means of a Rayleigh gas interferometer as made by Carl Zeiss.[1]

The use of this device enabled the progress of the reaction to be followed with ease under varying conditions of the experiment, and the values indicated agreed closely with those obtained by chemical analysis.

The effect of varying catalysts was investigated under varying pressures and with different velocities in the flow of gas. Of the catalysts tried the most effective were osmium and uranium. Uranium is advantageously used in the form of carbide, which in the presence of nitrogen and hydrogen changes into nitride. It was found that many substances have a poisoning influence on the catalyst, while a number of substances, such as alkali, alkaline earths, hydroxides, and salts, and many metals act as promoters and have the power of rendering catalysts in general more active.

Molybdenum and many of its compounds, also tungsten, either as metal, alloy, or nitrogen compound were also found to be very good catalysts.

While uranium compounds form (after osmium) the most effective catalysts, they are susceptible to deterioration through "poisoning" by the action of traces of carbon monoxide. Materials which are freer from this tendency though less active were found in several ferruginous mixtures, such as iron and potash, and some sodamide metal mixtures, and these have now been adopted in technical plants. In practice the most favourable yield of ammonia is obtained by finding the optimum temperature above which the declining equilibrium values more than counterbalance the advantage gained in the increased reaction velocity. A further factor to be considered is that at a certain temperature the catalyst begins to sinter and loses its efficiency. With increase of pressure, both the equilibrium ratio of ammonia and the reaction velocity are increased, but beyond a certain point this advantage is counterbalanced in practical operation by the increased cost in compressing the gases and the strengthening of the reaction vessel. With a given arrangement of catalytic surface, and in an apparatus through which the gases are circulated, the concentration of ammonia in the outflowing gases is regulated by the speed of circulation. The tables given below of results obtained by E. B. Maxted (loc. cit.) show a continuous increase in the total ammonia produced with increasing rate of circulation, though attended by a lower actual proportion in the excess of uncombined gases. As the difficulties in separating the ammonia, which

---

[1] Cf. Haber, "Zeit. für Elektrochem.," 1907, **13**, 460.

is most conveniently brought about by refrigeration, are increased with increasing dilution of the gas, it follows that in practice there is a limit to the advantage gained by increasing the rate of circulation.

The following results of laboratory experiments show the progress of this synthesis under different conditions.[1]

The values given for the yields refer, in all cases, to kilos per hour per litre of catalyst space.

TABLE I.
IRON-POTASH CATALYST AT 550° C. AND AT 150 ATMOSPHERES PRESSURE.

Time of Contact of Gas with Catalyst in Seconds.	Percentage of Ammonia Formed.	Yield of Ammonia in Kilos per Hour per Litre of Catalyst Space.
0·34	0·65	2·7
0·56	0·94	2·3
1·8	1·8	1·4
3·6	3·2	1·25
7·2	4·7	0·91
10·8	5·1	0·65
24·5	6·8	0·38

TABLE II.
IRON-POTASH CATALYST AT 550° C.

Velocity of Flow of Gas in Cubic Metres per Hour at N.T.P. per Litre of Catalyst Space.	50 Atmospheres.		150 Atmospheres.		180 Atmospheres.	
	Per Cent NH₃ Formed.	Yield in Kilos per Litre.	Per Cent NH₃ Formed.	Yield in Kilos per Litre.	Per Cent NH₃ Formed.	Yield in Kilos per Litre.
20	2·2	0·34	4·85	0·75	6·7	1·0
40	1·2	0·37	3·7	1·1	5·4	1·65
60	0·8	0·37	2·9	1·55	4·5	2·1
80	0·65	0·4	2·25	1·4	3·9	2·4
100	0·55	0·42	1·8	1·4	3·5	2·7

TABLE III.
CATALYST, IRON-POTASH. PRESSURE = 150 ATMOSPHERES.

Time of Contact in Seconds.	Temperature = 530° C.		Temperature = 580° C.	
	Per Cent NH₃.	Yield in Kilos per Litre per Hour.	Per Cent NH₃.	Yield in Kilos per Litre per Hour.
0·6	0·96	2·4	1·5	3·5
1·0	1·3	2·0	2·4	3·3
1·5	1·7	1·7	3·2	2·9
2·0	2·05	1·4	3·8	2·6

[1] E. B. Maxted, loc. cit.

CLAUDE PROCESS FOR AMMONIA SYNTHESIS.

A modification of the Haber process, in which use is made of higher pressures, has been brought into operation by G. Claude. It is stated [1] that, in industrial practice, the pressure employed is 1000 atms., with a temperature of the catalyst of 500° to 700° C. At 600° C., with a rate of flow of 100 c.m. per litre of catalyst space, the proportion of ammonia in the effluent gases is given as 25 per cent, corresponding to a combination of 40 per cent of the hydrogen and nitrogen. This yield is equivalent to 6 grams of ammonia (of 25 per cent content) per gram of catalyst per hour.

A cooling of the apparatus is arranged during the operation, the heat evolved being more than sufficient to maintain the desired temperature of the catalyst.

Economics of Synthetic Ammonia Processes. [2]—In technical operation the synthesis of ammonia from hydrogen and nitrogen is effected by passing a mixture of the gases in suitable proportions at high pressures over a catalyst heated to a definite temperature, removing the ammonia formed from the circulatory system by condensation or absorption, and passing the residual gases into the system again together with a supply of fresh gas mixture. The process thus consists mainly in the manufacture, compression, and circulation of very large volumes of gases. About 70,000 cubic feet of hydrogen and 23,000 cubic feet of nitrogen are required theoretically for every metric ton of ammonia produced.

GENERATION OF HYDROGEN ON A LARGE SCALE.

The main sources of hydrogen which come into consideration for the Haber process in this country are the following :—

(*a*) *By-product Hydrogen.*—Certain quantities of by-product hydrogen are available from the electrolytic alkali and cyanide industry, and from the fermentation process for making acetone which has been established in this country since the war. However, the quantity of hydrogen available from these sources is insufficient for the commercial manufacture of ammonia on a large scale.

(*b*) *The Electrolytic Process.*—This process yields hydrogen of a high degree of purity with practically pure oxygen as a by-product. The method is worked on a large scale in Germany by the Machinenfabrik Oerlikon, a solution of potassium carbonate being utilised as electrolyte. The hydrogen contains 1 per cent of oxygen, while the

[1] "J. du Four electrique," 1920, **28,** 5; "Chem. Age," 1920, **2,** 466; "Compt. rend.," 1921, **172,** 442.
[2] "Report of the Nitrogen Products Committee," p. 244.

oxygen produced at the same time contains 2 per cent hydrogen. The Heraeus Company in Hanau uses as electrolyte a 20 per cent solution of potassium hydrate maintained at 60° to 70° C. Both processes require an expenditure of 6 kw. hours per cubic metre of hydrogen, or an output of 5·8 cubic feet per kw. hour. At a cost of £4 per h.p. year, this amounts to 0·75d. per cubic metre of hydrogen (1s. 9d. per 1000 cubic feet), or 4½d. per lb. without allowing for the value of the oxygen.

(*c*) *The Steam-iron Process.*—This process involves the decomposition of steam by means of heated iron and the regeneration of the iron oxide that is formed by means of a gas such as water-gas, containing carbon monoxide. This process is largely used in connexion with the fat-hardening industry in this country, and yields hydrogen containing a small proportion of carbon monoxide as an impurity. It is estimated that under pre-war conditions the total cost of production with a large scale plant might be reduced to about 3s. per 1000 cubic feet. This process is worked in Germany by the Internationale Wasserstoff Aktien-Gesellschaft of Frankfort. Pyrites is found by the company to be the best material for use in the alternate operation of reduction and oxidation, as it retains its porous character.

(*d*) *The Linde Water-gas Process.*—In this process water gas is freed from dust and then compressed and treated with water under pressure in order to remove the carbon dioxide. The compressed gas is then passed into a liquefier in order to condense the carbon monoxide, which, in the case of moderately large plants, can be utilised as a power gas and is sufficient to generate all the power required. This process was employed by the Badische Company for their early Haber installation, but was subsequently abandoned for the catalytic method (*e*).

(*e*) *Catalytic Process.*—In this method carbon monoxide reacts with steam at a temperature of 400° to 500° C. in presence of a suitable catalyst, to give carbon dioxide and hydrogen. The reaction is exothermic, and the heat evolved can be arranged to suffice to cover losses by radiation and conduction. The composition of the cool gas after steam has been removed is approximately as follows : Hydrogen, 65 per cent, carbon dioxide 30 per cent, nitrogen 4 per cent, carbon monoxide 1 to 2 per cent. The carbon dioxide can be removed by treating the mixed gases with water under a pressure of about 30 atmospheres, and the traces of carbon monoxide still remaining by a further purifier containing ammoniacal cuprous chloride solution of such a strength as not to attack the iron vessel.

This process was patented by Mond and Langer as early as 1888, developed commercially by the Badische Company, and used on a large scale at their works at Ludwigshafen. From the data obtained from these works it is estimated that the total production costs, including interest and depreciation, would amount to 2s. 4d. per 1000 cubic feet.

Nitrogen.—All large scale nitrogen production is now conducted by means of the liquefaction and fractionation of air. The power requirements in a modern Claude type of plant amounts to about 9 kw. hours per 1000 cubic feet of nitrogen. If the cost of energy is taken at say 0·25d. per kw. hour, the cost of production, exclusive of interest on capital, would amount to about 5·37d. per 1000 cubic feet of nitrogen.

OPERATION OF THE HABER PROCESS IN GERMANY.[1]

The technical synthesis of ammonia on the lines indicated by Haber has been taken up by the Badische Anilin und Soda Fabrik, and the process was eventually brought to a commercial stage in 1913, and very largely applied during the war. The method originally used consisted in compressing a mixture of one volume of nitrogen and three volumes of hydrogen in presence of a catalyst, when under a pressure of 175 atmospheres, and a temperature of·nearly 550°, about 8 per cent by volume of ammonia is formed. The present working details of the process have been carefully guarded, and as far as is known, no commercial installation has been erected outside Germany. A modified form of the Haber process has, however, been developed by the General Chemical Company in the United States, and a plant erected on a large scale during the war by the U.S. Government at Sheffield, Alabama. The favourable results obtained with experimental units have also led recently to the formation of a company for its operation on a large scale in this country. The plant installed in Germany in 1913 had a capacity of 30,000 tons of ammonium sulphate annually, while the actual production is stated to have been 20,000 tons. This increased to 60,000 tons in 1914; 150,000 tons in 1915; and 300,000 in 1916; and the 1917 output at the Badische works at Oppau, near Ludwigshafen, is estimated to be equivalent to over 500,000 tons of ammonium sulphate.

Synthetic Ammonia Process in England.—A factory commenced operations early in 1918 for the manufacture of synthetic ammonia and the production of ammonium nitrate. The plant has now

[1] " Journ. du Four electrique," 1919, **28**, 15.

(1920) been taken over by Messrs. Brunner, Mond & Co., and, in a factory erected at Billingham-on-Tees, the manufacture of synthetic ammonia and its oxidation to nitric acid and ammonium nitrate has been undertaken for the production of explosives and fertilisers. Improvements in the plant are being embodied which have resulted from the research work carried out by the laboratory founded in 1918 by the Nitrogen Products Committee of the Ministry of Munitions.

The Haüsser (or Explosion) Process.[1]—In this process the union of atmospheric nitrogen and oxygen is brought about by the high temperature produced during the explosion of a combustible gas and air in a closed vessel. The small percentage of oxides of nitrogen which are formed can be recovered from the exhaust gases in the form of dilute nitric acid by absorption in water. A concentration of 0·5 to 0·6 per cent by volume of nitric oxide can be obtained in this method as compared with about 1·25 per cent in the case of the arc furnaces. For this purpose it is necessary to use a gas of high calorific value such as coke-oven gas.

A trial factory to operate this process was erected in 1913 at Heeringen, in Westphalia, the installation comprising stationary bombs having a capacity of 100 litres (3·53 cubic feet). The data obtained show a yield of 135 grams of HNO_3 per cubic metre of coke-oven gas having a calorific value of 4300 kg. calories per cubic metre (483 B.Th.U. per cubic foot), when operating at an initial compression of 5·5 to 6 atmospheres and by pre-heating to about 240° C. The net power requirements of such a plant, in which no use is made of the potential power of the explosions, but allowing for the steam-raising value of the exhaust gases, amounts to about 1 kw. year per metric ton of fixed nitrogen recovered in the form of dilute nitric acid. It was originally proposed to carry out the process with a gas engine, and to utilise the energy of the explosion for performing external work, and large scale trials in this direction have given some success. By enriching the air used in the explosion mixture with oxygen, a considerably increased yield of nitric oxide is obtained.

[1] Cf. " Report of the Nitrogen Products Committee," p. 60.

SECTION IX.

THE AMMONIA OXIDATION PROCESS.[1]

GASEOUS ammonia in the presence of air and suitable catalysts undergoes combustion with the formation of oxides of nitrogen and water. Though the final product of this reaction on proceeding to completion is nitrogen and water, yet by rapidly removing the mixture from the neighbourhood of the heated catalyst, nearly the whole of the nitrogen can be recovered in the form of the intermediate oxygen compound. The technical developments of the process were mainly due to the investigations carried out by Ostwald during the years 1900-2, but many important improvements and developments have since been made.

Technique of Process.—Ammonia is obtained by passing air through a coke tower down which ammonia liquor flows, or else directly in the gaseous form from the Haber process, or from autoclaves in which calcium cyanamide is decomposed. The gas is filtered, particular care being taken to remove any oxide of iron dust, which has a deleterious effect on the platinum catalyst, and after well mixing with air is passed into a tower-shaped converter, which widens towards the centre giving the form of a double cone. The oxidation takes place on the surface of a catalyst consisting of one or more gauzes of platinum wire of fine mesh. The reaction takes place during the very small interval of time in which the gases are in contact with the catalyst, and is represented by the equation

$$4NH_3 + 5O_2 = 4NO + 6H_2O + 220 \text{ kg. cals.}$$

The operation of the catalyst must be initiated by heating the gauze, either by means of a non-luminous gas flame, or electrically by a suitable current, conducted through leads attached to the opposite edges of the gauze. When the reaction begins, the gauze is main-

[1] " Report of the Nitrogen Products Committee," p. 264 ; J. R. Partington, " Journ. Soc. Chem. Ind.," 1918, 37, 337 ; "The Alkali Industry," J. R. Partington. London, 1919, Balliere, Tyndall & Cox.

tained at a red-heat by the heat of oxidation, and the conversion proceeds uninterruptedly as long as the supply of air and ammonia is maintained. The platinum gauze may last for about three months, after which it is desirable to replace it by a new gauze and to clean and refit. New gauzes are not very active at first, but acquire their full catalytic activity after a few hours' running. When no electrical heating is used an increase in the yield is obtained by the addition of one or more gauzes to the first one, placed close together so as to form a composite catalyst. It was found that with a single gauze, and supplying no heat beyond that generated in the chemical action, 90 per cent of the ammonia is recoverable as oxides of nitrogen with a flow rate not exceeding the production of 0·25 ton HNO_3 per square foot of catalytic area, per twenty-four hours. With two gauzes under the same conditions, this may be raised to 0·35 ton with the same efficiency, or to 1·5 tons with an efficiency of 85 per cent. With four gauzes the production attains 2 tons HNO_3 with an efficiency of about 85 per cent. It is found that the efficiency may be maintained at high flow rates by the application of additional heat, either by electrical heating of the gauze, or by pre-heating the mixture of ammonia and air, or the air alone, to about 350° C. With regard to the proportions of air and ammonia, the best mixture appears to be in the neighbourhood of the proportion required to give N_2O_3, when the gauze has a temperature of 650° to 700°. If too rich a mixture is used the gauze becomes too hot, and decomposition into elementary nitrogen results, whereas if too poor a mixture is used, the gauze is too cool and some ammonia passes through unoxidised.

Under established conditions, an output of 1·5 tons of nitric acid (HNO_3) per square foot of catalyst area per twenty-four hours, with an efficiency of 95 per cent, has regularly been attained.

Catalysts.—The commercial plants at present in operation rely upon the use of platinum catalysts. Base catalysts in the form of certain metallic oxides can be used with equal efficiency at a higher temperature.

Catalyst Poisons.—Ammonia of a reasonably high degree of purity is required when the process is carried out with platinum catalysts, and the presence of impurities such as compounds of sulphur and phosphorus, which poison the catalyst, must be avoided.

Recovery of Nitric Acid.—Under average technical conditions a concentration of about 10 per cent nitric oxide is obtained in this process as compared with 1 to 2 per cent in the case of the arc process. The system of absorption employed is similar to that adopted in connexion with arc furnace plants. Large towers constructed of granite or acid-resisting bricks are packed with blocks of stone, brick or hard coke, down which a stream of water is allowed to flow while the oxides of nitrogen pass upwards. In the first absorption tower dilute acid from a later tower is used as the solvent, and in the last tower an alkali solution is used to absorb the last traces of gas. In presence of air and water, the reactions undergone by the nitrogen peroxide are expressed by the equations

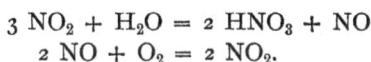

$$3 \ NO_2 + H_2O = 2 \ HNO_3 + NO$$
$$2 \ NO + O_2 = 2 \ NO_2.$$

The second requires time for its completion, and with a 10 per cent mixture of NO_2 and air $2\frac{1}{2}$ minutes contact in the towers should be allowed. The preliminary oxidation is effected by having the first tower empty and the gases are admitted to this after first cooling to 30°, together with secondary air (or oxygen) for the oxidation. An alternative method of recovering the oxides of nitrogen which has been proposed consists in removing the water vapour, adding sufficient air or oxygen to form NO_2, and liquefying this gas by cooling. The liquid peroxide could then be converted into nitric acid by treating with water and oxygen in an autoclave. In this way the bulky and expensive absorption towers would be dispensed with.

Output of Converter Units.—In the commercial developments of this process, comparatively small units have been found preferable to those having a large output. The units normally employed are capable of producing about 100 kg. of HNO_3 per day, or 30 metric tons per annum per unit.

Power Requirements.—As sufficient heat is evolved in the catalytic oxidation of ammonia to maintain the catalyst at the reaction temperature, the power requirements are small even when electrically heated catalysts are employed. Power is mainly necessary for the fans, blowers, and pumps used for circulating the gases and the solutions in the absorption towers. It is estimated that a power expenditure of 135 to 150 kw. hours per metric ton of nitric acid (calculated as 100 per cent), in the form of dilute acid would be adequate for a large-scale ammonia oxidation factory. These figures are equivalent to 0·07 to 0·08 kw. year of 8500 hours per metric ton of combined nitrogen in the form of dilute acid.

Application of Oxides of Nitrogen in Sulphuric Acid Manufacture.—The most profitable application which has so far been made of the ammonia oxidation process is to furnish the oxides of nitrogen needed in sulphuric acid manufacture by the lead-chamber process. For this purpose the oxides of nitrogen from the converter can be passed directly into the sulphuric acid chambers. The absorption towers required for recovering the nitric oxide are thus rendered unnecessary, and a few small converters will meet the requirements of a large sulphuric acid works. Converters have been extensively used for this purpose in Germany during the war, and application has been made of them in this connexion in this country.

Production in America of Nitric Acid by Catalytic Oxidation of Ammonia.

The whole of the calcium cyanamide manufactured at the plant of the Air Nitrates Corporation at Muscle Shoals, Alabama (cf. p. 138), is applied for the manufacture of nitric acid by catalytic oxidation of the ammonia generated in autoclaves. The ammonia is mixed with nine to ten times its volume of air and passed to the catalysers. Each catalysis chamber comprises four rows of 29 units each, separated at intervals of $4\frac{1}{2}$ feet. Ammonia is admitted from above through iron pipes to an aluminium vessel 1 foot 3 inches × 2 feet 3 inches × 5 feet high, at the base of which is a horizontal net of platinum gauze of 0·07 mm. wire, weighing 135 grm., and heated to 750° C., by an electric current. Each catalyser is supplied by a transformer of 8 kw., with 375 amps. at 21 volts. On passing through the gauze the ammonia is converted into oxides of nitrogen which can be observed by means of a small window at the side of the aluminium case. In order to cool down from 600° to 200° C. the gases are passed through aluminium tubes traversing a boiler, and then through a second series to cool to 30° C., when the nitric acid begins to form and is drained off into a well. The remaining gases, containing nitric oxide and nitrogen peroxide, are circulated along brick channels, when complete oxidation of the NO results, and the gas is brought into contact with a spray of water which is circulated in a series of towers in the opposite direction to the gas stream, so that the solution contains finally 50 per cent of acid. The acid is then neutralised by ammonia and crystalline ammonium nitrate obtained.

An efficiency of 90 to 95 per cent is obtained in the oxidation, with 5 to 6 lb. of ammonia per square foot of catalyst screen per hour. The catalyser is usually employed continuously for a year, by which time it loses about 5 per cent of its weight and is renewed. It has not been found necessary to submit the ammonia generated from the autoclaves to careful purification, no cases of poisoning of the catalyst having been experienced.

The plant installed at Muscle Shoals has a capacity corresponding to about 90,000 tons of HNO_3 per annum.

In view of the control obtainable with electrical heating, the cost of the energy consumed, amounting to about 1/3 of 1 per cent of the present market value of nitric acid, may be regarded as negligible.

Other Processes for the Catalytic Oxidation of Ammonia in the United States.[1]—Processes for the production of nitric acid from ammonia prepared synthetically, similar to the one at Muscle Shoals were, during the war, brought into operation at several other centres in the United States. A smaller plant was erected at Sheffield, Alabama, in conjunction with the synthetic ammonia plant, with a capacity of about 15,000 tons of HNO_3 per annum. The catalyser used in this case consists of a non-electrically heated multiple screen, of several layers of platinum gauze, which are welded together at points and rolled in the form of a cylinder. The ammonia mixture flows outwards through the screen at a rate several times as fast as with the electrically heated single screen. After the oxidation has been started by the external application of heat, the temperature is self-sustained from the heat of reaction.

A plant similar to the above is also said to be in operation by the Semet-Solvay Company, at Syracuse, New York. In addition to the above, the construction of a plant is said to have been undertaken at Indian Head, Maryland, by the United States Naval Department for fixing nitrogen by a modified Haber process, followed by oxidation of the ammonia to nitric acid, with a yield of 30,000 tons per annum.

Catalytic Oxidation of Nitric Oxide.—Considerable work is, with promising results, being carried out on the use of catalysers to hasten the conversion of the oxides of nitrogen obtained from the converters into nitric acid.

Manufacture of Nitric Acid in Germany.

The Ostwald process is the method which has been most largely used for the production of nitric acid in Germany. In 1909 a factory was equipped at Gerte, near Borkum, in Westphalia, for producing annually 2400 tons of nitric acid (53 per cent HNO_3). Shortly after the announcement of Ostwald's invention, F. Baeyer & Company of Elberfeld brought out a similar contact method[2] in which the oxides

[1] E. J. Pranke, "Metall. and Chem. Engineering," 1918, **19**, 395.
[2] Ger. Pat. 168,272.

of the heavier metals, copper, iron, etc., were employed as catalytic agents. A current of air mixed with 4 to 5 per cent of ammonia gave the best results, and the temperatures ranged from 600° to 750° C. This method does not appear to have been brought into operation on an industrial scale.

The ammonia oxidation process was installed shortly before the war in the Kayser works at Spandau, while a modification of the process, viz. that of Frank-Caro, was employed at the Badische Anilin und Soda Fabrik at Ludwigshafen, at the chemical works Griesheim-Elektron at Frankfort-on-Main, and at the Hochst works. The Frank-Caro converter contains an electrically heated single platinum gauze of wire 0·065 mm. diameter with 80 meshes to the linear inch. The annual production by the Frank-Caro process in Germany is estimated at 100,000 tons. Frank and Caro (Ger. Pat. 224,329) propose the use of cerium and similar oxides as contact substances, and more especially thoria, which is claimed to give the highest yield in oxidising ammonia.

A process of Kayser has also been largely applied in Germany. The original apparatus contained several platinum gauzes which were placed close together across a tube, and the air was pre-heated separately in a coke furnace to 300° C. before mixing with the ammonia. A very high flow-rate was employed, and it is claimed that a much larger output per unit area of catalyst is obtained than in other processes.

Ammonia Oxidation Process in England.—A works is in operation at Dagenham, Kent, which was erected during the war, for the production of nitric acid from by-product ammonia. The method used is the Ostwald process, the patents for which have been taken over and exploited by the Nitrogen Products and Carbide Company.

Economics of Ammonia Oxidation Process.—It is estimated that the cost of converting autoclave ammonia into concentrated nitric acid is just about equal to the cost of converting nitrate of soda to concentrated acid, while ammonia gas is a cheaper form of nitrogen than is nitrate of soda.

In the manufacture of sulphuric acid by the chamber process, the oxides of nitrogen employed are now being generally produced by this method of oxidising ammonia.

ESTIMATED PRODUCTION OF SYNTHETIC NITROGEN COMPOUNDS IN ALL COUNTRIES.[1]

Material.	Estimated Production in 1914. Metric Tons of Nitrogen.	Estimated Production in 1919. Metric Tons of Nitrogen.
Nitrate of soda	386,400	472,700
Sulphate of ammonia . . .	259,100	300,000
Arc process products	10,000	45,000[2]
Cyanamide products	28,200	327,300
Synthetic ammonia and miscellaneous	10,900	105,500
Total	694,600	1,250,500

FUEL REQUIREMENTS OF PROCESSES FOR SYNTHETIC PRODUCTION OF NITROGEN COMPOUNDS.[3]

Coal-fired Steam Power Plant Utilised.

Process, and Power Requirements in Kw. Years per Metric Ton of Combined Nitrogen in the Form of Product Specified.	Fuel Required per Metric Ton of Combined Nitrogen, with Large Scale Plant, and with Coal of 12,500 B.Th.U. per Lb.		
	For Power Plant. Tons.	For Process Operations. Tons.	Total. Tons.
Arc process. 8·41 kw. years of 8760 hours. Product: concentrated HNO_3	52·56	nil[4]	52·56
Calcium cyanamide process. 1·97 kw. years of 8400 to 8500 hours. Product: calcium cyanamide .	13·79	3·0[5]	16·79
Haber process. 0·42 kw. years of 8500 hours. Product: pure ammonia liquor (20 to 25 per cent NH_3) .	3·36	3·85[6]	7·21

[1] W. S. Landis, " Trans. Amer. Electrochem. Soc.," 1918, 34, 105.
[2] Estimate of E. Kilburn Scott, *ibid.*, p. 113.
[3] " Report of the Nitrogen Products Committee," p. 57.
[4] Waste heat of arc furnace gases utilised for concentration of acids.
[5] Anthracite and coal for carbide stage.
[6] Coke for hydrogen manufacture.

PRODUCTION COSTS OF NITRIC ACID BY DIFFERENT PROCESSES.[1]

£ per metric ton of combined nitrogen in the form of concentrated nitric acid. Factory costs of unpacked product exclusive of Royalties and on pre-war basis of charges.

Initial Process or Product.	Basis of Costs.	Total Cost per Metric Ton N₂, in Form of Concentrated Nitric Acid.
		£
By-product ammonia. *Via* ammonia oxidation process	Based on pre-war (average 1911-13) market price.	89·72
Haber process. *Via* ammonia oxidation process	" Probable."	44·15
Cyanamide process. *Via* ammonia oxidation process including cost of carbide, cyanamide, and ammonia stages .	With power (steam generated) at £3·75 per kw. year.	54·12
	With power at £3 per kw. year.	52·36
	With water power at £2 per kw. year.	49·95
Arc process.	With power (steam generated) at £3·75 per kw. year.	51·80
	With power at £3 per kw. year.	45·50
	With water power at £2 per kw. year.	37·08
Chile nitrate.	Based on pre-war (average 1911-13) market price.	100·48

FUTURE DEVELOPMENTS OF NITROGEN FIXATION INDUSTRY.

(a) *Arc Processes.*—In view of the large expenditure of power required for the synthesis of nitric acid by the arc process, and as water, air, and limestone constitute the whole of the raw materials required, the development of this industry may be expected to take place in more isolated and undeveloped centres where an abundance of cheap water power is available. In more developed centres, even where water power occurs, on account of its use for motive power, chemical and metallurgical industries and illumination, the amount of power being limited, its selling price is automatically brought to the level of the cost of power generated from fuel. Plans for the future development of the nitrate industry are accordingly directed to the acquisition of rights and the development of water powers in districts

---

[1] " Report of the Nitrogen Products Committee," Appendix V; cf. also C. L. Parsons, " Journ. Soc. Chem. Ind.," 1916, **36**, 1081.

isolated from any existing industrial development. The most note-worthy of such places now under consideration are in Iceland, Western Norway, Alaska, the Zambesi river, the Assuan Dam, the Eastern slopes of the Andes, and tidal power in this country.

In the case of all synthetic nitrogen processes, the scope offered is determined by the market price of Chile nitrate and by-product ammonia in comparison with the cost of power generation and operation of the process. In Scotland the cost of generating power on a large scale has been shown to be possible under pre-war conditions at a cost of £3 per kw. year (*vide* p. 412).

According to data given above (table p. 160), at this cost, nitric acid can be prepared by the arc process at a cost lower than one-half that required for the production of the acid from Chile nitrate, on the basis of pre-war market prices for the saltpetre.

Arc Processes as Loads for Intermittent Power.

The characteristics of the arc processes for the fixation of nitrogen with their low cost of supervision should make them suitable for favourable operation with an intermittent supply of power such as that obtainable during times of "off peak" loads at large supply stations, and more particularly that derived from tidal power (cf. p. 437).

(*b*) *Calcium Cyanamide.*—The position with the fixation of nitrogen as calcium cyanamide in regard to the power expenditure per unit weight of nitrogen is considerably more favourable than that of the arc process, but even in this case the cost of power generated from fuel can hardly be expected to enable the process to be operated economically except by the use of low-price power.

(*c*) *Haber Process.*—The consumption of power in the Haber process is of minor importance compared with the total engineering costs, so that the operation and situation of this manufacture will not be determined by the availability of cheap power.

SECTION X.

ELECTRIC SMELTING OF IRON-ORES.

In its application to the production of pig-iron, the electric furnace has introduced itself into a department of metallurgy which had already been developed to the highest state of efficiency by the use of fuel-heated methods. The modern blast furnace on account of this high degree of efficiency, might, *prima facie*, be regarded as an established process, which is the least likely to admit of the introduction of a system of such disputed economy as the electric furnace. In the reduction of iron-ore by coke in a large-size and modern blast-furnace, using pre-heated air for the blast, and utilising the calorific value of the furnace gases, it is estimated that in the smelting and reduction of the charge, a thermal efficiency of 70 per cent of the heat value of the coke is obtained. The iron-smelting industry has become situated in districts where iron-ore and suitable coke are available at a low price, and the conditions in these selected regions are certainly adverse to the competition of any alternative method of heating. This position, however, does not apply in the very large areas where fuel is scarce, but iron ore and electric energy, generated from water power, cheap. Moreover, the blast furnace, in spite of the high efficiency attained, has several disadvantages. On account of the large units and high shafts a special quality coke is required to withstand the weight of the column of charge ; wrong compositions of charge are difficult to correct when working on a large scale, and the nitrogen present in the blast has a deleterious effect by imparting brittleness to the iron.

There are two distinct directions where electric furnace processes have received successful commercial application in this connection :—

1. For the production of ordinary pig-iron in districts where furnace coke is scarce and electric power cheap. With the blast furnace the quantity of coke or charcoal used as fuel is about equal to the amount of pig-iron produced. In the electric furnace the carbon is used only for the reduction of the oxide, the heat being

supplied electrically. The consumption of carbon is consequently reduced to about one-third of the amount needed with the blast furnace. (On the basis of the production of carbon monoxide and dioxide in the ratio of 2 : 1, the amount of carbon theoretically necessary per metric ton of pig is 214 kilos.) Electric smelting processes are now operated extensively in Sweden and in California, and it is estimated that other factors being equal, electric smelting becomes cheaper than the blast furnace when the cost of 3 tons of fuel is higher than that of 1 kw. year of power.

2. For the production of a high-grade pig-iron, comparable with Swedish charcoal iron, by smelting selected ores with good quality coke or charcoal, which is added in the limited quantity needed for the chemical reduction, the heat being applied electrically.

It was in this direction that Stassano directed the first attempts which were made to establish commercially the electric smelting of iron ores, a wrought iron or mild steel being produced directly in one stage from ore. These attempts were afterwards abandoned in favour of the electrical production of steel from pig-iron, but within the last few years a considerable development has taken place in the production of a high-grade pig-iron (1) from high-grade ores and charcoal, and (2) a product of low phosphorus content from scrap steel and iron-ore. Several plants for these processes are now in successful operation in Sweden, California, France, and Canada.

The main advantages obtained by the electric furnace process for the smelting of iron-ores may be summarised as follows :—

1. Less contamination of the product with impurities from the coke or other form of carbon used, on account of the smaller quantity required.

2. Absence of contamination by nitrogen, through exclusion of air from the electric furnace.

It is well known that in the blast furnace the formation of cyanides occurs through reaction between carbon and nitrogen in presence of a basic slag (cf. p. 133), and concurrently the formation of nitride of iron occurs, which renders the iron brittle. This reaction is now known to be an important factor in determining the superiority of electrically produced iron and steel, in the preparation of which nitrogen is eliminated.

3. The possibility, through the higher temperature attainable, of employing a more basic slag, and thus securing a better removal of sulphur and phosphorus.

4. Exact control of the temperature in the reducing and melting zone.

11 *

5. The recovery of a gas from the smelting with a much higher content of carbon monoxide than is present in blast-furnace gas, which is highly diluted with nitrogen. The richer gas is thus better suited for utilisation by combustion, particularly when used in gas engines for power generation.

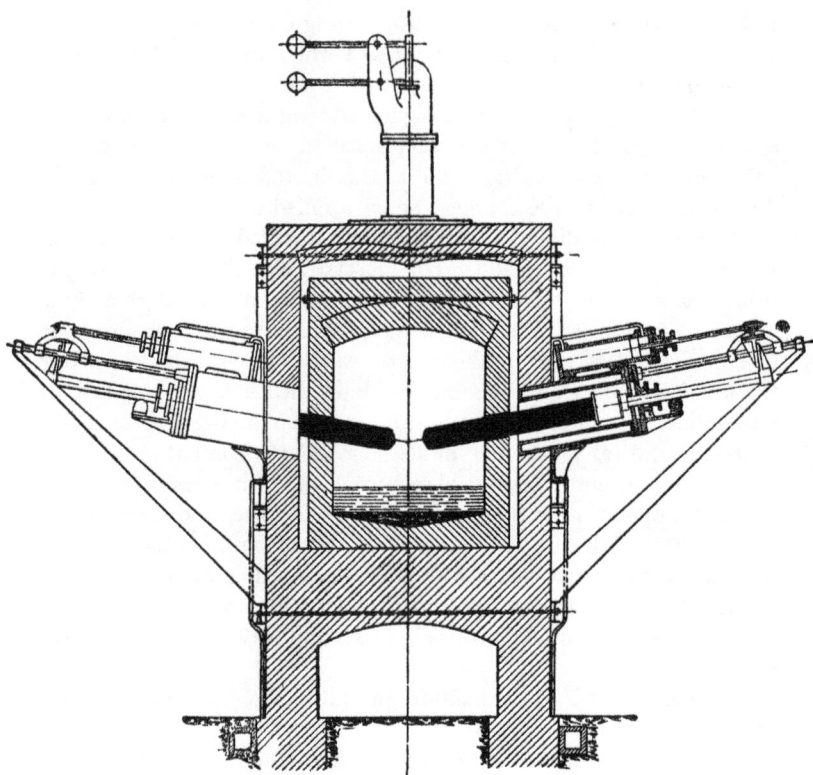

FIG. 106.

STASSANO'S FURNACES.

The original furnaces used by Stassano for smelting iron-ores consisted, as seen in Fig. 17, of a structure of similar form to the blast furnace, the tuyères being replaced by carbon electrodes arranged at a slight inclination to the horizontal position, allowing an arc of the "smothered" type to be formed in the body of the charge. The charge was admitted through the hopper at the head of the shaft, while the molten product collected in the space below the electrodes. The electrodes were surrounded by a water-cooled jacket, and could be regulated by a feeding screw. The gases from the reaction were led away through vents near the head of the shaft. However, with this method, the resistance of the charge was too high and the resulting iron had too high a carbon content. The method was finally replaced by a furnace in which the arc was formed above the surface of the charge, as seen in Fig. 106. This design of furnace was installed at Darfo (Lago d'Iseo), Lombardy, Italy, and later adopted at the Royal Smelting Works in association with the Arsenal, Turin.

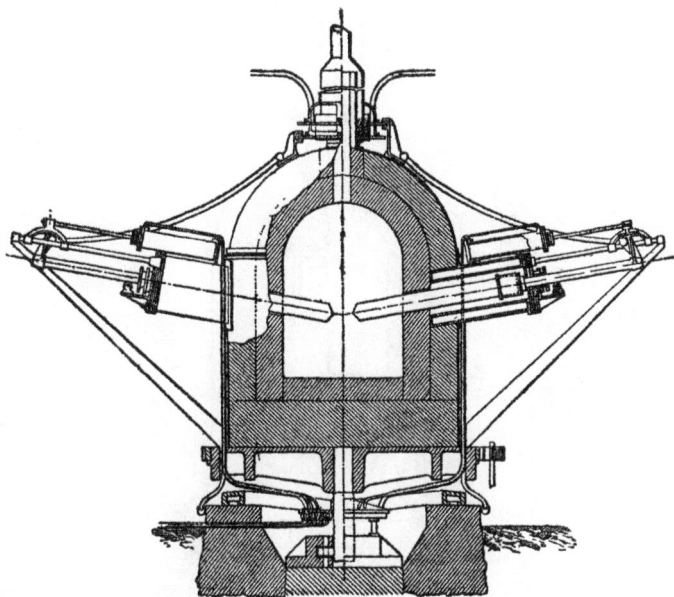

FIG. 107.

At Darfo the first furnace erected consumed 100 h.p. and employed an arc of 1000 amps. at 80 volts. A second furnace had a power consumption of 500 h.p. and employed 2000 amps. at 170 volts alternating current. Regulation of the electrodes was made by hand. In the 500 h.p. furnace a yield of 30 kg. of cast-iron was obtained from 70 kg. of ore in two hours. In the works at Turin, which were erected in 1903, a rotating type of furnace was installed. The construction of this is shown in the diagram in Figs. 107 and 108. Fig. 107 shows a vertical section through the two electrodes; and Fig. 108 a vertical section through the plane of the tapping spout and charging hopper. The vertical axis of the furnace is slightly inclined to the perpendicular, so that, on rotation, a mixing of the charge is brought about. Gases from the reaction pass out through a vent at the head of the enclosure. Two or three electrodes, spaced at equal intervals, pass through the refractory walls of the furnace and are inclined to the horizontal so that the carbon poles point downwards towards the charge. The electrodes and attached clamps are, for purposes of cooling, surrounded by a double-walled cylindrical jacket, through which water

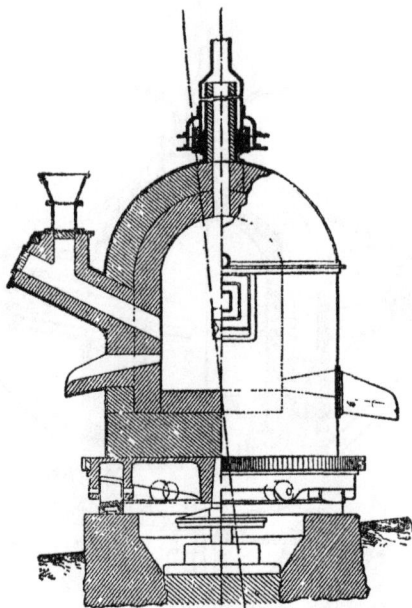

FIG. 108.

circulates. In Fig. 108 the funnel and shaft for admitting the charge are seen and openings at different levels for tapping off the metal and slag respectively. Movement of the electrodes is effected by hydraulic cylinders. The current is applied to the fixed rings near the base from each of which, by means of a movable brush contact, the current is led to the respective electrodes. Access of air during the smelting is precluded.

The Stassano furnace was originally applied for the smelting of Italian magnetite and hæmatite, containing 0·05 to 0·07 per cent sulphur, and 0·08 to 0·09 per cent phosphorus. Lime and silica were added as required for the formation of slag, and the necessary amount of carbon for the reduction. In some cases the ore was finely ground, and together with the other added materials in the form of powder was moulded into briquettes, by adding pitch and compressing in a hydraulic press, and smelted in this form. In this way loss of material through carrying away of the dust was avoided. The ore, after grinding, was, in some cases, concentrated magnetically.

For the production of one metric ton of iron or steel at Darfo, the following expenditure of materials and power was required: 1·6 tons of ore, 0·2 ton of slag-forming material, 0·25 ton of coke, and a consumption of electrodes of 12 kg. and electrical power of 4000 h.p. hours (0·34 kw. year). With the prevailing prices of materials (including power at 0·056d. per h.p. hour, or a total of 18·2s. per one ton of metal), the total cost of production was estimated at £4 9s., while the heat value of the carbon monoxide generated was equivalent to 14s., giving a net cost of £3 15s.

In some cases cast-iron and turnings were used as raw material in place of ore, thus giving a more favourable yield for the power consumed. In the earlier experiments a product of somewhat high carbon content was obtained, but later, by using a hearth of magnesia in place of graphite, the proportion of carbon was greatly reduced. Products of the following average composition were obtained:—

Fe = 99·68, Mn. = 0·09, Si = 0·03, S = 0·06, P = 0·02,
$\quad$ C = 0·10 per cent.

A typical chrome-steel was obtained of the composition

$\quad$ C = 1·51, Mn = 0·26, Cr = 1·22 per cent.

In the furnaces at Turin malleable iron was produced direct from ore, while steel for artillery uses was made from a mixture of ore and scrap metal and turnings, which were supplied from the arsenal.

The life of the refractory lining of the furnace, when in continuous operation, was thirty days.

HÉROULT FURNACE.

Experiments have also been successfully carried out by Héroult on the electric smelting of iron-ores, the type of furnace used being that which was designed for the production of steel, and is described on page 209.

KELLER FURNACE.

The Keller Furnace for the electric smelting of iron-ores was brought into operation about 1901 by the company Keller,

Leleux & Co. at Kerrousse (Morbihan), France. The procedure included the use of two furnaces, one for the production of pig-iron from the ore, and a second for refining the pig-iron. As seen in Fig. 109, the installation included a furnace with two shafts. A power of 1000 e.h.p. was employed. Heating is effected by means of arcs formed from the electrodes EE and the resistance of the charge. The hearth M of the furnace is slightly inclined, and shortly above it are openings in the furnace walls on opposite sides for the withdrawal of the slag and matte respectively. The charge of ore, carbon, and fluxing material C is admitted at the top of the shafts. The carbon monoxide evolved from the reaction is collected, and by burning employed to dry the raw materials, and in some cases

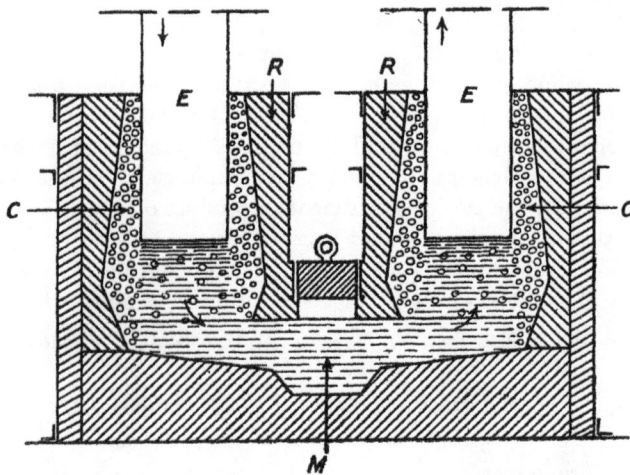

Fig. 109.

for a preliminary reduction of the ore. The tapped cast-iron is collected in a second or refining furnace, covered with slag-forming materials, on the surface of which an arc was formed from two suspended vertical electrodes. For the production of a ton of cast-iron by this process, a power expenditure of 1800 kw. hours, and for one ton of steel 2800 kw. hours is stated to be necessary.

Investigation of Canadian Commission of Electrothermic Processes for the Smelting of Iron-ores and Production of Steel.

In order to form an opinion as to the feasibility of introducing electric smelting of iron-ores as a commercial process in those provinces of Canada which lack coal for metallurgical coke, but are well supplied with water power and iron-ore deposits, the Canadian Government in 1903 appointed a Commission to investigate the

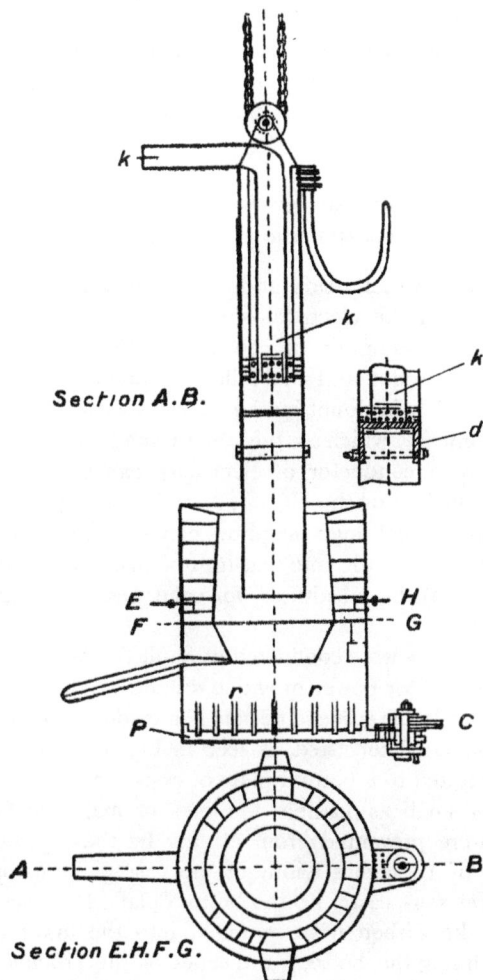

FIG. 110.

different electrothermic processes for the smelting of iron-ores and the making of steel in operation in Europe. Experiments on the reduction of iron-ores which were witnessed by the Commission were conducted by Dr. P. Héroult, at La Praz, France, and by

Messrs. Keller, Leleux & Co. at Livet, France. Héroult's experiment was only made to, show the possibility of smelting iron-ores, and no data in regard to output, etc., were obtained.

In the experiments of Messrs. Keller, Leleux & Co. the ore used was a porous hæmatite containing 3·21 per cent of manganese and only 0·02 per cent of sulphur, an ore, therefore, easily reduced and desulphurised. The consumption of electrical energy amounted to an average of about 0·350 e.h.p. year (365 days) per ton of cast-iron produced.

EXPERIMENTS CONDUCTED AT SAULT STE MARIE, CANADA, ON THE ELECTRIC SMELTING OF CANADIAN IRON-ORES.[1]

The Canadian Commission undertook the carrying out of a series of experiments on the electrical smelting of iron-ores, in order to determine the consumption of power with Canadian ore, the consumption of electrodes and the following points which were either not taken up or left in doubt by the Livet experiments :—

(1) If magnetite, which is the chief Canadian ore, and which is to some extent a conductor of electricity, can be successfully and economically smelted by the electrothermal process ; (2) If iron ores with a comparatively high sulphur content, but not containing manganese, can be made into pig-iron of marketable composition ; (3) If charcoal can be substituted for coke, and use thus made of a home product.

The experiments were conducted at Sault Ste Marie, in a building and with facilities for power provided by the Lake Superior Power Corporation, while the furnace design was made by Dr. P. Héroult.

The furnace used consisted, as seen in Fig. 110, of an iron casing ¼ inch thick, bolted to a bottom plate of cast-iron P. To render the inductance as small as possible, the lines of magnetic force in the iron casing were prevented from closing by the replacement of a vertical strip of 10 inches width of the casing by a copper plate. Rods of iron *rr* were cast into the bottom plate P to secure a good contact with the carbon paste rammed into the lower part of the furnace and lining the bottom and sides of the crucible, while the lining of the upper part of the furnace was made of firebricks.

The electrodes, manufactured by the Héroult process and imported from Sweden, were of 16 inches × 16 inches cross section by

---

[1] " Report on the Experiments at Sault Ste Marie, Ontario, under Government Auspices, in the Smelting of Canadian Iron-ores by the Electrothermic Process," E. Haanel, Ottawa, 1907.

6 feet long.   One end of the electrode was planed to fit into the steel shoe d, and held tight by means of wedges.   The steel shoe was riveted to four copper plates, two of which were strengthened on top with steel plates and were attached to a pulley.   A pipe k was put in the electrode holder, through which a current of air was circulated on to the holder.

The aluminium block into which the cables constituting the conductor C were cast was bolted to one of the copper plates.   The power applied was 200 kw. single-phase alternating current at a mean value of 5000 amps. at 40 volts, and a power factor of 91·9 per cent.

Smelting of Magnetite.—No difficulty was experienced in the smelting of magnetite on account of its conductivity, when using charcoal as a reducing agent.

Charcoal as a Reducing Agent.—Charcoal was found to be successful as a reducing agent, and did not require briquetting with the ore, but gave good results after crushing so as to pass a ¾ inch ring.   As some of the charcoal used was of very poor quality containing only 56 per cent carbon, and also on account of its consumption through access of air at the top of the furnace, a larger quantity was used per ton of product than with coke.

Content of Sulphur and Phosphorus.—By increasing the amount of limestone which, together with sand, was added to form the slag, the content of sulphur and phosphorus could be reduced to a very low degree, as shown in the detailed results given in the table below :—

RUN No. 14.[1]

Ore treated : Magnetite from Blairton Mine.
Reducing Agent : Charcoal.
Flux : { Limestone
          { Sand
Analysis of Raw Material.
  Blairton Ore.

	Per Cent.			Per Cent.	
SiO_2 .	. 6·60		MgO	. 5·50	
Fe_2O_3	. 60·74 } Fe = 55·85 per cent		MnO	. 0·13	
FeO .	. 17·18 }		P_2O_5	. 0·037 P = 0·016 per cent	
Al_2O_3	. 1·48		S .	. 0·57	
CaO .	. 2·84		CO_2 and undetermined .	4·923	

Loc. cit., pp. 53-9.

Proportions in Charge :—

Ore .	.	400 lb.
Charcoal .		125 ,,
Limestone		25 ,,
Sand	.	6 ,,

Analysis of Slag.

SiO_2	.	36·16
Al_2O_3	.	18·21
CaO	.	23·14
MgO	.	20·44
S .	.	2·00
FeO	.	0·42
P_2O_5	.	0·018

Analysis of Iron Produced :—
Cast No. 80, Grey Iron
Total carbon 3·73

Si	.	. 3·53
S	.	. 0·042
P	.	. 0034

Yield, 11,989 lb. pig-iron = 3·62 tons per h.p. year.
Mean current, 5000 amps. Mean volts on furnace, 35·8.
Power factor, 0·91. Mean kilowatts, 161.
Electrode Consumption.—The consumption of electrodes amounted on an average to 18 lb. per ton of pig-iron.

PRODUCTION OF FERRO-NICKEL PIG.

Ore treated : Roasted Pyrrhotite.
Reducing Agent : Charcoal.
Flux : Limestone.

Analysis of Raw Material.

(Roasted Pyrrhotite.)

SiO_2.	. 10·96		MgO	.	3·53
Al_2O_3	. 3·31		S .	.	1·56
Fe_2O_3	. 65·43		P .	. 0·016	Fe = 45·8
CaO .	. 3·92		Cu .	.	0·41
			Ni ·	.	2·23

Proportions in Charge.

Ore .	.	400 lb.
Charcoal .		110 ,,
Limestone		50 ,,

Analysis of Slag.

SiO_2	.	16·44
Al_2O_3	.	13·86
CaO	.	53·25
MgO	.	8·80
S .	.	5·28
Fe .	.	0·65
Cu .	.	trace
Ni .	.	trace

Analysis of Iron Produced :—
Total carbon 3·38

Si	.	. 4·50
S	.	. 0·006
P	.	. 0·037
Cu	.	. 0·87
Ni	.	. 4·12

Yield, 7336 lb. Ferro-nickel = 2·6 tons per h.p. year.

Results and Conclusions of Experiments.—The general results of the above experiments are summarised as follows :—

1. Magnetite ores can be smelted as economically as hæmatites by the electrothermic process.

2. Ores of high sulphur content can be made into pig-iron containing only a few thousandths of 1 per cent of sulphur.

3. The silicon content can be varied as required for the class of pig to be produced.

4. Charcoal, which can be cheaply produced from mill refuse, or wood which could not otherwise be utilised, and peat-coke can be substituted for coke without being briquetted with the ore.

5. A ferro-nickel pig can be produced practically free from sulphur and of fine quality from roasted nickeliferous pyrrhotite.

6. Titaniferous iron-ores containing up to 5 per cent can be successfully treated by the electrothermic process.

Remarks.—The main importance of the results achieved in the above experiments is due to the fact that many of the Canadian magnetites are too high in sulphur to be smelted in the blast furnace and consequently have so far been of no commercial value. The above experiments have shown, on the other hand, that the best of pig-iron can be made from ores which contain as high as 1·5 per cent of sulphur. With regard to the availability of water power, it was estimated that many sources of such power exist in Ontario and Quebec, surrounded by iron-ore fields, which could be developed to furnish an electric horse-power year for from $4·50 to $6·00. With the present advance which has been made in the transmission of electric energy, batteries of electric furnaces could be set up at various iron-ore deposits, which could be supplied with electric energy from some centrally located water power, thus effecting a saving of the transportation costs of the ore from the mine to the furnace.

FERRO-NICKEL PIG.

The plant at Sault Ste Marie used in the above experiments was afterwards purchased by the Lake Superior Corporation and applied for the semi-commercial production of ferro-nickel pig. For this purpose roasted nickeliferous pyrrhotite, containing about 2 per cent S, was smelted together with charcoal and limestone. The average composition of the product was 2·75 per cent Si, 0·01 per cent S, 0·03 per cent P, 4 per cent Ni, and 0·8 per cent Cu.

The consumption of materials for 1 short ton of pig was 1500 lb. limestone, 1200 lb. charcoal, and 40 lb. of electrodes.

SMELTING OF IRON-ORE AT WELLAND, ONTARIO.

On account of the success obtained with the experimental plant at Sault Ste Marie, the commercial manufacture of pig-iron by electric smelting was undertaken at Welland, Ontario, on a site facing the Welland Canal. The installation consisted of cne 3000 h.p. furnace, with a production of about 35 tons of pig-iron per day. The electric smelting of iron-ore was also undertaken in California (p. 187).

ELECTRIC SMELTING OF IRON-ORE IN SWEDEN.[1]

In Sweden, where the conditions relating to the iron industry are, in the main respects, identical with those existing in several of the provinces in Canada, the importance of electric smelting processes has been fully realised. The work carried out at Sault Ste Marie has been continued in Sweden, mainly by Messrs. Grönwall, Lindblad, and Stalhane, who have successfully applied, on a large scale, the commercial production of pig-iron by electric smelting. Experiments on a large scale were first undertaken at the Domnarfvet iron works, with the co-operation of the owners of large iron-ore deposits in Sweden, and later a special company was formed, known as the "Aktiebolaget Elektrometall," to which the patent rights were assigned.

The construction of the first electric shaft furnace was begun in 1906, and the points systematically investigated were :—

1. The construction and operation of electric furnaces.

2. The conductivity and other characteristics of materials at high temperatures.

3. The quality of the refractory lining materials.

4. The most suitable manner of designing and constructing the masonry of the furnaces.

5. Different methods of supplying the current, and the use of various contact devices.

Electrodes.—With large scale furnaces, in distinction from the early type of Héroult-Haanel, the use of an electrode in the base of the furnace is found disadvantageous, and in all types which have

---

[1] "Report on the Investigation of an Electric Shaft at Domnarfvet, Sweden," E. Haanel, Department of Mines, Ottawa, 1909; "Report on Electrothermic Smelting of Iron-ores in Sweden," A. Stansfield, *ibid.*, 1915; Helfenstein, "Iron and Coal Trades Review," 1914, **88**, 505 ; J. Härden, "Met. and Chem. Engineering," 1914, **12**, 82, 223, 444; J. Orten-Boving, "The Iron Age," 1914, **93**, 1268.

been subsequently developed, the electrodes are all arranged in a central position above the smelting hearth.

The use of an electrode in the base of the furnace leads to a weakening of the construction, and there are, moreover, great technical difficulties in conducting very big currents away from the floor of a furnace. The heating of the base of the furnace means a loss of energy, and has a very injurious effect on this most important part of the furnace.

Phase Displacement.—A difficulty which is encountered when applying very big currents on adjacent electrodes is that of phase displacement, due to the inductive e.m.f. which increases in proportion to the size of the units. As explained above (p. 39), the existence of this factor necessitates the provision of a larger generator plant in order to produce a given amount of power.

Merits of Open and Closed Types of Furnaces.—The use of an open top facilitates the feeding-in and inspection of the charge and is in general use with smaller sizes of furnaces, and with nearly all furnaces for the manufacture of calcium carbide and ferro-alloys. The disadvantages of the open type of furnace, however, are in the combustion of the furnace gases which takes place in the free space above, the heat being injurious to the conductors and electrode connexions. Further, the attendants are exposed to the heat and smoke emitted. These disadvantages increase in proportion to the size of the furnace, and may be obviated by adopting the following features :—

1. A durable furnace cover or roof.
2. Utilisation of the gas.
3. Continuous charging.

Advantages of Large Units.—A saving of space is effected by the use of large units, as one large furnace requires a much smaller space than several small ones.

The capital expenditure per h.p. is considerably smaller with a large unit than with a small one. The amount of materials required in the construction of a large unit is, per h.p., only a fraction of that in small furnaces. The larger and closed furnaces require fewer attendants for a given output. Utilisation of the evolved gases is only possible in the case of large furnaces.

Nature of Ore Employed.—It is important to use a high-grade ore for electric smelting, as the content of iron in the charge has a large influence on the electrical efficiency. If coke is used as the reducing material, it is necessary, on account of the higher content of sulphur and phosphorus, to work with a basic slag in order to obtain a pure material. The quantity of lime to be added is, in consequence, larger. It is advisable to use a mixed charge of ores of acid and basic nature so as to avoid a larger addition of lime.

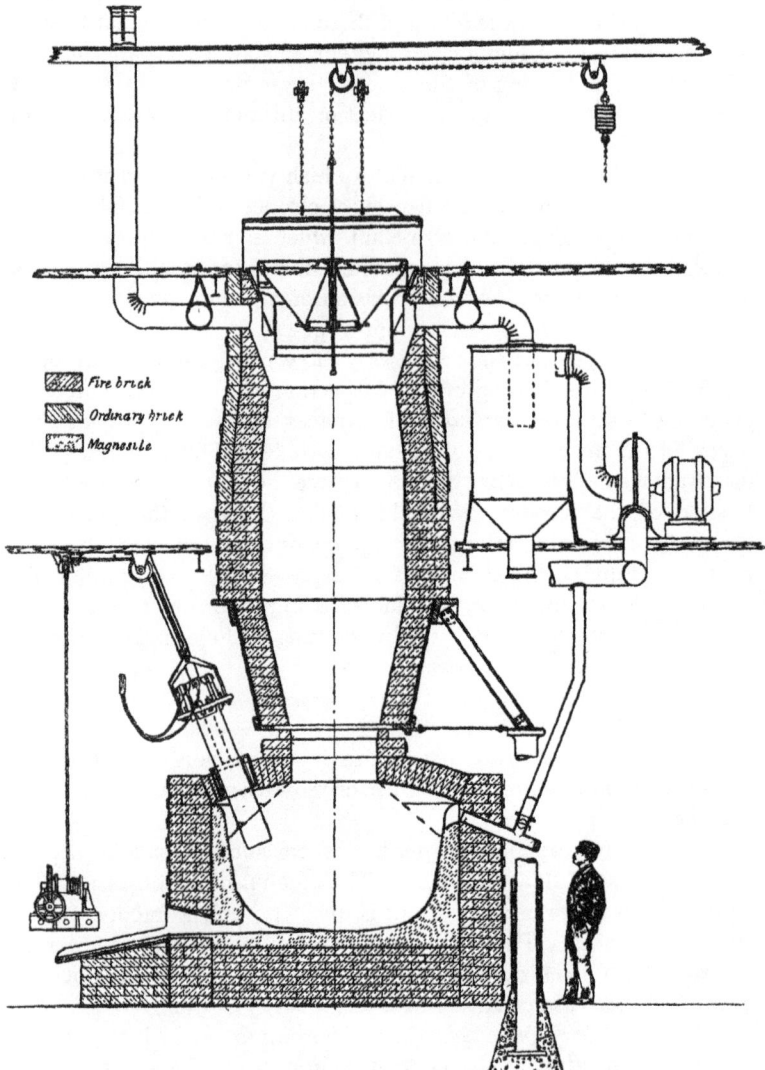

Fire brick

Ordinary brick

Magnesite

FIG. 111.

Form of Carbon Used for Reduction.—When coke is used as the reducing material, the disadvantages obtained are the low resistance, which necessitates the use o₁ lower voltage and consequently higher current. If it is attempted to increase the voltage by raising the electrodes, the formation of arcs and local overheating result. It is, moreover, difficult with coke to obtain a satisfactory slag, this being either too viscous or containing too high a percentage of metal. These difficulties may be overcome by using correspondingly larger electrodes and allowing for a higher power consumption. With coke the zone of reaction is more localised in the immediate neighbourhood of the electrodes where the main heating occurs, whereas with charcoal a heating zone is obtained of larger area and more uniformity of temperature.

ELEKTROMETALL FURNACE.

The Elektrometall furnace which was finally devised is illustrated in Fig. 111, and consists essentially of a smelting chamber (or crucible), in which the charge is finally melted down by electrical heat, and a shaft in which the charge is heated and the ore partly reduced, by heat rising from the crucible. The heat is carried up by gases produced by the reduction of the ore, and this action is assisted by returning some gas from the furnace top to the crucible, so as to increase the stream of gases up the shaft. The lower part of the shaft, down which the charge descends, has the form of a truncated cone, which serves to direct the charge into the crucible in such a manner that the descending charge does not come into contact with the lining at the point where the electrodes enter the furnace. The charge on falling through the circular aperture at the base of the shaft into the free space of the crucible assumes a definite angle, viz. $50°$ to $55°$ to the vertical shown in Fig. 111, by dotted lines, and thus leaves a free air space between the surface of the charge and the lining at the top of the crucible. In earlier furnaces even when the electrodes were cooled by water jackets the temperature of the brickwork in close proximity to the electrodes became very great, and led to a rapid destruction of the most refractory lining materials, and also to loss of power by the leakage of current through the conducting lining. The reduction of the iron ore in the crucible during the progress of the smelting leads to the formation of a mixture of carbon dioxide and monoxide in varying proportions in accordance with the equation

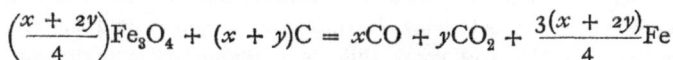

$$\left(\frac{x + 2y}{4}\right)Fe_3O_4 + (x + y)C = xCO + yCO_2 + \frac{3(x + 2y)}{4}Fe$$

On ascending the heated column of charge in the shaft, reaction

12

takes place between the carbon monoxide and iron oxide, giving iron and carbon dioxide. The gases reach the head of the furnace at a low temperature, and with an increased ratio of carbon dioxide to monoxide. By means of an arrangement of gas pipes as shown in Fig. 111, the gas is withdrawn from the head of the furnace and passes through dust catchers, then through pipes where it meets with a spray of water, through a centrifugal fan, also supplied with a water spray, and finally through a separating chamber for the removal of the entangled water. The washed gas is supplied to six tuyères entering under the furnace arch and between adjacent electrodes. The excess of gas evolved during the smelting escapes from the dust-collecting chamber through a pipe extending above the roof of the building and, in absence of any further utilisation, is allowed to burn. The circulation of the gas in this manner, in addition to utilising some of the carbon monoxide for the preliminary reduction of the ore, also serves the important purpose of cooling the roof of the crucible. This cooling action is due in part to the absorption of heat, which is later imparted to the charge, and in part to the decomposition of the carbon dioxide and water vapour in the gas in contact with the incandescent carbon in the crucible.

The operation of the furnace depends very largely on the gas circulation. Increasing the circulation raises the temperature in the shaft, facilitates the reduction of the ore, raises the percentage of carbon dioxide in the escaping gases, and increases the economy both of electric power and of fuel. On the other hand, the circulation increases the consumption of electrodes as these are attacked by the carbon dioxide in the circulating gas.

The crucible is circular, of large capacity so as to serve as a mixer and balance small irregularities in the charge, and is provided with one tapping-hole from which both the slag and the metal are withdrawn. The metal and slag are separated by a dam as they flow, the metal being cast into pigs or taken in a ladle to the steel furnace. The crucible is lined with firebrick like the ordinary blast furnace, and not, as in the earlier furnaces, with magnesite. The stack of the furnace is constructed in a steel shell, and is supported on steel beams, independent of the crucible.

The arch of the crucible is constructed of fireclay bricks, and is the least substantial part of the furnace, but repairs to it can be effected without great delay, by introducing some cold ore-charge into the crucible, and using this as a temporary support for the new brickwork.

Electrodes and their Adjustment.—The electrodes are circular, about 2 feet in diameter, and 4 or 5 feet long, and can be attached end to end by moulded carbon nipples, which are screwed into threaded holes in the ends of the electrodes (cf. p. 342). In this way, as the electrode becomes short, use is made of the remnant by attaching it to the end of a new electrode.

The electrode holders consist of two inclined guides, between which the electrodes lie, supported by guide rollers. At the bottom of the guides is a water-cooled collar built into the furnace arch. This collar is packed around the electrode with asbestos, so that the gas does not escape from the furnace. Electrical contact with the carbon is made by means of a water-cooled ring, consisting essentially of a number of metal blocks, forming a flexible collar that can be tightened around the electrode, and by means of a short piece of flexible cable, each block is connected to one of the copper bus-bars. The electrodes do not, as a rule, need moving more than once in two or three days. The larger furnaces have six electrodes, supplied with three-phase current from three transformers. Each transformer is connected to two diametrically opposite electrodes, so that the electric current tends to pass between these instead of between adjacent electrodes. The voltage of each transformer can be regulated separately by means of tappings on the primary windings and a nearly constant power can thus be supplied to each pair of electrodes in spite of changes in the electrical resistance between them. The regulation of the electric power is thus effected by changing the voltage of the supply and not by moving the electrodes up and down. When through consumption of the electrode the voltage between it and the material in the furnace exceeds a certain limit the electrode is fed further in.

The gas escaping from the Swedish furnaces is not fully utilised at present, but it will probably be employed for heating open-hearth furnaces, or for similar purposes, and this will represent an important economy in the operation of the furnace.

Dimensions of Furnace.—The dimensions of a 2500 h.p. unit at Trollhättan are given[1] as

> Total height, 54 feet.
> Outside diameter of crucible, 17 feet.
> Outside diameter of shaft, 10½ feet.

Commercial Efficiency of the Elektrometall Furnace.—The steadiness of the furnace load is, as usual, of first importance in determining

[1] O. Frick, " Met. and Chem. Engineering," 1911, **9,** 631.

the commercial efficiency, since the cost of power is mainly based on the maximum demand. It may be assumed with this furnace that, while in operation, the average power utilised is about 92 per cent of the maximum, and deducting further 2 per cent for lights and motors, the average utilised during operation amounts to 90 per cent when measured on the low tension side, and excluding losses in the transformer. In Sweden it is estimated that the stoppages during the holidays, when the furnaces must be banked up, amount to 2½ per cent of the time, and the stoppages for repairs, relinings, and other causes, bring this up to 8 or 9 per cent of the whole time. The power utilised by the furnace amounts, consequently, to some 82 or 83 per cent of the gross amount paid for.

Power Consumption.—The figures given refer to the power actually used in the furnace (and transformer) while making 1 ton of pig-iron. From ores of 58 to 60 per cent iron, basic Bessemer iron needs 2245 kw. hours while acid open-hearth iron needs 2116; 'or an average figure of 2200 kw. hours. From ores of 50 per cent iron, acid open-hearth iron requires an average of 2500 kw. hours. To make grey iron for foundry use would necessitate a decided increase in the power consumption. Correcting the above figures in the ratio 100/82 to allow for periods of stoppage we get 2700 and ·3050 kw. hours, or 0·41 and 0·47 h.p. years respectively as the amount of electrical energy actually expended in order to obtain 1 ton of such iron from a 60 per cent and a 50 per cent iron ore.

Consumption of Charcoal.—This is given as 315 to 357 kg. per ton of iron produced in the case of white pig-iron and for grey pig-iron 370 kg. The consumption of charcoal depends on the nature of the ore (hæmatite needing more than magnetite) and on the ratio of carbon dioxide to monoxide in the furnace gases, the more carbon dioxide in the gases, the less the consumption of charcoal. For this reason the circulation of the gases through the furnace reduces the consumption of the charcoal but, at the same time, gives a by-product gas of lower calorific value. The circulation of the gases also increases the loss of electrodes, and furnaces having a rapid circulation show a greater electrode loss than those with a slow circulation.

By-product Gas Recovery.—The records show a variation in the percentage of carbon dioxide in the gases ranging from about 8 to 35 per cent depending upon the kind of ore and flux as well as on the speed of the gas circulation.

The sensible heat carried out of the furnace by the escaping gases is unimportant, owing to the low temperature of the furnace top.

Collecting the results of all the calculations, it would appear that (1) without circulation the escaping gases have a heat value about equal to the net heat requirements of the furnace; and (2) with gas circulation about one-fourth the value of the escaping gas is utilised in the furnace, thus saving about 11 per cent of the carbon and 7 per cent of the electrical energy.

HEAT DISTRIBUTION IN ELEKTROMETALL FURNACE.

Calculated by Messrs. Leffler & Nystrom.[1]

Results of Operation from 9 April to 18 May, 1911 (Trollhättan).
Ore, about five parts magnetite to three parts hæmatite (55·9 per cent Fe).
Product, open-hearth pig-iron, 15·1 tons daily of composition :—

Fe 95·35 per cent	Si 1·00 per cent	S 0·020 per cent
C 3·52 ,, ,,	Mn 0·89 ,, ,,	P 0·013 ,, ,,

Electrical Supply.

Phase I, 13,564 amps., 75 volts ;   Working time 851·7 hours
  „ II, 11,817 „ 83 „   Standing „   68·3 „
      1717 kilowatts.   ──────
                              Total . 920·0 „

For One Ton of Pig-iron.

Supplies.	By-products.
1725 kg. ore	455 kg. slag.
48·5 ,, raw limestone . .	738 ,, dry gases at 64° C.
92·5 ,, burnt lime . . .	
396·7 ,, charcoal . . .	$\Big\{$ CO₂ 8·77 per cent CH₄ 3·18 per cent $\Big\}$
4·99 ,, electrodes . . .	CO 81·57 ,, ,, H₂ 6·48 ,, ,,
2481 kilowatt hours . . .	

Heat Distribution.

Reduction of Fe, Si, etc.	1,722,155 cals.	
Oxidation of carbon	882,863 ,,	
Net reduction Fe, etc., by C .	839,292 = 39·46 per cent.	
Fusion of pig-iron . . .	302,400 = 14·22 ,, ,,	
„ „ slag	227,500 = 10·70 ,, ,,	
Decomposing limestone . .	27,391 = 1·29 ,, ,,	
Evaporating water . . .	38,189 = 1·80 ,, ,,	
Sensible heat of gases . .	10,361 = 0·49 ,, ,,	

Total heat utilised 1,445,133 = 1686 kw. hours = 67·96 per cent.

Transformer losses	68 kw. hours =	2·74 per cent.
Losses in low-tension conductors . .	95 ,, ,, =	3·83 ,, ,,
„ „ cooling water . . .	163 ,, ,, =	6·57 ,, ,,
„ by radiation, etc. (by difference) .	469 ,, ,, =	18·90 ,, ,,

Kilowatt hours supplied . . 2481   = 100·00 ,, ,,

Potential energy in gases, 1,623,000 cals. = 1890 kw. hours = 76·2 ,, ,,

[1] A. Stansfield, loc. cit., p. 19.

ELEKTROMETALL FURNACES IN OPERATION OR IN COURSE OF CONSTRUCTION IN SWEDEN IN 1915 WITH NATURE OF ORE AND PRODUCT.[1]

At Domnarfvet two furnaces of 3000 h.p. and one of 6000 h.p. are in operation.   Phosphoric hæmatite ore is smelted with charcoal and fluxes, producing a phosphoric pig-iron, which is converted into steel by the basic Bessemer process.   Non-phosphoric magnetite ores (about 50 per cent iron in the charge) are smelted (at present in the Helfenstein furnace, cf. below) with charcoal to make a high quality pig-iron.   This is used in the acid open-hearth furnace to make a high quality steel for export.

2. At Södersfors, three furnaces of 3000 h.p.   The furnaces are intended for the production of a special quality of iron which has hitherto been made here in charcoal blast furnaces from the Dannemora ores.

2a. At Ljusne, one furnace of 3000 h.p.

3. At Hagfors, three furnaces of 3400 h.p. in operation.

3a. At Nykroppa, two furnaces of 3400 h.p.

4. At Trolhattan, one furnace of 2000 h.p. in operation and one of 3000 h.p. building.

HELFENSTEIN FURNACE.

The Helfenstein furnace was originally designed for the manufacture of calcium carbide and ferrosilicon, but has now been applied on a large scale for the smelting of iron-ore.   This type of furnace consists of a large rectangular smelting chamber, above which are suspended the three vertical electrodes each consisting of a bundle of carbon blocks of rectangular cross section which enter through openings in the roof of the furnace and project downwards into the charge.   The roof is constructed by providing water-cooled partitions which are placed between the electrodes, and which serve the double purpose of supporting the arches and providing spaces which serve as charging hoppers.   The electrodes are suspended in these hoppers and surrounded by the charging material.   Underneath the roof the partitions form gasification chambers for the collection of the gases.   The sealing of the opening around the electrode is effected by the charge itself and enables the furnace to be always kept under a gas pressure.   The furnace at Domnarfvet, which was brought into operation in 1913, was designed for a capacity of 10,000 to 12,000 h.p., and the conducting cables were dimensioned

[1] Cf. p. 175.

accordingly. However, through the high frequency of the available current supply, which is of sixty-two periods, the phase displacement is too high when taking a load of 10,000 h.p., and it has accordingly been decided to employ from 6000 to 8000 h.p. only. With this load 26,000 amps. are used on each electrode bundle and 120 volts between the phases, while the phase displacement does not exceed 0·8. The maximum output of the furnace is 65 tons per twenty-four hours, and 1200 to 1600 tons per month. With charcoal, the consumption of energy per ton of pig-iron is 2000 kw. hours. The consumption of charcoal is 6 to 8 cwt. and consumption of electrodes 15½ lb.

The value of the furnace gases is 2800 to 3000 cals. per cubic metre of gas.

According to Oesterreich[1] the results obtained with this furnace during a week in June, 1913, were as follows :—

Maximum output in 24 hours, 65 tons.
Consumption of materials per ton of pig-iron from 60 per cent ore.

(a) *Charcoal as reducing material.*

Energy consumed	2170 kw. hrs.
Charcoal (70 per cent C.) . .	380 kg.
Electrodes	5 kg. (11·0 lb.).

(b) *Coke as reducing material, working experimentally.*

Energy consumed	2600 to 2700 kw. hrs.
Coke	310 to 330 kg.
Electrodes	4 kg. (8·8 lb.).

Power Factor.—With a load of 10,000 h.p. the value obtained for cos ϕ was 0·75, and at 6600 to 8000 h.p. the value was 0·8.

The results in respect of the consumption of energy and of charcoal are seen to be less favourable than those with the Elektrometall furnace, while the consumption of electrodes is about the same. The advantages claimed for the Helfenstein furnace, however, are :—

1. Lower capital expenditure (on a pre-war basis, 55s. 6d. per h.p. for the whole plant compared with 89s. per h.p. for the Elektro-metall plant).

2. By-product gases have a higher calorific value (2600 to 3000 cals. per cubic metre compared with 2300 cals. for the Elektro-metall).

3. Additional materials for improving the charge can be added close to the melting zone and quickly made operative, while in a furnace with a high shaft it may take hours before such admixtures are able to reach the melting zone.

[1] "Stahl u. Eisen," 1916, **36**, 1059 ; " Met. and Chem. Engineering," 1917, **16**, 509.

4. Pulverised ore can be employed without the working of the furnace being hampered.

5. The furnace allows of coke being used as the reducing material

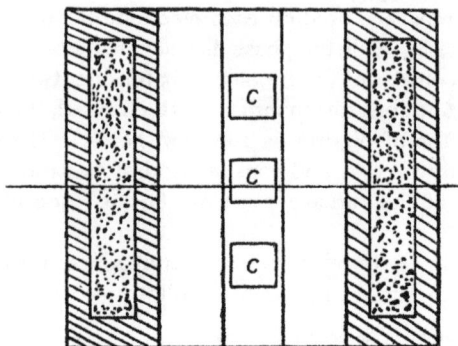

FIG. 112.

though, in this case, the voltage is lower than with charcoal. In the case of coke the voltage amounts to 50 to 55 per phase, while a current density of 5 to 6 amps. per square centimetre of electrodes

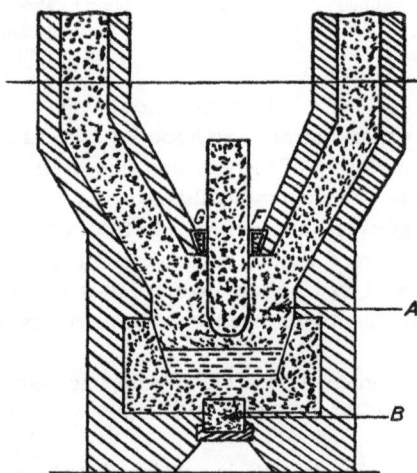

FIG. 113.

is required. With charcoal the furnace can be easily worked with 70 volts per phase, and a current density of 2 to 3 amps. per square centimetre of electrode. On account of the higher current entailed when working with coke it becomes necessary to reduce the load on the furnace to only 5000 to 5500 h.p., but with this load there are no difficulties experienced.

The gases from the Helfenstein furnace at Domnarfvet are now employed in the operation of open-hearth steel furnaces.

THE TINFOS FURNACE.

The Tinfos furnace (see Figs. 112 and 113) consists of a long rectangular smelting chamber A, having an inclined chute along each long side to supply the ore charge. The roof between these chutes is carried by means of two water-cooled beams F and G, running the length of the furnace and placed just far enough apart to admit the rectangular electrodes C, of which there are three connected to one pole of the supply. The opposing electrode B lies in the bottom of the furnace, and is covered with a bed of rammed coke.

In Norway the Tinfos Jernverk A/S. are working at Notodden three single-phase furnaces, which are operated together in order to utilise the three-phase supply. Each furnace employs about 1600 h.p. (or 1200 kw.) at about 35 to 55 volts, 50 cycles. During the smelting the gases pass up the chutes, thus heating the ore charge to some extent, but no circulation is used, and the gases are not employed at present in any way outside the furnace. The furnace is of interest, because as in the case of the Helfenstein furnace, coke is used instead of charcoal as the reducing agent. For obtaining a large output of pig-iron of moderate quality, it will often happen that coke is preferable to charcoal. The Elektrometall furnace has, so far, not proved satisfactory when using coke, and moreover is disadvantageous when using very powdery ores, owing to the charge in the shaft becoming too compact to allow the passage of the gases.

The use of a single-phase current and a bottom electrode in the Tinfos furnace ensures the current being led through the charge on to the metal beneath and leads to a more certain distribution of the heat throughout the charge. The disadvantages of this furnace, however, are the necessity of working three units simultaneously, and the use of a considerably lower voltage.

GENERAL RESULTS OF ELECTRIC SMELTING OF IRON-ORE IN SWEDEN.

According to Prof. J. W. Richards,[1] pig-iron was, in 1920, being produced in Sweden in electric furnaces from charcoal at a cost of $5 per ton less than their own blast-furnace pig-iron.

---

[1] " Met. and Chem. Engineering," 1920, **22**, 61.

At Hagsfors,[1] 100 miles west of Domnarfvet, there are now five electric pig-iron furnaces in operation, and electrodes up to 30 inches diameter are in use. The present cost of water power is said to average $10 to $12 per h.p. year.

It is estimated[2] that the production of cast-iron by electric smelting of ore in Sweden amounted, in 1918, to 100,000 tons per annum, mostly produced by smelting with wood-charcoal. The electric furnace production of cast-iron accordingly represents about one-eighth of the total production of this metal in Sweden, and the proportion will be considerably increased when installations in the course of erection at Trollhättan and Porjus are completed.

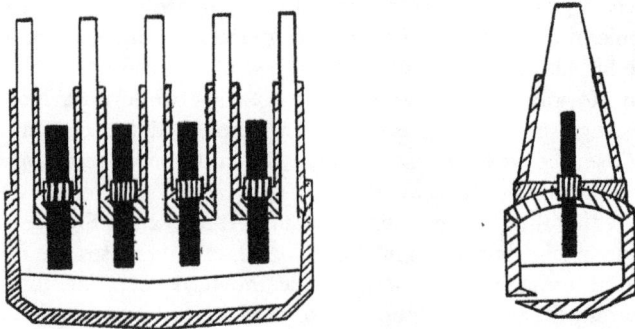

FIG. 114.

NOBLE ELECTRIC STEEL COMPANY, AT HÉROULT, CALIFORNIA, U.S.A.[3]

The plant of the Noble Electric Steel Company is situated on the banks of the Pitt River, at Héroult, Shasta County, California. The plant consists of one 2000 kw. and one 3000 kw. iron furnace of a modified Helfenstein model, one 2000 kw. furnace of the shaft type (not now operated), one 2000 kw. steel furnace of the tilting type not now in operation, and charcoal retorts, lime kilns, etc.

In order to meet the local demand, it is necessary to produce a foundry iron, or a soft high-silicon iron (with 2 to 3 per cent Si). A deposit of magnetite is available of very high grade. A good limestone is also obtained locally.

Operations at these works were commenced in 1907 by H. H. Noble, who erected a three-phase furnace of 1500 kw. capacity. A

_____

[1] Cf. p. 183.　　　　　　[2] " J. du Four electrique," 1919, **28**, 17.
[3] " Met. and Chem. Engineering," 1913, **11**, 16, 383.

shaft-type furnace was introduced in 1909, but later abandoned in favour of a 2000 kw. three-phase furnace of the Helfenstein model, having four electrodes delta-connected and suspended between five charging stacks. The first furnace was brought into operation in 1911, and has been working since. A companion furnace of 3000 kw. was built later. This type of furnace, as illustrated in Fig. 114, was designed by R. E. Frickey and consists of a rectangular steel shell, 28 feet long and 10 feet wide, lined with standard furnace brick, and surmounted by five charging stacks, 18 feet high. Between the stacks the top of the furnace is arched over and through the centre of these arches the electrodes penetrate vertically into the charge. The electrodes used are of Acheson graphite, 12 inches in diameter, with screw threaded connexions at the ends for joining on a new electrode. No utilisation has been so far made of the furnace gases.

Electrical Supply.—Power is supplied by the Northern California Power Company at 60,000 volts, three-phase, and by means of three 750 kilovolt-ampere transformers, converted into three-phase currents at 40 to 80 volts.

Furnace Charge.—The composition of the ore averages 67 to 68 per cent iron, while the silica is sometimes as low as 2·5 per cent, the P 0·011 per cent, and the S 0·021 per cent. For a pig-iron containing 2 to 3 per cent silicon, it is consequently necessary to add silica in the form of quartz. The requisite amount of lime or limestone is added so as to give a slag containing about 47 per cent of silica. Charcoal is generally used as the reducing agent. It is found that with coke the electrical conductivity is so good that much of the current passes between the electrodes in the upper part of the furnace, thus giving too high a temperature near the roof of the furnace, and too low a temperature at the bottom. Further, with coke, less intimate contact with the ore is obtained, resulting in a slower reduction of the ore, a lowered content of silicon in the iron, and an increased power consumption per ton of product. Mixtures of coke and charcoal have given good efficiencies however.

Gases Produced.—The gases evolved contain from 5 to 9 per cent CO_2, 56 to 70 per cent CO, 12 to 16 per cent CH_4, and 0·8 to 1·2 per cent H_2.

Composition of Iron.—The average composition of pig-iron produced for making steel castings, is as follows :—

Si, 2·88 per cent; C, 3·47 per cent; S, 0·028 per cent; P, 0·031 per cent.

Furnace Efficiency.—The efficiency of this type of furnace increases considerably with increase of load. With the use of 3000 kw. a power consumption of 2200 kw. hours per ton of pig is obtained, and a consumption of charcoal of 400 kg.

ELECTRICAL CONNEXIONS FOR FOUR-ELECTRODE FURNACE.[1]

The transformer connexions for a four-electrode furnace can be arranged as illustrated in Fig. 115. The primary is connected to a three-phase supply which may be arranged with either star or delta-connexions. The windings *ab*, *bc*, and *ca* as shown in the figure are arranged as delta-connexions. The secondary windings are shown at PQ, RS, and MO. Wires from the terminals 1, 2, 3, and 4 lead to the electrodes. From the centre of RS a wire is taken to the neutral point N. The four independent phases are given by PVRN,

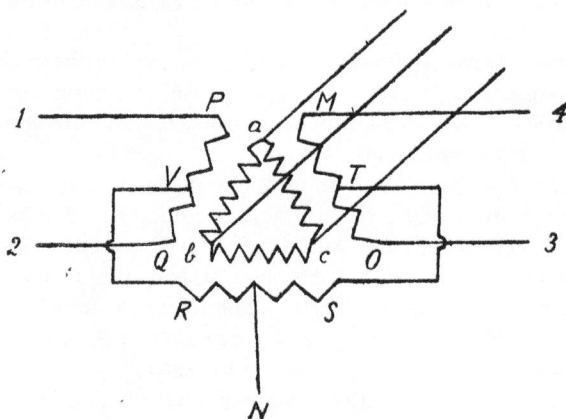

FIG. 115.

QVRN, OTSN, and MTSN. The windings can be so proportioned that the four vectors or electrode voltages are all equal. When equal currents are taken through the electrodes, three equal currents are taken from the primary lines. The current flowing through the neutral N is equal to the vector sum of the currents in the four electrodes, and this is approximately equal to that in one electrode. There is accordingly obtained a four-phase, five-wire system in which the conductors are similar in cross-section and connexions.

The neutral point is made the hearth of the furnace. It follows from the nature of the windings that fluctuations on any electrode, such as those caused by an interruption of the arc, are distributed

[1] J. Bibby, "Iron and Coal Trades Review," 1918, **97**, 719.

amongst the three primary mains. On commencing to heat, when the hearth is non-conductive, the furnace works as a four-phase, four-wire system and the unbalancing on the primary mains is not excessive.

COMPARISON OF DIFFERENT TYPES OF FURNACES.[1]

The distinguishing characteristics of furnaces such as the Helfenstein type which are not provided with shafts are :—

1. The absence of preliminary reduction which occurs in the shaft.

2. The absence of circulation of gas.

3. The possibility that is offered by this type of using calcined limestone and ore in a finely-divided condition. Reduction of metal in the non-shaft type is first brought about in contact with the electrodes, and this would account for the fact of the furnace being adapted for the production of high silicon metal.

The objection to the use of calcined lime in the Swedish furnaces is that it increases the proportion of pulverised material in the charge through being more fragile, and the presence of finely-divided material impedes the circulation of the gases.

CHAPLET FURNACE FOR THE REDUCTION OF IRON-ORE.[2]

The electrical reduction of iron-ore is carried out by the "Société, La Néo-Metallurgie" in France by a type of furnace known as that of the Chaplet-Néo-Metallurgie. This consists of an arc furnace in which the current is introduced by one or several vertical electrodes as illustrated at E in Fig. 116, while the second terminal is connected to the base of the furnace, through the electrode F. The reaction chamber is of circular form. During operation the slag-forming materials are first introduced, then the mixture of mineral and carbon in suitable proportions. Fusion and reduction of the charge takes place in the neighbourhood of the arc, and the metal produced passes through the slag and collects on the hearth. As in the case of the blast-furnace, there are two reaction zones, an outer one in which the oxide is reduced and an inner one of higher temperature in which the charge is fused. The materials can be employed in powder form, either direct or after agglomerating or briquetting. Of the different varieties of carbon, anthracite was found to give the most favourable yield for a given power consumption, though either coke

---

[1] Cf. D. A. Lyon, "Rev. d'Electrochimie," 1913, **7**, 210 ; "Met. and Chem. Engineering," 1913, **11**, 15.

[2] Arnou, "Rev. de Metallurgie," 1910, **7**, 1190.

or charcoal can be used. With charcoal, the higher resistance offered enabled the application of a higher voltage. Magnetite was found to give a better yield than hæmatite.

Nature of Product.—A mild steel can be produced in one operation by this furnace from ore of the following composition :—

C	0·08 to 0·11	per cent.
Mn	0·09 ,, 0·46	,,
Si	0·02 ,, 0·19	,,
S	0·01 ,, 0·02	,,
P	traces	,,

This product was obtained without the addition of slag-forming materials or refining compounds. Tests on the physical properties of these products have shown the material to be of a similar grade to Swedish steels.

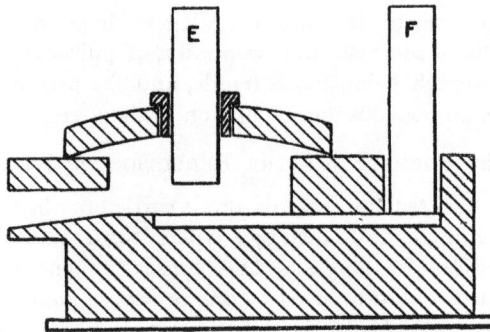

FIG. 116.

The power consumption in experiments which were mostly carried out with a furnace of 120 kw. had the following values per ton of soft iron or steel :—

        3430 kw. hours with a unit of 120 kw.

        2600 ,,   ,,   ,,   ,,   ,,   ,, 200 ,,

In a furnace of 120 kw., one ton of soft steel required the following power expenditure in the case of the following materials :—

Magnetite and charcoal	3150 kw. hours.	
,, ,, anthracite	3050 ,,	,,
Hæmatite and charcoal	3430 ,,	,,
,, ,, anthracite	3100 ,,	,,
Siderose and ,, ...	4000 ,,	,,

The theoretical power consumption is estimated to be 2000 to

2050 kw. hours per ton, which gives an efficiency of about 80 per cent in the case of the 200 kw. unit.

Consumption of Carbon.—With hæmatite in powder form 360 kg. of charcoal or 270 kg. anthracite are needed per ton of product, while magnetite requires 310 kg. of charcoal or 260 kg. anthracite.

Consumption of Electrodes.—In normal operation this varies from 25 to 35 kg. per ton of product.

The Chaplet furnace has been applied for the manufacture of ferro-alloys (cf. p. 276).

Cost of Plant for the Electric Smelting of Iron-Ores in Canada.— An estimate based on the results obtained in Sweden has been made by Prof. A. Stansfield[1] of the cost of operating a plant in Canada for the electric smelting of iron ores with the following results :—

With Rennerfeld or Elektrometall furnaces the total inclusive cost of installation of three furnaces of 4000 h.p., including building and all accessories, will amount to $360,000.

The cost of production of 1 metric ton (2204 lb.) of grey (high silicon) pig-iron from coke is estimated as follows :—

1600 kilos of ore at $4·60 per ton	$6·40
130 ,, ,, lime ,, $15·00 ,, ,,	1·95
400 ,, ,, coke ,, $5·00	2·00
0·42 e.h.p. year (2700 kw. hours at $10·00 per h.p. year)	4·20
9 kilos of electrodes at $85 per ton	0·77
Labour and engineering	1·20
Office and organisation	0·50
Repairs	0·60
Depreciation (6 per cent of $360,000) for yearly output	0·72
General expenses	0·50
Petty Charges	0·66
Total (exclusive of royalty)	$19·50

Output.—The daily output for one 4000 h.p. furnace is as follows :—

	Tons daily.
White (charcoal) iron	29·6
Grey (charcoal) iron	27·4
White (coke) iron	28·8
Grey (coke) iron	26·1

Analysis of Costs.—The cost of electric energy, while undoubtedly a very important factor, is not even the largest single item, and in the above estimates only amounts to about 20 per cent of the whole

[1] " Cassier's Magazine," 1916, **49**, 97.

costs. In each case the ore represents a large expense, and in making charcoal iron the fuel for reduction costs as much as the electric power.

Use of Waste Gases.—The waste gases can be used for heating an open-hearth furnace, and would have a heating power (per ton of iron) equal to that of the producer gas obtainable from about ¼ ton of coal.

Development of Electric Iron and Steel Furnaces in Norway.[1]

Experiments on the electric smelting of iron-ore were begun in 1910 at Notodden, with a furnace of 500 kw. capacity. After successful trials with coke as reduction material, the *Tinfos Jernwerk* was formed. Three furnaces were installed, as detailed above (p. 186). The furnaces now operate very regularly, and produce either white or grey pig-iron. Some of the manganese of the ore is reduced giving a product which competes with charcoal-iron. The energy consumed per ton of product is 2700 kw. hours for white pig, and 3000 kw. hours for grey pig from ore containing 44 per cent iron.

A furnace of the same type has also been erected at Ulefos, using phosphoric ore from the Faehn deposits and producing foundry iron, which is remelted in a Rennerfeld furnace and made into stove castings.

Steel Furnaces.—The most important installation for making and refining steel is the *Stavanger Staalverk*, where for treating steel which has been previously refined in an open-hearth furnace a 5 ton Ròchling-Rodenhauser single-phase induction furnace was installed in 1913. During the war the plant was considerably extended and, on account of the scarcity of coal, the operation of the open-hearth stage has been discontinued and cold charges applied in the induction furnaces.

Other electric steel furnaces which have been installed in Norway since 1914 include a small Röchling-Rodenhauser furnace at Eureka, Christiania, where acid-resisting steel is made; at Hamar, Hougsund, and Drammen, cast steel is made in Rennerfeld furnaces, and at Kongsberg and Naes tool steel is made in the same type of furnace.

"Synthetic Cast-Iron."[2]

The production of a high-grade cast-iron from scrap steel and turnings is a process which was applied to a very large extent during

---

[1] H. Styri, " Trans. Amer. Electrochem. Soc.," 1918, **32**, 129.
[2] Ch. A. Keller, " Trans. Amer. Electrochem. Soc.," 1920, **37**, 17.

the war. The largest development of this work was made by M. Ch. A. Keller at the ferro-alloy works at Livet (Isère), France.

The cast-iron was produced to supply the special quality of metal needed for shell construction, while the scrap steel was obtained from munitions plants. The process involves the melting of scrap steel, recarburising and desulphurising, together in some cases with dephosphorising and increasing the silicon content, and is brought about in a Keller type of steel furnace with conducting hearth (cf. p. 221). In a typical procedure, steel turnings of the percentage composition: Silicon, 0·44; manganese, 0·55; sulphur, 0·07, was transformed into a white cast-iron containing: Si, 0·52; Mn, 0·48; S, trace; C, 3·55. The main advantages obtained by this system over the blast furnace are that the melting is accompanied by desulphurisation, and losses of manganese and other alloyed metals by oxidation are avoided, while the method permits the preparation of any desired type of cast-iron.

Ordinary coke or charcoal in a fine state of division, so as to make good contact with the steel turnings, is used together with materials to form a basic slag for the purpose of desulphurisation, which also serves to increase the resistance of the charge. The upper part of the charge is heated by the gases produced lower in the furnace. In a furnace of 2500 kw. capacity, producing 80 to 100 tons per day, the power consumption has been reduced to 675 kw. and the consumption of electrodes to 6 kg. per ton of cast-iron.

When dephosphorising treatment is also undertaken, the power consumption is increased to about 750 kw. hour per ton. Though primarily developed to utilise turnings and scrap metal obtained in projectile factories and for the manufacture of shells, this system is now receiving an extended application for industrial requirements. The main directions in which synthetic cast-iron is being applied is for the preparation of malleable castings of high-silicon content, the production of castings of extra mechanical strength, the preparation of special qualities of cast-iron, such as those containing nickel, chromium, and titanium. In the preparation of this class of metal the absence of dissolved gases, which is ensured in the electric furnace, is of particular benefit.

In 1914 the manufacture of malleable iron was begun at Livet for the preparation of shells, the product having the following percentage composition: C, 2·90; Si, 1·75; Mn, 0·50; S, trace; P, 0·05. Beginning with a 2-ton electric furnace, there were in operation at Livet in 1916 five furnaces of 2000 kw. and one of 2500 kw. capacity, with a total capacity of 300 tons of cast-iron daily. In 1918 there was brought into operation a further water-power development at Vernes on the Romanche where 7000 h.p. was harnessed, while in the same year there was in the course of development a water power of 7000 h.p. at Bâton.

In 1916 a national artillery factory was erected at Nanterre, utilising power from the Paris plant, and in 1917 300 tons of synthetic cast-iron were manufactured daily. In 1918 a total of 15,000 kw. had been brought into operation at this factory. Furnaces of 1650 kw. were used, as illustrated in Plate IX, and the expenditure of materials per metric ton of product was as follows :—

Steel turnings used	1113 kg.
Coke	89 to 95 kg.
Electrodes	6 to 10 kg.
Kw. hours consumed ;	815.

ELECTRIC PIG-IRON IN CANADA.[1]

At the present time no pig-iron is being made from ore in the electric furnace in Canada. A low-phosphorus pig-iron is, however, being prepared by melting steel scrap in electric furnaces at Orillia, St. Catherines, and Collingwood, Ontario, and Shawinigan Falls, Quebec, as well as a number of places in the United States. The furnaces are constructed with a carbon base and lining of silica bricks, and provided with one tapping hole and two charging doors. The charge employed consists of steel turnings containing some manganese, charcoal, coke or coal, ferro-silicon, and lime. The iron produced contains up to 3 per cent carbon and about 0·02 per cent sulphur and 0·03 per cent phosphorus.

[1] R. Turnbull, " Trans. Amer. Electrochem. Soc.," 1917, **32**, 119; 1919, **34**, 143 ; " Metall. and Chem. Engineering," 1919, **20**, 178.

PLATE IX.—KELLER FURNACES—NATIONAL ARTILLERY FACTORY, NANTERRE.

(Ch. A. Keller.)

SECTION XI.

ELECTRIC STEEL FURNACES.

DIRECTION OF DEVELOPMENT.

THE application of the electric furnace for the production of steel has developed in two main directions :—

1. *To Replace or Supplement the Open-Hearth or Converter Process.*

In this connexion the electric furnace is used for the production of a steel intermediate between the "ordinary" and "high-grade" or crucible steel qualities. In some of these cases electrical heating serves merely to replace the use of fuel such as gas. The economy of electric furnaces as an alternative to fuel-heated methods in this connexion comes into consideration in localities and under circumstances where fuel is scarce and electrical power cheap. Thus in Norway, during the war, the open-hearth system has, in some cases, been actually replaced by electric furnaces. In other districts where such extreme conditions do not apply, as in Scandinavia, it has been found advantageous to adopt a combined process in which, for the production of steel in bulk, a preliminary refining is conducted in open-hearth furnaces. In these instances carbon and phosphorus are removed by the use of an oxidising slag containing iron oxide, and the liquid steel is then transferred from the open-hearth to the electric furnace for deoxidation, desulphurisation, and for the additions of alloys and adjustments of carbon. Here the electric furnace has a distinct advantage, through the use of a higher temperature, the maintenance of a reducing atmosphere, the protection of the metal from furnace gases, and the provision of a slag which has powerful reducing properties and contains a high percentage of lime.

An installation of this nature was first made in the United States by the Halcomb Steel Company and has been in operation since 1905. A similar installation was made at the Richard Lindenberg works in Germany at about the same time.[1] Other plants soon followed in

---

[1] J. A. Mathews, " Trans. Amer. Electrochem. Soc.," 1917, 31, 43.

different countries and, in these instances, the electric furnace has become established in centres which have been selected as being the most favourable for the operation of fuel-heated furnaces, and where electric power is generated from steam or blast-furnace gas. The development of this combined system has mainly arisen from the belief in its metallurgical soundness, and the further belief that electricity could not with economy be applied for the first melting. This latter view does not, however, at present apply, as on account of the present cost of large-scale power supply, and the electrical efficiency now attained, cold-smelting has been proved feasible in many localities, and is the method which has almost exclusively been adopted in France and England.

It was considered by Girod [1] that a great disadvantage attending the use of a duplex system as compared with electric refining from a cold charge was due to the following facts. Bessemer and open-hearth steel at the high temperature of melting dissolve metallic oxides, oxides of carbon, nitrogen and hydrogen to such a degree that their removal afterwards in the electric furnace is of doubtful possibility unless the charge is allowed to cool off and solidify, followed by remelting.

The manufacture of steel for rails in America is now being largely conducted by this combined process, in which the electric furnace is used for the finishing treatment. This development has been largely stimulated on account of the continual raising in the standards specified for steel for this purpose.

A clear advantage gained by the electric furnace is that it enables the production of a uniformly high quality of steel to be obtained from low-grade materials.

The electric furnace has been adopted on a large scale in America in conjunction with the Bessemer Converter (*vide* p. 217), whereby Bessemer steel is subjected to a further treatment in a Héroult type of furnace. This use of the electric furnace in conjunction with the Bessemer Converter is of significance as promising to lead to the preservation of the latter process which, in order to meet the present more exacting requirements in the quality of steel, is tending to be supplanted by the open-hearth process.

2. *Replacement of Crucible Process.*

The second direction of development of the electric furnace is in the production of high-grade steels of the quality of those obtained

---

[1] " Trans. Amer. Electrochem. Soc.," 1909, **15**, 245.

by the crucible process, and for alloy-steels. The electric furnace enables the production of steel of a quality now generally admitted to be equal to that obtained in the crucible furnace. The advantage offered by the former method is that while the crucible process is limited to the production of small batches of metal, with a consequent lack of uniformity between different lots, the electric furnace gives larger masses of metal of necessarily uniform composition. The economy of operation here is decidedly in favour of the electrical practice both in the saving in the cost of crucibles, in the greater efficiency in the application of heat, and in the labour costs. The crucible process is, moreover, impracticable for the production of steel in the quantities which can be provided with the electric furnace.

For the production of special alloy steels the main advantage obtained with the electric furnace is the diminished loss of alloy metal by oxidation. In preparing alloy steels from the open-hearth product, on account of the serious loss by oxidation the addition of alloy metal can only be made while the metal is in the ladle, which, as in the case of the crucible process, is liable to lead to lack of uniformity.

At the present time the main application which has been made of the electric furnace is to the special class of steels required in automobile and aeroplane construction. In addition to these purposes, the requirements for munitions of war gave a great impetus to the industry. It is estimated[1] that in the United States the total power consumed in electric steel furnaces in operation in 1917 amounted to 150,000 kw., and the output to 1,250,000 gross tons of ingots and castings. This output amounts to about eight times that of crucible steel, and one-eighth that of Bessemer production.

Electric steel promises to find its main market in the more exacting requirements of steel for structural and tensile purposes, as in connexion with automobile and aeroplane parts. Further, for these applications, crucible steels are hardly applicable because of the difficulties attending the manufacture of alloy steels of very low or medium carbon content. Moreover, the tonnage now demanded is far beyond what could be met by crucible steels, and the electric furnace product has to some extent created its own market.

REMELTING OF ALLOY STEEL.

A further important application of the electric furnace is for the remelting of scrap alloy steel which can be brought about with

---

[1] J. A. Mathews, " Trans. Amer. Electrochem. Soc.," 1917, **31**, 49.

minimum loss through oxidation of the valuable alloy metal. In armour-plate manufacture, for instance, about 50 per cent of the steel from the furnace remains as scrap after the formation of the plate, and is returned to the furnace for remelting. With the ordinary furnace the chromium is thereby lost by oxidation, and at the same time a slag produced that is very pasty, infusible, and difficult to handle.

Decentralising Influence of Electric Furnaces.

A notable feature in England is that the application of electric furnaces is causing a decentralisation of the steel industry. Formerly iron and steel scrap were forwarded to Sheffield for the manufacture of steel, but at present, partly on account of the transport difficulty which arose during the war, and partly to take advantage of local sources of cheap power such as from blast-furnace gas, electric processes now in operation are very widely distributed in this country (cf. p. 272).

Economy of Electric Furnaces.

Comparing electric with fuel-heated furnaces generally, the main advantage obtained is that electric furnaces allow the use of more impure raw materials, while producing the highest quality steel.

A further important factor which will determine the extent to which the electric furnace will supplant the fuel-heated processes is that of the relative costs of fuel and electric power. In the open-hearth furnace starting with cold scrap, the consumption of fuel in the form of bituminous coal may be taken at from 600 to 700 lb. per ton of steel produced. In an electric furnace the expenditure of power may be taken from 700 to 800 kw. hours per ton of steel to give a product refined to a somewhat higher degree than open-hearth steel. In the crucible process the amount of fuel required in the form of coke may be taken as $1\frac{1}{2}$ to 2 tons per ton of steel, while in an induction furnace, starting with the same materials and giving a similar product, the power expenditure will be from 600 to 700 kw. hours.

It is seen, therefore, that from the point of view of the cost of fuel and power, the electric furnace will only possess the advantage in economy under circumstances where fuel is dear in comparison with electric power, or else when applied for the final treatment of a previously melted and partially refined charge. The crucible process on this basis cannot, on the grounds of economy, be justified under any circumstances.

Other factors which determine the relative merits of fuel and electric furnaces are the labour costs, relative costs of upkeep of crucibles and furnaces, and the advantages gained in working with larger masses of metal. All of these factors are very markedly in favour of the electric furnace. An important advantage which in many cases has determined the adoption of electric furnaces is that, unlike the open-hearth process, the electric furnace will produce steel from scrap metal alone without the addition of pig-iron.

ADVANTAGES OF ELECTRIC FURNACE OPERATION IN THE PRODUCTION OF STEEL.

The advantages of the electric furnace production of steel may be briefly summarised as follows :—

1. On a scale comparable with open-hearth treatment and with comparable costs of operation a product can be obtained of a similar grade to the best crucible steel.

2. High-grade electric steel can usually be produced for castings at an average cost which, for equivalent tonnage, is less than that of the ordinary commercial grades of open-hearth steel or converter steel.

3. In comparison with the crucible process, electric furnace operation enables the production of a high-grade product starting from impure materials, the cost of the electrical method is considerably lower, and the process is carried out with much larger units than in the case of the crucible process.

4. In the electric furnace alloy additions may be made in the furnace itself rather than in the ladle, which increases the assimilation, diffusion, and homogeneity.

5. With alloy steels made in the electric furnace the loss of alloyed metal through oxidation is considerably less than in fuel-heated furnaces.

6. For steel foundries, metal can be prepared from scrap metal alone, without the addition of pig-iron. The steel can be taken to a higher temperature than in open-hearth practice which confers a considerable advantage with small and medium castings, and in these cases the process is cheaper to operate.

TYPES OF ELECTRIC STEEL FURNACES.

The essential points which are of main importance in the design of electric steel furnaces are the following ·—

(*a*) In order to obviate the use of tapping holes provision for teeming the charge should be made by means of a tilting device.

(*b*) The slag to be readily removable, which is usually arranged by the provision of side doors.

(*c*) The whole of the hearth to be capable of being readily inspected and repaired, by the provision of suitable doors.

(*d*) The furnace to be reasonably gas-tight in order to control the atmosphere.

(*e*) The roof to be removable so as to enable rapid replacement.

(*f*) During the progress of the heating of the metal an efficient circulation should be imparted to the metal so as to ensure uniformity of composition.

The types of furnaces used in the production of steel may be classified under the following headings :—

1. The direct arc or combination of direct arc and resistance.

2. The radiation arc.

3. The induction furnace.

1. In the types in the first category, the heating is applied by the formation of an arc between one or more carbon electrodes suspended vertically above the bath, and the surface of the charge, or covering slag. In this type, according to the thickness of the layer of slag, more or less heat is generated by its resistance, thus combining the principles of arc and resistance heating. With a single-phase current, the connexions are made by two similar electrodes arranged side by side, or by one vertical electrode and the base of the furnace which makes contact with the metal charge, either through metal rods embedded in the hearth of the furnace, or by making use of a conducting refractory. In the latter case, the resistance thus offered to the passage of the current serves as an additional source of heat. With a three-phase current, three vertical electrodes can be similarly arranged in place of two, or the circuit can be completed by means of two vertical electrodes and the conducting hearth of the furnace. With a two-phase current, three terminals are also used, one serving as a common return for the two phases.

Advantages of System using Conducting Hearth and those of System using Series Carbon Electrodes.—The advantages obtained when using a carbon-free base electrode as one of the current terminals are as follows :—

(1) The current is caused to pass through the metal bath, instead of, in the case of series electrodes, passing along the surface of the

metal. In the former system the so-called "electrification" of the mass of metal is considered to give a product of more homogeneous quality through the better distribution of heat which is obtained. This is due to the generation of a certain amount of heat by the passage of the current through the mass of metal, and a circulating effect produced by the magnetic field.

(2) When starting with a cold charge of scrap metal a particular advantage is obtained through the higher resistance when using a base electrode. This facilitates the "lighting up" of the furnace, while with series electrodes above the bath a short circuit is generally obtained through the top layers of metal.

(3) The roof is more easily preserved.

On the other hand the advantages obtained by the series carbon electrode system are as follows :—

(1) The electrical connexions are all above the furnace and more accessible.

(2) A higher voltage is given by the use of two electrodes in series, and thus, for a given power expenditure, a smaller carrying capacity of the leads has to be provided for.

(3) The induction effect of the current is less, and consequently the power factor is higher. With a base electrode, whereby the current all passes in one direction, the iron casing of the furnace, unless divided by a strip of non-magnetic metal, forms a magnetic circuit which leads to the generation of secondary currents.

2. *The Radiation Arc.*—In this system the main example of which is the Stassano furnace, heating is brought about by means of an arc between electrodes arranged horizontally or obliquely above the surface of the charge. The electro-magnetic field produces a certain deflection of the arc downwards towards the contents of the bath. In the Stassano process, a circulation of the bath is brought about by rotating the whole furnace which is mounted obliquely above the axis of rotation. The main advantage gained in this system is the avoidance of contact of the carbon electrodes with the furnace charge, and the production of low carbon steels can thus be readily obtained.

3. *The Induction Furnace.*—In the usual type of induction furnace the crucible holding the metal under treatment in the form of a ring or loop constitutes the single-turn secondary of a transformer. The iron core or yoke of the transformer passes through and interlinks the primary coil with the single-turn secondary, and thus provides the necessary close coupling between the two.

The main advantages obtained by this type of furnace are the absence of contamination of the metal by carbon, and in the better exclusion of furnace gases from the metal; the furnace is conveniently adapted for the introduction of the current, enabling the application of a high potential circuit without the use of transformers or copper cables of large cross section; fluctuations in the current do not occur as in other types, and a much steadier load is therefore offered to the power supply.

One disadvantage, however, in the original type of furnace, containing a ring of metal of uniform section, is that the temperature could not be taken to the same high degree as in the direct-arc type, and the powerful reducing action which is accompanied by the formation of calcium carbide is thus not secured. On account of this lower temperature the furnace could not be applied for carrying out any extensive refining, but rather a melting which required the use of pure materials in order to yield a high grade of steel. However, in more recent types of the induction furnace attempts are made to overcome this limitation by arranging a narrowing of the channel of a portion of the circuit, or by other means causing an increased resistance and higher temperature to be obtained locally. A good circulation of the metal is in all cases ensured in the induction type of furnace through the operation of electro-magnetic forces.

CHARACTERISTIC FEATURES OF ELECTRIC STEEL FURNACES.

To summarise the main characteristics of the different types of steel furnaces, the difference in the functioning of the electric furnace from that of the usual fuel-heated processes is determined by the following two factors :—

1. The control in the electric furnace of the surrounding atmosphere, whereby contact of the metal and slag with oxygen and nitrogen can be avoided, and

2. The production in electric furnaces of higher temperatures, which modifies in a large degree the chemical reactions taking place in the slag and between the slag and the metal. In the direct-arc furnaces, such as those of Héroult and Girod, a zone of very high temperature is produced in the slag immediately below the electrodes, while a circulation of the main mass of metal is induced by the local high temperature and the electro-magnetic influence of the current. In the radiation-arc furnaces such as that of Stassano, a more widely distributed but less intense heat is imparted to the metal, and circulation is ensured by the rotation of the furnace. In the Kjellin type of induction furnace a high temperature is produced uniformly

throughout the metal; active circulation is ensured by the rotary magnetic field set up by the large currents; absence of carbon from electrodes and all access of air is precluded. In other types of induction furnaces temperature differences are produced by special means, such as reducing the cross section of the bath at certain places, or by additional "pole plates" as in the Röchling-Rodenhauser furnace.

COMPARATIVE FEATURES OF DIFFERENT TYPES OF ELECTRIC FURNACES.

With the direct-arc furnaces, on account of the higher temperatures obtained, the most complete refining from impure materials can be effected. Amongst themselves the differences in various designs relate mainly to facilities for the introduction of the current, and in carrying out repairs and replacements, in the power efficiency, and in the cost of repairs and maintenance. Through the disposition of the electrodes, which determines the nature of the electro-magnetic currents induced, important advantages are in some cases claimed in the securing of a more effective circulation of the metal during treatment.

With the radiation-arc furnace the heat efficiency is not as a rule as high as in the direct arc, and the heating and deterioration of the furnace roof is greater. However, in avoiding direct contact of the carbon electrodes with the metal or slag, a metal of lower carbon content can generally be obtained, and other contamination from the electrodes avoided.

With the induction furnace of the Kjellin type the high temperature of the direct arc is not obtained, so that, while with the latter, special quality steels can be produced from impure materials, with the Kjellin furnace, as with the crucible process, a high-grade product can only be obtained by using pure raw materials such as Swedish irons or Styrian charcoal pig-iron. Apart from this limitation of temperature, a certain advantage appears to be obtained by the induction furnace over other electric furnaces, in the more complete exclusion of air from the metal during treatment and consequent absence of nitrogen from the product.

CHEMICAL REACTIONS IN ELECTRIC STEEL FURNACES.[1]

(a) *Oxidising Treatment. Removal of Phosphorus.*—The operation of the electric steel furnace begins, as in the case of the open hearth,

[1] R. Amberg, "Electrochem. and Metall. Ind.," 1909, **7**, 115; "Report of Eighth International Congress of Applied Chemistry," New York, 1912; Field, "Trans. Faraday Soc.," 1917, **13** (1), 1.

by the addition of pig-iron and scrap steel, or scrap steel alone. This charge is melted down or admitted to the furnace in a previously fused condition, the slag-forming materials, consisting of lime, silica, and iron oxide in the form of ore or scale, are added, and the first or oxidising stage of the refining is carried out. During this stage carbon is oxidised, and phosphorus eliminated to a larger or smaller degree by the formation of phosphate which passes into the slag.

The ferric oxide added to the slag is first reduced to ferrous oxide, which partly dissolves in the metal according to the temperature and the coefficient of division of FeO between the two phases. If the slag is saturated with FeO the latter can assume its full concentration in the metal bath according to the solubility curve given

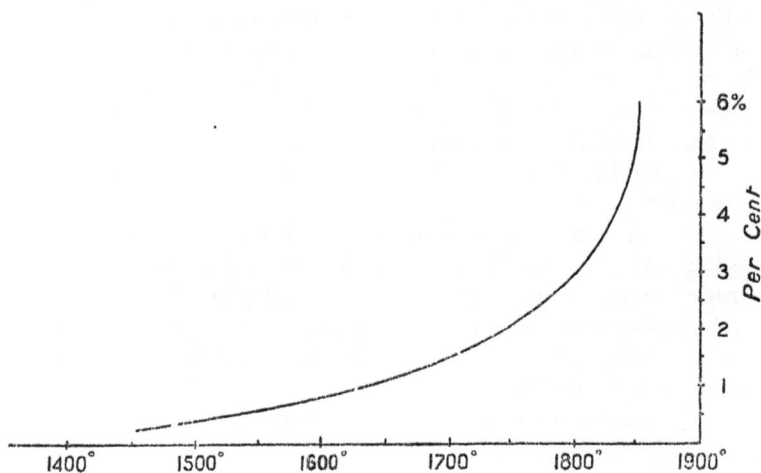

FIG. 117.

by Eichoff (Fig. 117). An essential difference in these reactions in the open-hearth and electric furnaces respectively is that with the latter, when well constructed, practically all the oxygen has to be supplied in the form of iron oxide, whereas with the open-hearth, an unlimited supply of oxygen is available from the air admitted, from which the slag continuously replaces the amounts transferred to the steel. At this period of oxidation the basic slags of an arc furnace and an induction furnace show a material difference; the dephosphorisation is accomplished more quickly in the comparatively colder slag of the induction furnace than under the higher heat of the arc. It is on this account that the application of electric furnaces in conjunction with the open-hearth process has developed very largely. In this combined method the initial refining is carried as far as possible in the fuel-heated furnace, and the charge is admitted in the molten condition to the electric furnace for finishing operations, for which it is especially fitted, and which cannot be performed in the open-hearth.

The procedure of the first stage of refining does not then differ essentially from that of the open-hearth furnace. The CaO in the slag will amount to about 40 per cent, iron and manganese oxides together to 26 to 29 per cent, and the P_2O_5 may amount to 4 per cent.

In the oxidising period sulphur is removed only to a small extent in the open-hearth furnace, but to a larger degree in the electric furnace, especially in presence of manganese. Apparently a small quantity of sulphur dioxide forms and disappears with the gases, as the partial pressure of oxygen and of sulphur dioxide in the atmosphere of the electric furnace is smaller than in the open-hearth.

This sulphur dioxide would appear to be derived from the reaction

$$FeS + 2\ FeO = 3\ Fe + SO_2.$$

(*b*) *Reducing Treatment. Removal of Sulphur.*—The final refining stage is conducted in the presence of a reducing slag. In this stage the amount of oxygen in the slag is gradually diminished, until any phosphate present begins to be reduced. The danger of this element re-entering the bath is obviated by removing the old slag and adding a fresh one, or, alternatively, by bringing about the formation of a compound of the probable formula Ca_3P_2, which is not reabsorbed by the steel phase.

In the molten steel sulphur is chemically combined with iron and manganese. These sulphides are soluble both in the steel bath and in the slag. According to the distribution law of Nernst relating to two immiscible solvents, such as molten iron and slag, the ratio of the concentration of the different substances which are soluble in the two media is a constant for any given temperature, provided the dissolved substance has the same molecular weight in the two solvents. Thus, in the case of iron sulphide,

$$\frac{c\ .\ \text{FeS (in slag)}}{c\ .\ \text{FeS (in metal)}} = k,$$

where k is a constant for any given temperature and composition of slag. A similar relation holds for manganese sulphide. The value of k or the quantity of sulphides passing into the slag will be the greater the higher the temperature, and the greater the basicity of the slag and its content of lime and manganous oxide.

Considering a case where manganese is assumed to be absent, the reactions in the slag will be determined by the following relationship :—

$$FeS + CaO \rightleftarrows CaS + FeO \qquad . \qquad . \qquad . \quad (1)$$

Unlike iron sulphide, calcium sulphide is insoluble in iron.

The concentration of lime, being very large, may be considered to be constant, so that from the law of mass action the amount of FeS present in the slag will be proportional to the product of CaS and FeO in the slag, and according to the distribution ratio k in slag and metal the amount of sulphur as FeS dissolved in the metal will be proportional to the same product.

However, during the furnace treatment, a factor which leads to the continuous diminution of the iron oxide present is the progress of the reaction expressed by the equation

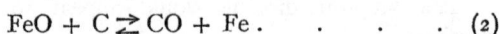

$$FeO + C \rightleftarrows CO + Fe . \qquad . \qquad . \quad (2)$$

which, by removal of carbon monoxide, proceeds completely in the direction from left to right, and thus leads to a similar completion of reaction (1). The equations (1) and (2) may be combined in the form of

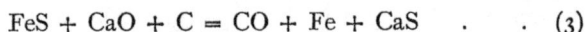

$$FeS + CaO + C = CO + Fe + CaS \qquad . \qquad . \quad (3)$$

Apart from the equilibrium values, the removal of the sulphur depends greatly on the velocity with which the substances participating in the reaction are brought to the contact surface and carried away from it, i.e. on the velocity of diffusion and mechanical convection. The reaction is therefore facilitated by every condition which makes the slag less viscous, by an increase in the area of the contact surface between the two phases, and by rapid renewal of the contact surface.

In addition to the above, two other phases which have to be taken into consideration are the atmosphere above the bath, and one or more solids in the hearth of the furnace.

In addition to equation (3) at the high temperature of the direct arc furnace, a reaction is brought about through the presence of calcium carbide in accordance with the equation

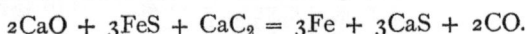

$$2CaO + 3FeS + CaC_2 = 3Fe + 3CaS + 2CO.$$

At the lower temperature of the induction furnace, the following reaction is produced by the addition of ferro-silicon :—

$$2FeS + 2CaO + Si = 2Fe + 2CaS + SiO_2.$$

The reaction at the basis of these processes is the change of FeS or MnS into CaS and appears to lead to an equilibrium, which is displaced by a small addition of metallic oxides towards the CaO side, viz. from right to left in the equation

$$CaO + Fe(Mn)S \rightleftarrows CaS + Fe(Mn)O,$$

while from left to right is the underlying principle of reactions (1) to (3).

Formation of Silicon Sulphide.—A further reaction which has been established in the desulphurisation of iron in presence of silicon is the formation of silicon sulphide, in accordance with the equation

$$2FeS + FeSi = 3Fe + SiS_2.[1]$$

Influence of Sulphur Phosphorus, and Gases in Steel.—The thorough removal of sulphur and phosphorus is of importance in the making of both ingots and castings. Sulphur which is present as sulphide of iron forms an eutectic which divides the grains of metal, so that the strength of the structure is to a large extent limited by the strength of this eutectic. Further, as the iron sulphide and phosphide eutectics have a lower melting-point than the metal, they cause the metal to yield or tear across the joint during solidification.

Gases become insoluble in the steel as it cools, and separate out in the form of blow-holes. Some of these become filled in with compounds of iron with sulphur and phosphorus which are forced in during the solidification of the metal.

With steel that is free from sulphur the metal is ductile on solidification and has not, therefore, the same tendency to develop cracks and fissures. The sulphur present can, with advantage, be combined with manganese when the sulphide of manganese formed is less deleterious than the iron compound, but with steel made in the electric furnace there is less need for such a remedy. A thorough removal of sulphur is also an indication that deoxidation of the metal has been brought about.

(c) *Deoxidation.*—The deoxidation treatment is conducted by the addition of carbonaceous material to the slag, followed generally by the addition of ferro-silicon or ferro-manganese. Carbon is then introduced to give the necessary content of this element to the steel, and other ferro-alloys for the preparation of special alloy steels.

Removal of Hydrogen.—According to Coussergues,[2] in the direct-arc electric steel furnaces, a reaction takes place between the calcium

---

[1] Cf. W. Fielding, " Trans. Faraday Soc.," 1909, 5 (1), 110.
[2] " Rev. de Metallurgie," 1909, 6, 589.

carbide formed in the slag and dissolved hydrogen from the steel in accordance with the reaction

$$CaC_2 + 2H = Ca + C_2H_2.$$

Composition of Slags.—Acid slags become more viscous with an increase of the silica content, and basic slags have a higher melting-point with increasing amounts of CaO. In the electric furnace a slag can be used containing over 75 per cent of lime without the use of fluorspar. This mineral increases the fluidity of any given slag and is largely applied for this purpose. Its presence also enables a lower proportion of silica to be used. For some special purposes slags are used containing as little as 2·1 per cent SiO_2.

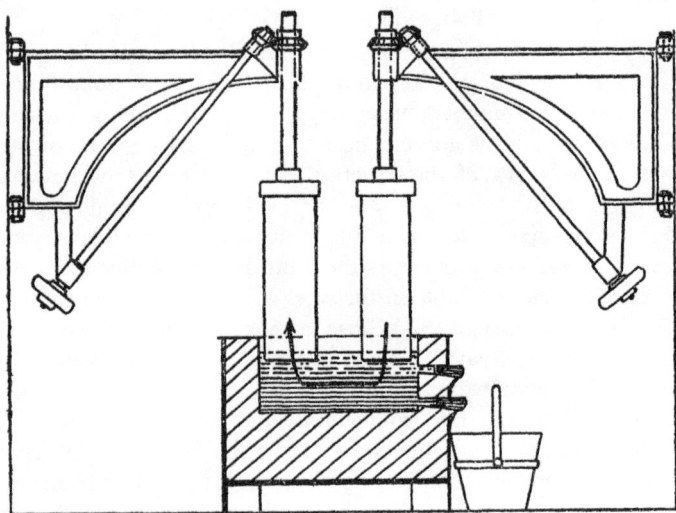

FIG. 118.

Carbon added to the slag and that derived from the electrodes may form silicon carbide, calcium silicide, and calcium carbide. The latter compound is recognised by the development of acetylene from a cooled sample, and it is only after this carbide has been formed to some extent that the deoxidation of the charge can be relied upon as being complete.

THE HÉROULT FURNACE.[1]

The Héroult furnace is operated by forming arcs between the surface of the charge and carbon electrodes which are suspended

[1] "The Electrician." 1918, **81**, 588. (The writer is also indebted to Messrs. Campbell, Gifford & Waite, London, for data supplied.)

vertically over the furnace and connected to the different poles of
an alternating current supply. Formerly only single-phase current
was used together with two electrodes joined in series, but the

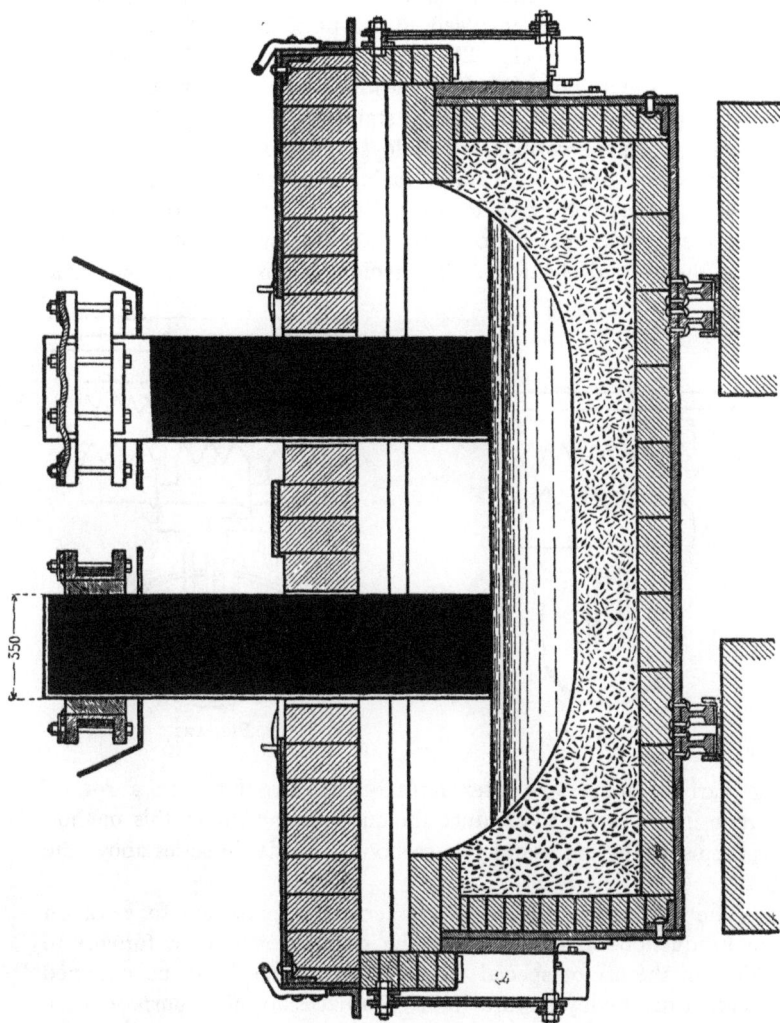

FIG. 119.

larger furnaces now in use are operated with three-phase current
and contain three electrodes. In all cases when the furnace
charge is molten, the current, as shown in Fig. 118, passes in the
form of an arc from the electrode to the surface of the slag,

14

through the slag to the metal, and horizontally through the surface layers of metal to the adjacent electrode, where a second arc is formed. A zone of very high temperature is thus produced in the slag in the neighbourhood of the electrodes, and rapidly brings about the chemical reactions involved in the purification of the iron and its conversion into steel. The work of Héroult on the production of steel in the electric furnace followed as an outcome of the aluminium process which had been established in 1887. The possibility of producing various ferro-alloys was shown in a similar type of furnace consisting of a metallic casing of crucible form, the bottom of which is carbon-lined to form one pole, whilst the movable carbon electrode making contact with the surface of the charge forms the second electrode. To avoid contamination of the metal

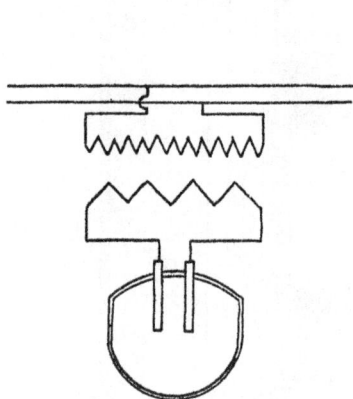

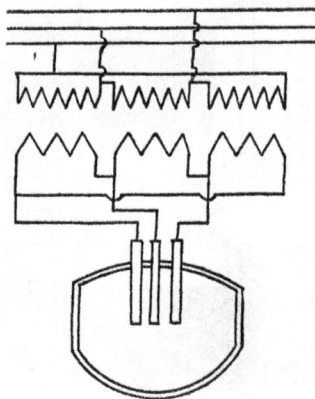

FIG. 120.   FIG. 121.

the carbon lining was later replaced by chromite with a rod of carbon in the centre to conduct the current, and finally this method was replaced by the use of two carbon electrodes in series above the bath.

The production of low-carbon ferro-chromium led to work on the production of steel and to the establishment of a furnace in which, by the use of special slags, high-grade steel can be obtained directly from highly impure iron. Furnaces for this purpose were brought into operation at La Praz and Froges in France at the aluminium works of P. Héroult (cf. p. 356), and in 1906 the process was applied at Remscheid, in Germany. The type of furnace used is as illustrated in Fig. 119 which shows the metal bath and covering layer of slag.

The transformer connexions for the operation of a furnace of two electrodes from a high-tension single-phase supply are illustrated in Fig. 120, and those for a three-phase furnace in which both transformer windings are delta-connected, in Fig. 121.

A diagram of the furnace as at present constructed for the use of three-phase current is shown in Fig. 122, and a view of the furnace during teeming in Plate X. When in operation, the lower surface of the electrode is maintained at a distance of 18 inches (45 cm.) above the layer of the slag, and contamination of the metal by the carbon is minimised while a high voltage, amounting to from 80 to 110 volts, is maintained.

Lining of Furnace.—A suitable lining is given by a mixture of a good quality magnesite, mixed with basic slag and tar as a binding material. Burnt dolomite is also found to give good results.

The life of the lining varies from three months to one year, minor repairs being carried out after each heat by throwing in magnesite or dolomite.

Roof of Furnace.—The roof is generally composed of dolomite, magnesite, or silica bricks, and usually requires renewal once a month.

Electrodes.—With furnaces of two tons capacity and less, graphite electrodes are generally used and again in the largest sizes, where it is necessary to employ electrode bundles on account of the difficulty of preparing rods of very large diameter. For furnace of intermediate sizes the electrodes are generally of amorphous carbon.

Water Cooling.—Water-coolers are placed around the electrodes on the roof.

Electrode Consumption.—When working with cold scrap metal, the consumption of electrodes of amorphous carbon is usually from 30 to 40 lb. per ton, and those of graphite 12 to 16 lb. per ton.

Acid Linings.—In cases where refining is not required, such as in the melting of turnings of high-grade metal, an acid lining, such as silica may be used together with an acid slag, and many advantages thus gained.

Power Consumption.—With a furnace of 5 tons capacity, starting from cold scrap, the power required for melting and refining one ton of steel varies from 650 to 800 kw. hours, according to the nature of the scrap and the material to be made. When starting with molten metal the power required for recarburising, desulphurising, and deoxidation of the steel amounts in a furnace of 5 tons capacity to from 140 to 180 kw. hours.

Electrical Connexions.—The Héroult furnaces, as designed for three-phase operation and supplied by means of transformers from a

14 *

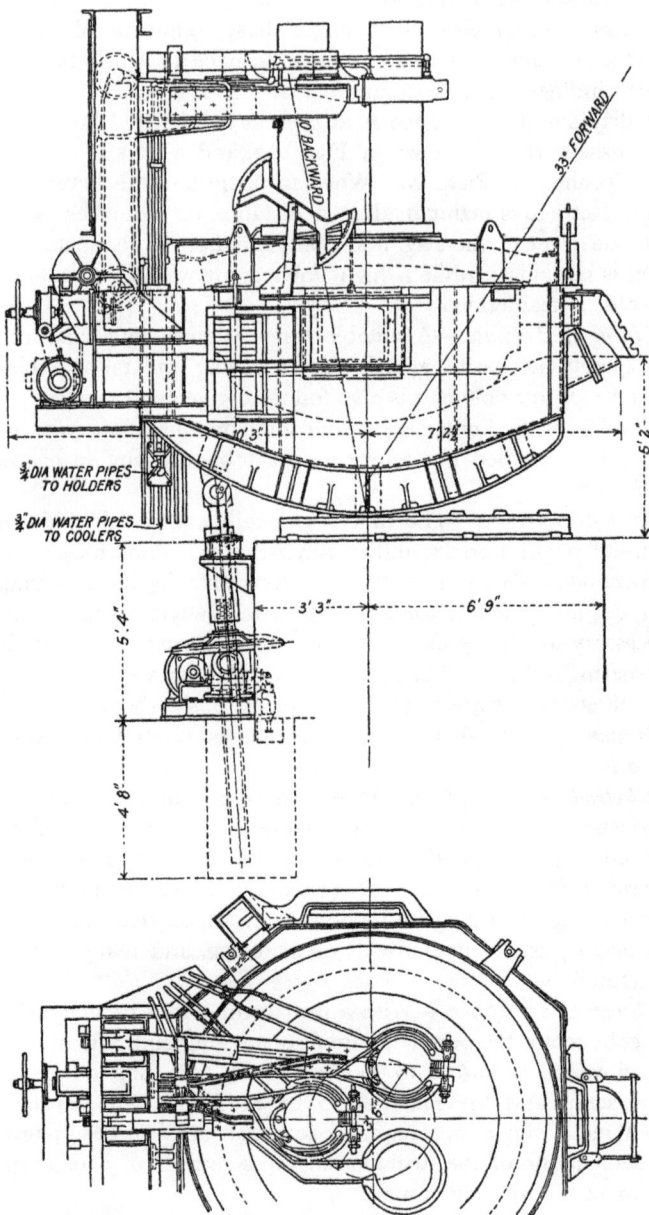

FIG. 122.

PLATE X.—HÉROULT STEEL FURNACE.

(Messrs. Campbell, Gifford & Waite.)

high-tension circuit, are generally controlled by two switches, a principal high-tension switch connecting with the supply circuit, and a change voltage switch which is interlocked with this main switch so that the change of voltage can only be made when the supply is cut off. The transformer connexions are generally arranged in star form on the high-tension side and delta on the low-tension side, thus dividing any momentary load on any single low-tension phase between two high-tension phases and causing less unbalance on the supply lines. The usual line voltages at full load for basic working are 90 volts and 75 volts, the former for rapid melting and the second lower voltage for use during the refining period. The method of changing voltages is illustrated by the diagram in Fig. 123 which shows the connexions as follows :—

1. High-tension switch.
2. Change voltage switch.
3. Current transformers for trip coils in case of overload.
4. Primary transformer windings.
5. Secondary transformer windings.
6 and 7. Low-tension star-connexion and reactance coils.
7 and 8. Low-tension mesh connexion.
9. Voltage indicating lamps.
10. Hearth neutral point for lamps.
11. Wattmeter voltmeter connexions.
12. Current transformers.
13. Ammeters.
14 and 15. Current coil connexions to wattmeters.
16. Low-tension cables.
17. Electrodes.
18. Furnace hearth.

When working with acid hearth and slag a much higher arc voltage is required owing to the greater resistance of the acid slag. In order to be able to change the voltage to convert the furnace from basic to acid working without delay and without loss of power, the switch 7 is employed to change the low-voltage connexion from delta to star. This raises the low-tension line voltage from 75 to 128 volts and, at the same time, introduces a supplementary reactance in the circuit which protects the system from the greater fluctuations consequent on working at a higher voltage.

Working of the Furnace.—In the case of cold scrap, after admitting to the furnace by the side doors, the electrodes are lowered on to the charge and the switch closed. An electric lamp is connected to each circuit between the electrode and the hearth, and remains lighted until the electrode touches the charge when it becomes extinguished. An indication is thus given when contact of each electrode is made with the charge. The arc which is formed with the surface of the charge causes a pool of metal to form under

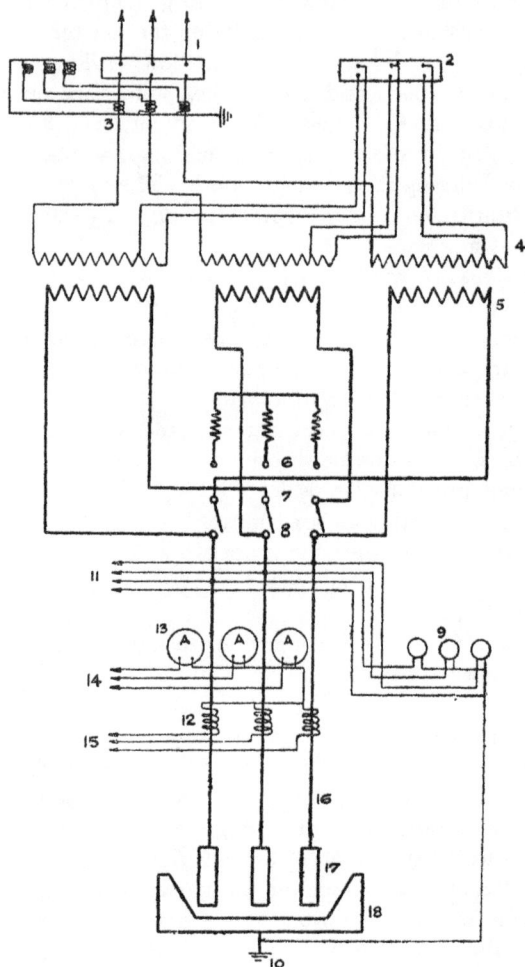

FIG. 123.

each electrode followed eventually by the melting of the whole of the metal and slag. The iron-ore and lime form an oxidising slag which removes carbon, manganese, silicon, and phosphorus. When the desired degree of refining has been obtained, the furnace is tilted slightly forward and the slag scraped off. Carefully-sized anthracite is then thrown on the surface to bring the bath up to the requisite carbon content. A second slag, consisting of lime and fluor-spar is then added. This, after a further addition of carbon dust, is partly converted into calcium carbide and forms a white slag which removes the sulphur as calcium sulphide.

For the manufacture of alloy steel the necessary additions are then made in the furnace, and when the steel is at the desired temperature, it is poured into the ladle by tilting the furnace. The bottom is then repaired and the furnace is ready for the next heat.

Capacity of Furnaces and Allied Data.—The relation between the capacity of Héroult furnaces, power consumption, and other data are given in the following table :—

Capacity of Furnace (Tons).	Power Consumption. K.V.A.	Nature, Number, and Diameter of Electrodes.	Duration of Heat (Hours).	Remarks.
0·5	200 to 400 single-phase	Graphite, 2, 5½ ins.	1¼ to 1½ (with 400 K.V.A.)	Suitable for high-speed steel tools or very light castings.
1·5 to 2	450 to 600 three-phase	Graphite, 3, 6 ins.	3 to 4	Suitable for tool and special steels and small castings.
3 to 3·5	600 to 900 three-phase	Graphite, 3, 7 to 8 ins. Amorph.-carbon, 3, 14 to 16 ins.	4 to 6	Suitable for alloy and carbon steels, and medium-size castings.
6 to 7	1200 to 1800 three-phase	Amorph.-carbon, 3, 18 to 20 ins.	5 to 8	Most widely-used size, suitable for alloy or carbon steels.
10 to 12	1800 to 2400		6 (with cold charge) 2½ (with molten charge)	Suitable for alloy and carbon steels and for refining molten steel from open-hearth furnaces.
15	2250	Amorph.-carbon, 3, 24 ins. Graphite, 3, 3 × 8 ins.	1¼ to 1¾ (with molten charge)	Used for refining molten steel from open-hearth or Bessemer Converters.
25	3750	—	—	,, ,,

15-ton Furnace.[1]—Héroult furnaces of 15 tons capacity have been applied in the United States at the South Chicago Works of the Illinois Steel Company, and at the American Steel and Wire Company at Worcester, Mass. The former plant is used for refining molten metal from an acid-lined Bessemer Converter, and the latter for

[1] " Met. and Chem. Engineering," 1910, **8**, 179.

metal from the basic open hearth, the object being to produce a higher grade of steel than that given by the open hearth at a slightly advanced cost.

In its earlier applications the Héroult furnace had been adopted essentially as a substitute for the crucible steel process for the manufacture of products such as high-grade tool steels, but in these larger plants which were established at South Chicago in 1909, and at Worcester in 1910, it entered into a much wider field of usefulness, being applied for large tonnage products such as rails, axles, and wire.

Up to this time (1910) the only two other electric furnace installations in America were at the Halcomb Steel Company, Syracuse, N.Y., and at the Firth Stirling Steel Company, McKeesport, Pa.

In the process at Chicago, the metal, after treatment in the Bessemer Converter, is submitted in the electric furnace to a treatment which involves both desulphurisation and dephosphorisation. The weight of each charge varies from 23,000 to 29,000 lb. The duration of treatment of each charge is $1\frac{1}{4}$ to $1\frac{3}{4}$ hours, and the energy consumption from 100 to 200 kw. hours per metric ton of metal, depending largely on the amount of phosphorus to be removed.

The furnace is operated by three-phase supply at 25 cycles, delta-connected for 100 volts, and employs three electrodes. The hearth is lined with dolomite and the roof composed of silica bricks. The consumption of electrodes is given, both in the case of graphite and amorphous carbon, as 6 lb. per ton of steel.

Considerable difficulty is experienced in the production of electrodes of amorphous carbon of the size demanded (24 inches diameter). Graphite electrodes, each of which consists of a bundle of three rods of 8 inches diameter are more generally used. A water-cooled jacket is provided around each electrode. The power factor obtained is from 0·8 to 0·9.

Electric Power Supply.—The electrical energy is generated by means of gas engines operated with blast-furnace gas at an estimated cost of $\frac{1}{2}$ cent (0·25d.) per kw. hour actually consumed.

Analysis of Product.—In the case of axle steel, the composition of the product obtained is as follows :—

> C, 0·35 to 0·45
> S, 0·028 to 0·032
> P, 0·026 to 0·039
> Mn, 0·34 to 0·49

Different batches of metal show a marked uniformity in composition.

Triplex Process of Steel Production.[1]—The method which, after

---

[1] T. W. Robinson, " Met. and Chem. Engineering," 1918, **19**, 15.

numerous experiments, has been finally adopted for the production of steel at the Chicago works of the Illinois Steel Company consists in the following triplex process. Molten Bessemer-blown steel or

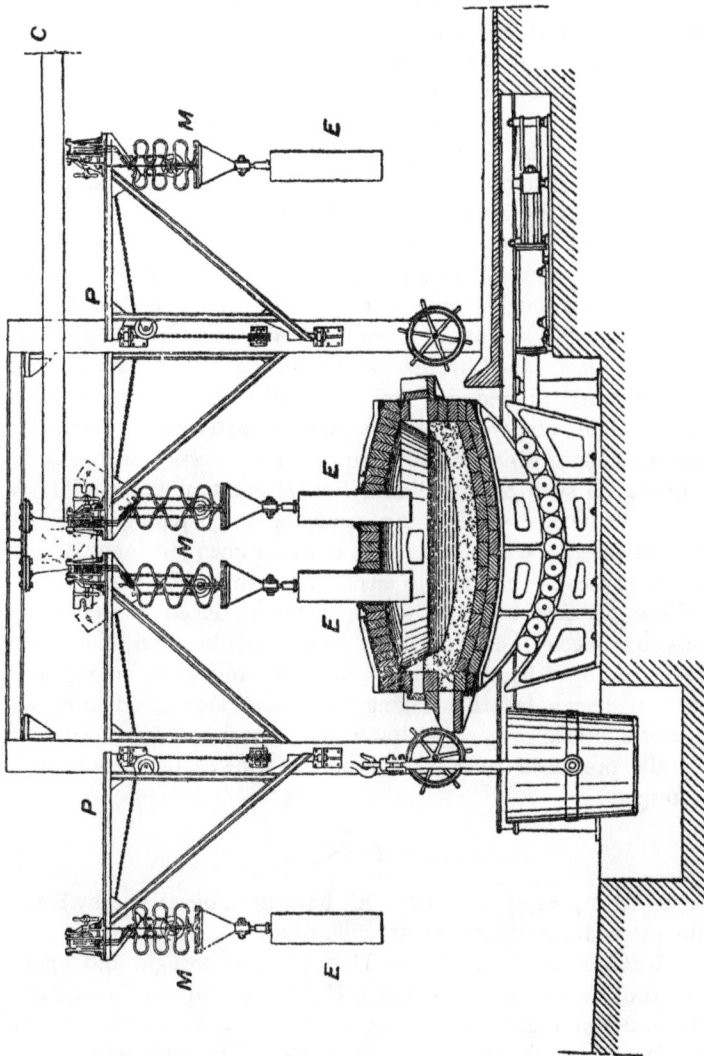

FIG. 124.

molten pig-iron together with cold alloy-steel scrap is transferred to a tilting open-hearth furnace, and after a refining treatment, which includes dephosphorisation, in this furnace, the metal is transferred to the electric furnace.

The present electric plant contains three 25-ton Héroult furnaces, each equipped with transformers of 3750 k.v.a. capacity. The output of the plant varies somewhat with the character of the steel made, but may be placed approximately at 12,000 tons per month. This, with the output of the two old furnaces, gives a capacity of 16,000 to 17,000 tons per month.

With four furnaces operating on the triplex process twenty-four-hour load factors of 75 or 80 per cent are not unusual.

In addition to the above equipment, ten 30-ton electric furnaces have recently been installed for the triplex process with steel previously treated in Bessemer and open-hearth furnaces.[1]

The furnaces ordinarily are operated with only a reducing slag, and care is taken at all times to see that such conditions obtain as will most thoroughly and quickly effect complete deoxidation. After the steel is thoroughly "dead-melted" and the reactions are complete as determined by the careful testing of the slag and the metal, the current is reduced until a proper pouring temperature is obtained. The temperature of the metal is determined pyrometrically, a procedure which is more particularly necessary with electrically produced steel than ordinarily, on account of the liability to excessive temperatures and the tendency to "piping" through absence of gases. With alloy steels, to avoid unequal ingot strains, special measures are adopted to ensure uniform solidification.

Nature of Steel Produced.—The metal produced in the above process has been found distinctly more ductile than either the Bessemer or open-hearth product, whereas resistance to repeated impacts, deflexion before breaking blow, and elongation after last blow before breaking were, at lower temperatures, greater with the electrically produced steel, but at 60° F. showed no advantage in these respects.

KELLER FURNACES.[2]

The main types of furnaces which have been developed by Keller for the preparation of steel are the following :—

1. *Multiple Electrode Type.*—This type was brought into operation in 1905, at the steel works of J. Holtzer, at Unieux (Loire), and, as illustrated in Figs. 124 and 125, consists of four movable vertical electrodes E, which serve for the introduction and return of a

---

[1] R. M. Keeney, "Chem. and Metall. Engineering," Nov. 17, 1920; "Jour. Soc. Chem. Ind.," 1921, 40, 24.

[2] "Trans. Amer. Electrochem. Soc.," 1909, 15, 87.

single-phase current, each pole being formed of two electrodes in parallel. The electrodes are admitted through four square openings (two of which are shown in Fig. 124) in the roof of the furnace, and each electrode forms at its base a heating zone in which the temperature can be regulated. The furnace bath is independent

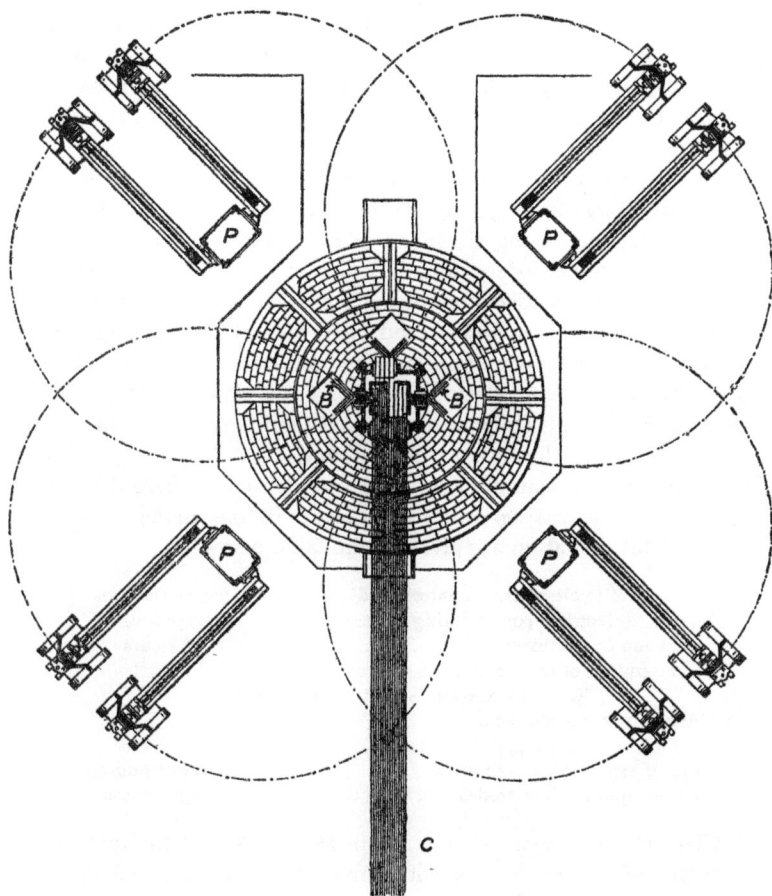

Fig. 125.

of the electrode connexions, and after removing the electrodes, can be tilted forwards for the purpose of teeming or backwards for pouring the slag. The electrodes are raised or lowered by means of the folding flexible conductors M, and are removed from above the furnace by means of the jointed arms which are attached to the pivoted support at P. The mechanism provided enables the electrodes to be raised or lowered either separately, in pairs of the same polarity, or the four electrodes simultaneously. By this means the voltage can be equalised and the power evenly distributed over the four electrodes. The bus-bars C which lead the current to the furnace consist of a group of twenty interlaced and closely-spaced copper bars, each being 10 inches wide and 0·2 inches thick, the phases being alternately adjacent to each other. The leads are brought directly over the centre of the furnace and connected to a central block from which four electric circuits radiate as at B, each carrying two connectors for each electrode.

The jointed arms carry at the point where the electrodes are suspended fittings for engaging the bus-bar terminals. The method employed for interlacing the leads reduces to a minimum the induction or reactance effect, and it is claimed that a power factor of $\cos \phi = 0·97$ is obtained when using a current of 12,000 amps.

The furnace bath is circular and lined with a hearth of magnesite. By means of the roller supports, the furnace body can be tilted either backwards or forwards for pouring the slag or metal. This furnace is applied at Unieux for the final refining of steel which has been treated in a Martin open-hearth furnace. The charge is admitted in a molten condition to the electric furnace where deoxidising treatment is carried out and alloy additions made. Typical results of this operation are given in the table below :—

Weight of (molten) charge introduced . .	7·5 metric tons
Average electrical power during treatment .	750 kw.
Duration of treatment	2¾ hours
Composition of charge added,	
C, 0·15 ; S, 0·06 ; and P, 0·007 per cent.	
Analysis of steel formed,	
C, 0·443 ; S, 0·009 ; P, 0·008 per cent.	
Power expenditure per ton	275 kw. hours
Consumption of electrodes	10 kg. per ton

Three-Phase Furnace.—It is proposed by Keller to apply the above type of furnace for use with three-phase current, making use of three electrodes. In the case of a star-connected supply, it is proposed to use a conducting hearth connected to the neutral point of the supply.

2. *Conducting-Hearth Type.*—In this system, as illustrated in Fig. 126, the hearth of the furnace is conducting and connected to one of the poles of power supply. For this purpose iron bars 1 inch to 1¼ inches diameter, regularly spaced about 1 inch to 1¼ inches

apart, are placed vertically and attached to a metallic plate at the
bottom, the groups of bars thus covering the entire area upon which
the liquid bath will rest. The spaces between the bars are filled by
ramming in a basic material such as magnesite. There is thus obtained a compact mass or block consisting of iron and refractory
material, of which the metallic sections are good conductors when
cold; when heated the refractory material also rapidly becomes a

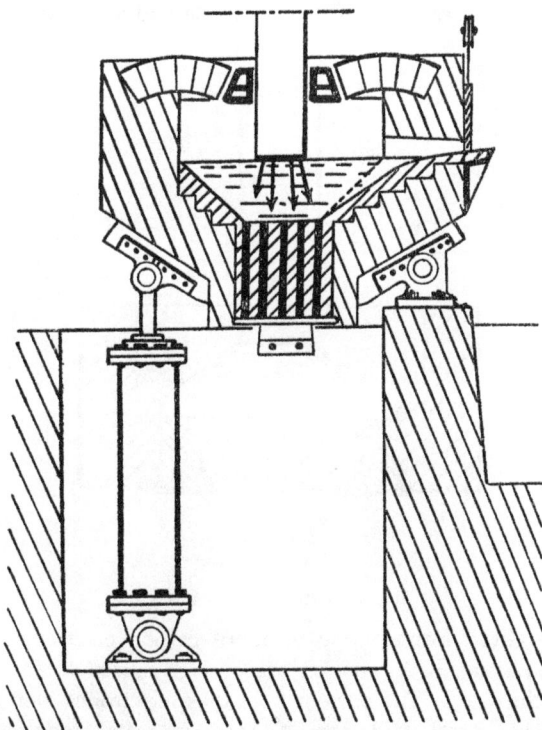

Fig. 126.

conductor. The whole of this base is surrounded by a metallic
casing serving as an envelope which may be cooled by a current of
water. The lower plate fastened to the iron bars is connected to
one of the poles of the power supply.

The furnace is covered by an arched roof traversed by the vertical
electrode, the height of which can be regulated either by hand or
automatic adjustment.

When changing an electrode, in order to complete this in the shortest interval of time, the electrode is connected to the end of a swinging arm, so that it can be replaced by another previously attached to the end of a similar arm, which enables the second electrode to be transferred to the position formerly occupied by the first.

This type of furnace is suitable for operation with several electrodes joined in parallel to one pole or connected with the different phases of a polyphase circuit.

Plate XI gives a view of a single-phase Keller steel furnace in operation containing one vertical electrode and a conducting hearth electrode.

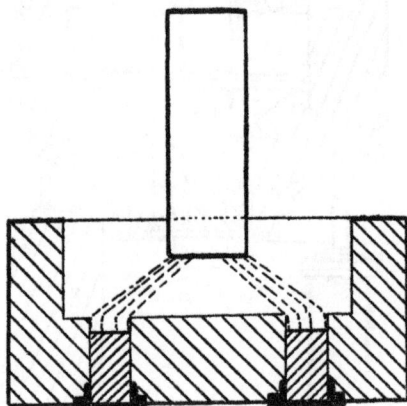

Fig. 127.

THE GIROD FURNACE.[1]

In the Girod furnace one or more carbon electrodes are suspended above the bath, and an arc or arcs caused to form on the surface of the slag. The current from these electrodes then traverses the mass of metal and returns through metal rods embedded in the hearth of the furnace. The rods are spaced radially near the periphery of the hearth so that the current passing from the top electrodes through the metal will proceed along definite paths, and by the electro-magnetic influence will cause a rotation of the molten metal in directions at right angles to the path of current flow, and so bring about a circulation of the bath. The circulation obtained is such that a current of metal can be observed passing upwards next to the walls, inwards towards the centre of

[1] " Electrochem. and Metall. Ind.," 1908, 6, 452 ; 1911, 9, 581.

PLATE XI.—KELLER STEEL FURNACE.

(Ch. A. Keller.)

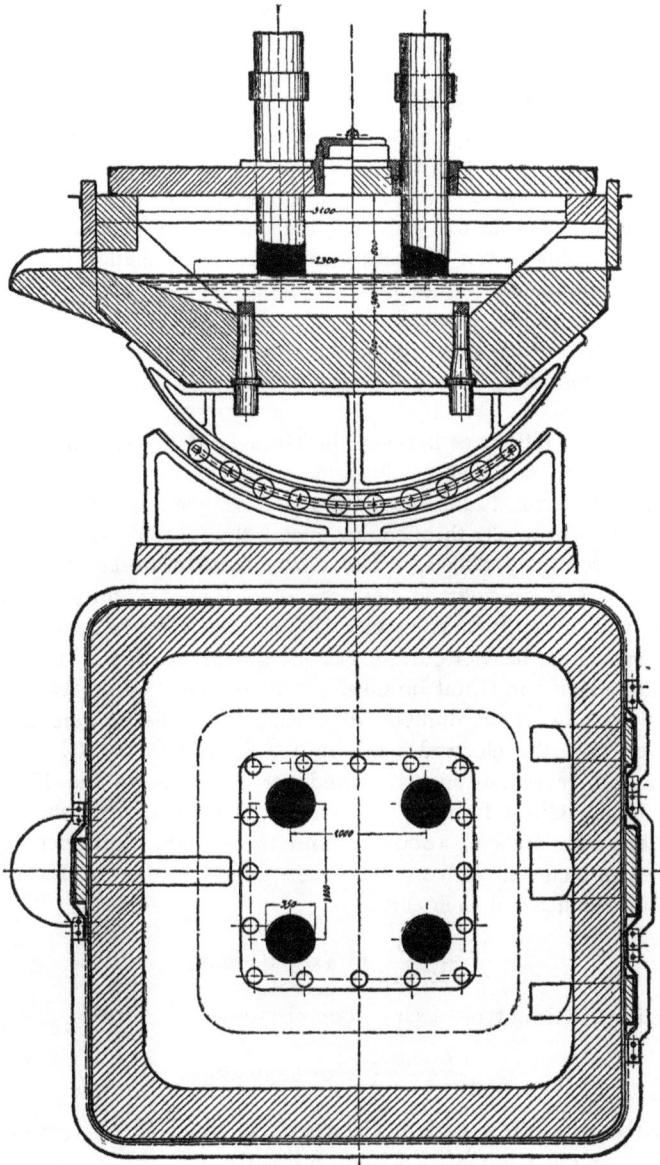

Fig. 128.

the furnace, and then downwards.   A diagram of a vertical section showing the principle of the electrode arrangement is seen in Fig. 127. These rods connecting the charge with the conductors are of soft iron, and in order to avoid overheating and melting, the ends projecting through the base of the furnace are water cooled as shown in Fig. 131.   The water cooling also prolongs the life of the refractory lining.   The efficiency of the circulation increases with the acuteness of the angle formed between the lines representing the axis of the electrodes and the path of the current respectively in the bath.   For this reason the bath is arranged to be shallow in proportion to the diameter of the base of the furnace.   Furnaces of small capacity (2·5 tons) generally operate with one carbon electrode, while those of 12·5 tons capacity have usually four upper electrodes joined in parallel.   Fig. 128 shows a plan and section of a 12-ton furnace.

The main difference between the Girod and the Héroult furnace arises from the fact that in the former the current passes obliquely through the metal to the base of the furnace, whereas with the latter it passes horizontally through the metal, thus leading to differences in the circulatory effects.   The Girod furnace works at a lower voltage (about 70 volts in the case of a 1000 kw. furnace), compared with about 110 volts with the Héroult.   In furnaces of the same capacity a heavier current is consequently carried by the leads in the case of the Girod furnace, but in furnaces of the two types containing the same number of carbon electrodes, where in the Héroult type the electrodes are in series, and in the Girod type in parallel, more current is carried per electrode in the Héroult than in the Girod furnace.

Thus with units of 1000 kw., and two carbon electrodes using single-phase current with a power factor of 0·8, the electrodes in the Héroult furnace will each carry

$$\frac{1,000,000}{0\cdot8 \times 110} \text{ or } 11,320 \text{ amps.,}$$

while in the Girod type, each carbon electrode will carry

$$\frac{1,000,000}{0\cdot8 \times 2 \times 70} \text{ or } 8850 \text{ amps.}$$

Similarly, with three carbon electrodes, in the Héroult type using three-phase current, each electrode will carry

$$\frac{1,000,000}{0\cdot8 \times 1\cdot75 \times 110} \text{ or } 6500 \text{ amps.,}$$

while with the Girod type of the same capacity, but using single-phase current, with the three electrodes in parallel, and assuming the same power factor, will take

$$\frac{1,000,000}{0\cdot8 \times 70 \times 3} \text{ or } 5960 \text{ amps.}$$

It is claimed that with the Girod furnace the heating of the charge is more equalised than in the case of the Héroult type, and a decided advantage gained when first applying the heating to a charge of cold scrap metal. In this case on account of the disposition of the lower electrode, the carbon electrode can be pressed on to the metal without causing a short circuit, and the metal is evenly heated and brought to the melting-point. The power fluctuations during operation, though considerable, are said to be less than in the case of the Héroult furnace. The power and electrode consumption of the two furnaces appears to be about the same.

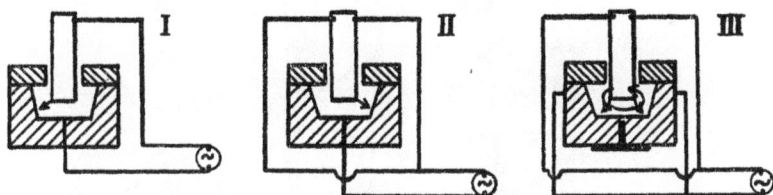

FIG. 129.

With the 10 to 12-ton Girod furnace the depth of metal in the bath is about 12 inches. The electrodes during operation are adjusted automatically by means of Thury regulators so as to give a definite current.

Arrangement of Electrical Conductors.—On account of the reactance effects given when using single-phase alternating current, the arrangement of the electrical conductors is of considerable importance. There are three methods in use for conducting the current to the carbon electrode at the top and the steel electrode at the bottom, employed in the Girod furnace. In method 1 (Fig. 129) the shortest path from the power supply to the carbon and steel electrodes has been chosen; all of the cables are on the side of the furnace which faces the source of supply. In method 2 the cable to the carbon electrode is divided into two sections which run parallel to each other. The steel electrode is connected by the shortest path with the supply. In method 3 the current is conducted to the carbon electrode in a manner similar to method 2, but while in methods 1 and 2 the bottom steel electrode is insulated from the furnace body itself (only the metal bath being in the electric circuit), the steel electrodes are here electrically connected with the furnace body. The arrangement of the conductors is symmetrical around the furnace, and the magnetic fluxes, formed by the iron shell, are reduced to a minimum. In this way the magnetic field around the carbon electrode and the heating effect of the arc become uniform, and the current passes through the charge in a more

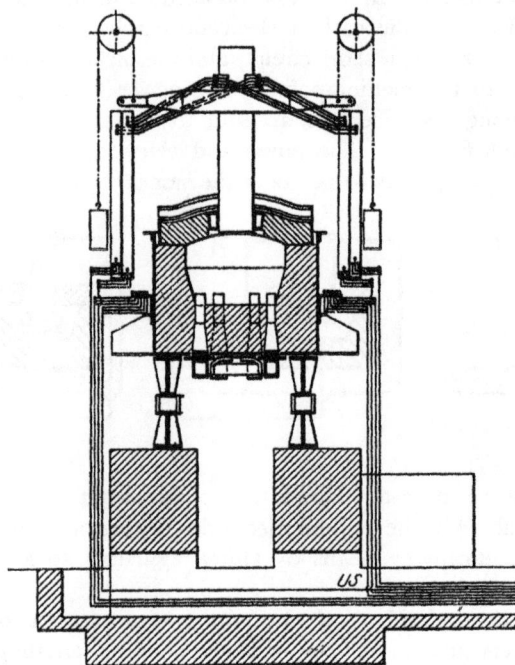

Fig. 130.

regular order. With methods 1 and 2 the electric arc is deflected towards the side marked with the arrow, which results in a non-uniform heating of the steel bath and a rapid destruction of that side of the furnace which is exposed to the excessive direct heat of the arc. In method 3, though the length of cable and attending resistance losses are greater, the furnace economy is greater and the main advantages obtained are as follows :—

1. The electric arc circling about the periphery of the carbon

electrode causes a strong agitation of slag and metal which accelerates the speed of reaction between the slag and the iron bath.

2. The roof and furnace walls receive a more uniform radiation from the arc and last longer.

3. The consumption of carbon electrodes is more uniform.

In a typical Girod furnace the metal in the bath is in electrical connexion with six steel electrodes passing through the hearth of the furnace, and electrically connected with each other and with the furnace body by means of a copper plate under the base of the furnace. The current is conducted to the furnace by means of eight parallel bus-bars of copper which are alternately of opposite polarity. Currents of opposite polarity are conducted along alternate electrodes

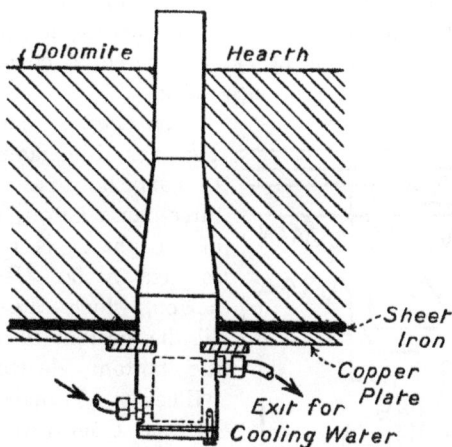

FIG. 131.

on both sides of the furnace to the pivoting point, where each bus-bar is connected by flexible copper bars or cables either with the furnace body or with the carbon electrode (cf. Fig. 130).

Hearth Electrodes.—The ratio of the cross section of the steel electrodes which are embedded in the bottom of the furnace chamber to the rest of the bottom area is 1 : 16. A renewal of the lining of the hearth does not become necessary until it is required to repair the walls. After 120 heats, it is necessary to re-line the walls and repair the hearth-bottom. This is filled up again to the original level since it has meantime corroded away to a depth of about 2 inches at the centre where the highest temperature is reached.

The water-cooling of the hearth electrode is limited to that

15 *

portion of the soft steel bars which project below the furnace body (*vide* Fig. 131). The loss of heat which follows from the water-cooling has been estimated in the case of a 265 kw. unit to amount, during an operation of 130 minutes, to 8721 kg. cals.

Since 1 coulomb = 0·239 cal., the above heat corresponds to

$$\frac{8721}{0\text{·}239 \times 60 \times 60} = 10\text{·}1 \text{ kw. hours.}$$

The total expenditure of energy during this period amounted to 1004 kw. hours. Hence the loss of heat due to cooling the bottom electrode is 1·01 per cent of the total energy consumption.

The heat loss due to the cooling of the carbon electrode and the arched roof, which was brought about by means of a cooling ring through which water was circulated was found to amount to 3·65 per cent of the total energy expended. The heat loss in cooling the steel electrode and the amount of water required is consequently very much less than for the carbon electrode.

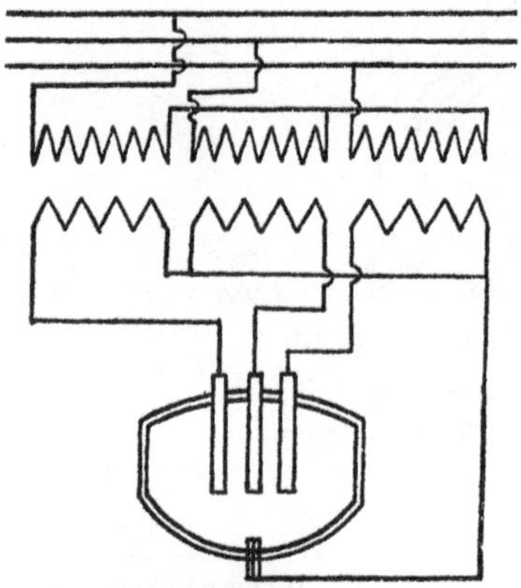

Transformer Connexions.—Fig. 132 illustrates the system of connexions as applied to a three-phase Girod furnace. One of the phases is reversed so as to give a 60° Y- (star-) connexion, which causes practically all the current to pass through the bottom electrode. (From "Electric Furnaces," General Electric Company, Schenectady, N.Y., U.S.A.)

Installations of Girod Furnaces.[1]—The Girod furnace was first brought into operation at the Société anonyme Electro-metal-

FIG. 132.

lurgique Procédés Paul Girod, at Ugine, Savoie, France, where in 1903 a water power was developed and a works installed for the manufacture of ferro-alloys (*vide* p. 275). A separate company, known as the Compagnie des Forges et Aciéries Electriques Paul Girod, was formed later, and a works built near that of the parent company from which electric power is supplied. The furnaces in use are of 2 tons, and 10 to 12 tons capacity.

An electrode factory is operated by the same company, and

[1] W. Borchers, "Eng. and Mining Journ.," 1909, **88**, 1113.

electrodes made of square or round cross section up to 13 inches diameter (350 mm.) These are made by finely grinding retort carbon, petroleum coke, and anthracite coal, kneading the mixture with tar, pressing in moulds, and baking in stoves.

The operation of the furnaces at Ugine consists in feeding in a charge of iron or steel scrap together with 10 per cent of its weight of oxide of iron in the form of ore or scale, and 4 per cent of lime. The charge is heated for about five hours when the oxide is mostly reduced. The slag is then skimmed off and a second charge of lime and iron oxide added. The slag is in some cases renewed a third time, and finally the metal is heated with a charge of lime alone. Alloys, such as ferro-manganese-silicon, are then added to complete the deoxidation. For the preparation of carbon steels Swedish char-

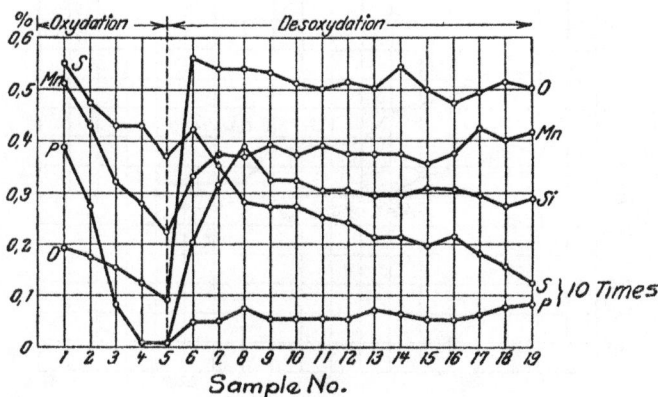

FIG. 133.

coal iron or high carbon iron is added and, for special alloy steels, ferro-nickel, ferro-tungsten, or other alloys are added.

When beginning with cold materials the total duration of the furnace treatment for each batch of metal is about eight hours. The power required to produce 1 ton of steel in a 1000 to 1200 kw. furnace is from 800 to 900 kw. hours, and in a small furnace from 900 to 1000 kw. hours. The consumption of electrodes is from 12 to 15 kg. per ton of steel. Typical analyses of samples taken at different stages during the refining are shown in the diagram in Fig. 133.

When beginning with a molten charge which has undergone a preliminary refining in an open-hearth furnace, each charge can be treated in about two and a half hours.

The lining of the furnace will withstand at least 80 charges, but
cavities are formed in the side walls through the action of slag at
shorter intervals of time and are repaired with calcined dolomite.
The bottom of the furnace will withstand from 120 to 160 charges
when the corrosion amounts to about 10 cm.   The roof requires
renewing after 20 to 25 charges.

The energy consumption for a given output varies considerably
with the weight of the charge and decreases in large furnaces.   The
minimum economical weight of charge is about 3 tons.   Beginning
with a liquid charge with a furnace of 2100 kg., a power expenditure
of 430 kw. hours is required to produce 1 ton of steel of medium

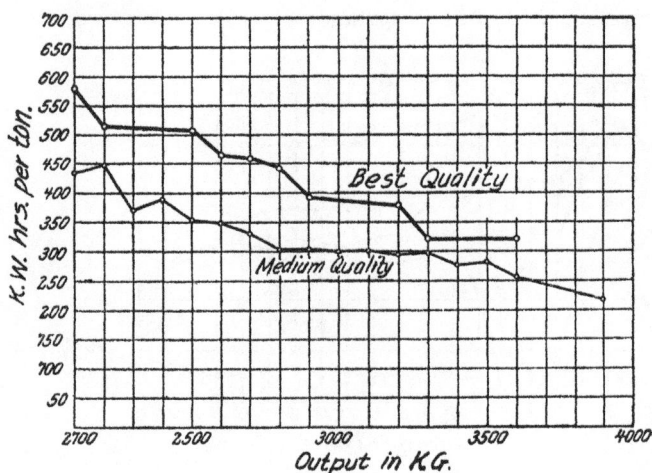

FIG. 134.

quality, and 580 kw. hours for the best quality.   With a unit of
3600 kg. 250 kw. hours has been found necessary per ton of medium
quality steel and 325 kw. hours per ton of best quality steel.   Under
very favourable conditions it has been found possible to refine 3654
kg. of steel, including oxidation, deoxidation, and desulphurisation,
with an energy consumption of 158 kw. hours per ton. (cf. Fig. 134).

At Ugine there are in operation at present (1920), nine Girod
furnaces with a total capacity of 96 tons and an output of 70,000
tons per annum.

10-*ton Girod Steel Furnace.*[1]—A 10-ton Girod steel furnace has
been in operation at the Bethlehem Steel Company's plant in South

[1] C. A. Buck, " Trans. Amer. Electrochem. Soc.," 1917, **31**, 81.

Bethlehem, Pa., U.S.A., since 1916. The furnace is cylindrical in form, about 15 feet in diameter, and 5 feet outside depth, and contains a charging door and pouring spout directly opposite. The furnace rests on heavy cast-iron rockers running on rollers which enables it to be tilted either towards the charging side, whereby the slag can be discharged through a notch in the charging door, or else towards the pouring side to pour the molten metal into a ladle of $12\frac{1}{2}$ tons capacity, placed in a pit in front of the furnace. The bottom has a thickness of 20 inches, and is made of burned dolomite well rammed in with tar. The metal bath is about 16 inches thick. Water-cooled electrical connexions are made with the base of the furnace and consist of fourteen soft steel poles about $3\frac{1}{2}$ inches diameter, the lower ends being water-cooled and the upper ends extending through the furnace hearth to the metal bath. The roof, which is arched, consists of silica bricks 9 inches in thickness, and insulated from the magnesite brick of the hearth wall by means of asbestos plates. The three electrodes of amorphous carbon, 17 inches in diameter, are held in water-cooled copper holders and pass into the furnace through openings in the roof. Adjustment of the electrodes is made so as to be about 4 inches above the surface of the bath when the arc is formed. About 1500 k.v.a. of three-phase alternating current, 25 cycles, and 65 to 80 volts is employed, each electrode receiving one phase of the three-phase current, the conducting hearth of the shell acting as the neutral point. The furnace has so far been mainly applied for the manufacture of alloy steels, and simple carbon steels ranging from 0·08 to 1·30 per cent carbon. For this purpose scrap steel and turnings are added together with lime and the charge heated, when phosphorus is removed and sulphur and carbon reduced by the oxidising slag to the required degree. When the oxidation has proceeded far enough to furnish a test piece which will bend through 180° without breaking, the oxidising slag is removed, recarburiser added, followed by the de-oxidising slag of powdered petroleum coke, silica sand, lime, and fluorspar. This slag dissolves ferrous oxide which has formed on the surface of the bath. After the lapse of a sufficient interval, varying amounts of ferro-silicon, ferro-manganese, etc., are added to bring the slag to the point where it becomes white and quickly disintegrates into a fine white powder. These alloys act very energetically on the oxides in the bath, forming a very fluid slag which easily rises to the surface. During this period when the slag is completely de-oxidised and is very basic, the desulphurisation, which was incomplete during the first stage of the operation proceeds further, and is rapidly completed at the time of tapping.   Products have been obtained containing below 0·010 P and 0·015 S.

The current is generated by gas engines driving 2500 kw. generators, producing three-phase alternating current of 6600 volts at 25 cycles.

THE "ELECTRO-METALS" FURNACE.[1]

The "Electro-Metals" furnace which was developed by Grönwall, Lindblad, and Stalhane employs the two-phase three-wire system, whereby two-phase current is taken to two upper electrodes arranged to form an arc with the surface of the charge.   One electrode is connected to each phase, and the common return current is taken through a fixed electrode beneath the hearth.   The two-phase current is obtained from any three-phase supply by means of two static transformers with Scott-connexions (cf. Fig. 135).   With this system, when the voltages are equal in the three primary mains, the voltages induced between each upper electrode and the return electrode are equal, and when equal currents are flowing through the upper electrodes, which are automatically regulated to accomplish this, equal currents are being drawn from the three-phase primary mains, so that so long as the regulation of the electrodes is good, the current demand from the high-tension mains will be balanced.

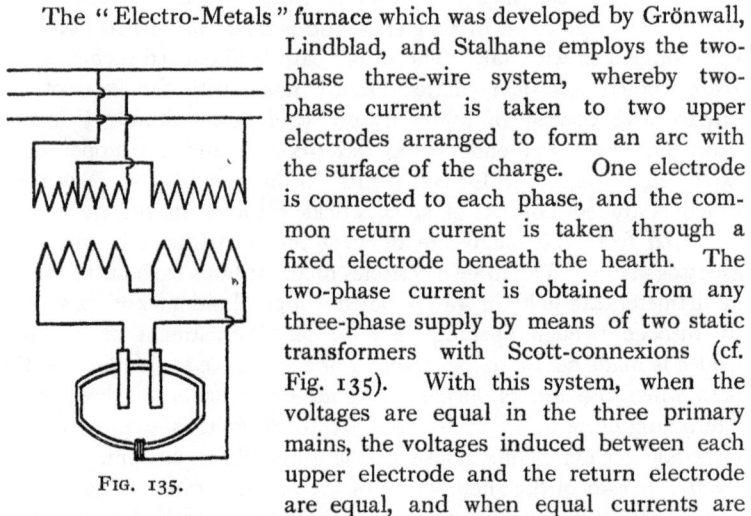

FIG. 135.

The application of a three-phase circuit directly to the furnace, with one phase on one of the upper electrodes, the second phase on the second electrode, and the third phase on the conducting hearth, is undesirable because of the unequal resistances that would result between the phases.   To obtain a balanced system with a three-phase supply, the impedances between any two of the phases must be equal.   In the "Electro-Metals" furnace the resistance between the two upper electrodes is equal to that of two arcs, whereas the resistance between each of the upper electrodes and the lower

[1] "Cassiers' Magazine," 1917, 52, 51; "Electrical Review," London, 1913, 73, 155.   (The writer is also indebted to Messrs. Electro-Metals, Ltd., London, for data supplied.)

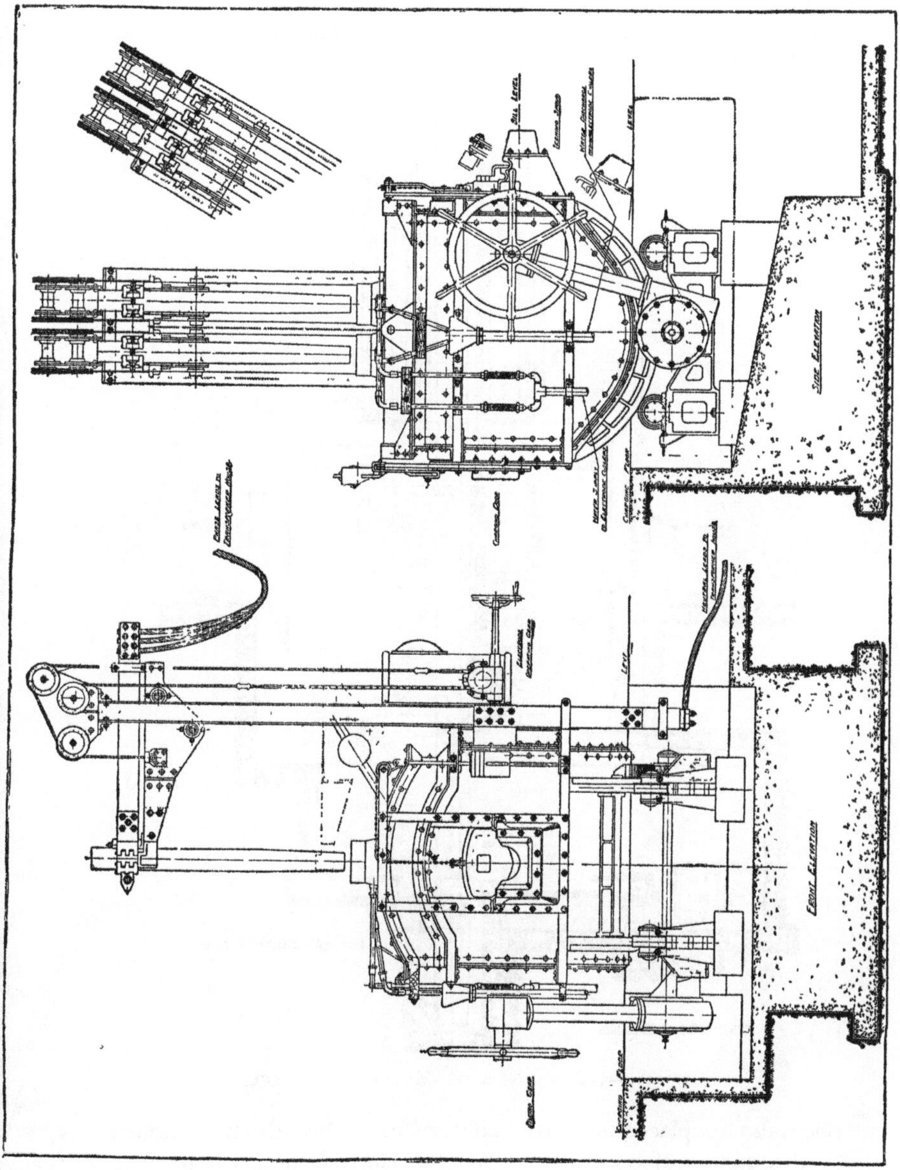

FIG. 136.—½ TON "ELECTRO-METALS" FURNACE.

electrode is equal to that of one arc plus a negligible constant resistance of the hearth, so that an accurately balanced three-phase
system is not possible on this type of furnace.   The two-phase system
on the other hand remains balanced for all loads if the two upper

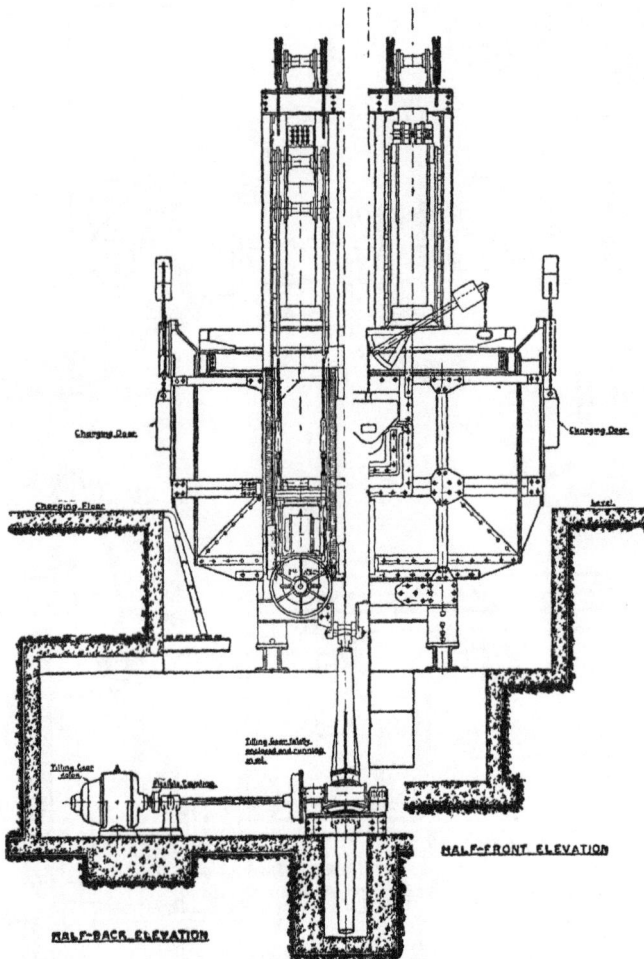

FIG. 137.—5-TON "ELECTRO-METALS" FURNACE.

electrodes are placed in correct adjustment.   The advantage gained
by this system of operation is that the electric current flows in a
horizontal direction in the bath of metal between the two phases of
the two upper electrodes and in a vertical direction between each of

the two upper electrodes and the neutral return electrode in the hearth, so that the bath of metal may be considered as being a movable conductor in a rotating magnetic field which causes a circulation of the metal in a vertical plane. The advantages which are

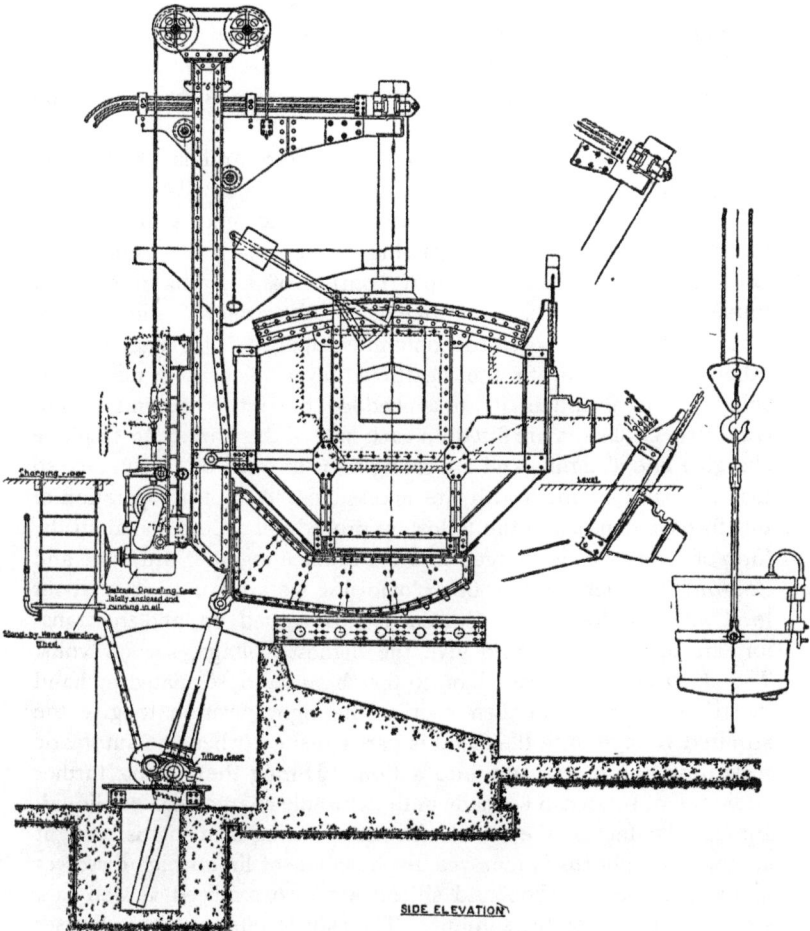

SIDE ELEVATION

FIG. 138.—5-TON " ELECTRO-METALS " FURNACE.

claimed to result from the circulation obtained in this type of furnace consist of the following :—

1. Preventing superheating and promoting uniformity of temperature throughout the bath.

2. Causing each portion of the metal to be brought into contact with the slag, thus facilitating the necessary reactions.

3. Ensuring the assimilation and diffusion of added alloys, which, like tungsten, may have a specific gravity considerably different from that of steel.

4. Radiation of heat to the roof is less intense than in the case of furnaces having only two upper electrodes.

Figs. 136 to 138 show diagrams of $\frac{1}{2}$ and 5-ton furnaces respectively. Plate XII gives a view of a $7\frac{1}{2}$-ton furnace. This is a standard size used for the manufacture of high-class steels from common scrap.

The furnace body is supported on rockers and is tilted by an electric motor through gearing and a thrust-rod. The electrodes are adjusted automatically by means of electric motors and suitable gearing, so that the current passing through each electrode has a definite desired value. The roof consists of a strong rectangular framework which acts as abutments for the arch of silica bricks. The hearth is composed of dolomite supported by magnesite or other bricks. Beneath the hearth a layer of carbon mixture is suitably placed, and in it are embedded the return neutral current conductors. Doors are fitted on each side of the furnace to facilitate charging and slagging, and a teeming door in front of the furnace so that all parts of the hearth are accessible. A typical operation of this furnace consists in the following procedure. Cold metal in the form of scrap, etc., is charged into the furnace together with lime and iron-ore, hammer scale, or other oxidising material until it is about level with the door sills. Current is now turned on, and the transformers adjusted so as to give the highest voltage, say 90 volts. The electrodes are lowered on to the charge and regulated by hand to strike the arc, and then kept at the right position to give the required voltage until the metal is partly melted, when the automatic regulation can be brought into action. During the melting further additions of scrap can be made until the whole charge becomes liquid. Intimate contact of the metal with the oxidising elements is brought about. Phosphorus is removed as phosphate of lime which dissolves in the slag, and carbon and silicon are also oxidised as well as a small proportion of the sulphur. The slag is removed, recarburiser added, and a new basic slag charged, consisting of lime and fluorspar. Deoxidation and desulphurisation are then brought about in the usual way (*vide* p. 215). At the completion of the reactions the slag is removed, and any alloy additions can then be made.

The furnaces consume about 600 kw. hours of power per ton of steel produced, while the heating extends over an interval of four hours.

PLATE XII.—"ELECTRO-METALS" FURNACE (7½ TONS CAPACITY).

PLATE XIII.—"ELECTRO-METALS" FURNACE.

" Electro-Metals " furnaces of larger capacity are constructed with four carbon electrodes which are connected to the transformers as shown in Fig. 139 (from " Electric Furnaces," General Electric Company, N.Y., U.S.A.).   By using two banks of transformers, various connexions can be made so as to vary the amount of current passing through the bottom electrode.   Plate XIII gives a view of one of these furnaces.

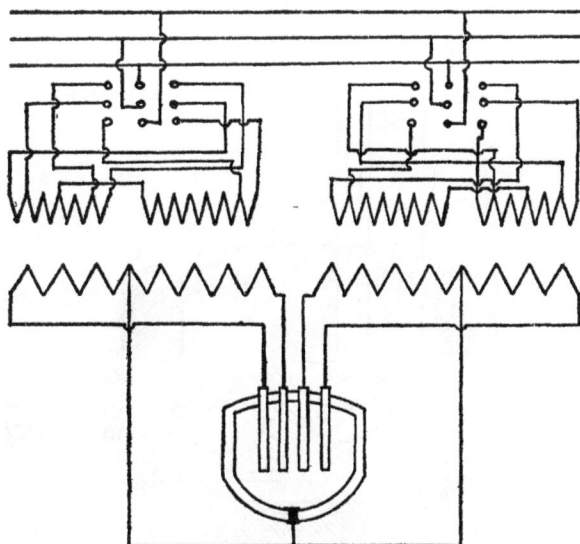

FIG. 139.

BOOTH-HALL FURNACE.[1]

The Booth-Hall furnace embodies the use of a vertical arc, a hearth which becomes electrically conducting when heated, and the use of an auxiliary electrode which acts as a return for the current until the hearth becomes heated and conducting.   The furnace is made for three, two, or single-phase circuits.   Fig. 140 shows a longitudinal section of a two-phase furnace with the auxiliary electrode resting upon the charge.   Arcs are shown passing between the main electrode and the charge, and a pool of metal forming on the hearth of the furnace.

[1] " Trans. Amer. Electrochem. Soc.," 1919, **33**, 247.

Fig. 141 shows diagrammatically the arrangement of main and auxiliary electrodes, the latter in black, in single, two and three-phase furnaces. With a three-phase supply, star-connected, the hearth of the furnace and the auxiliary electrode are connected to the neutral point.

In starting operation on a cold charge, the auxiliary electrode is lowered until it rests on top of the charge, and the arc is then formed between the charge and the main electrode or electrodes. The auxiliary electrode, which acts as a return for the current, is arranged

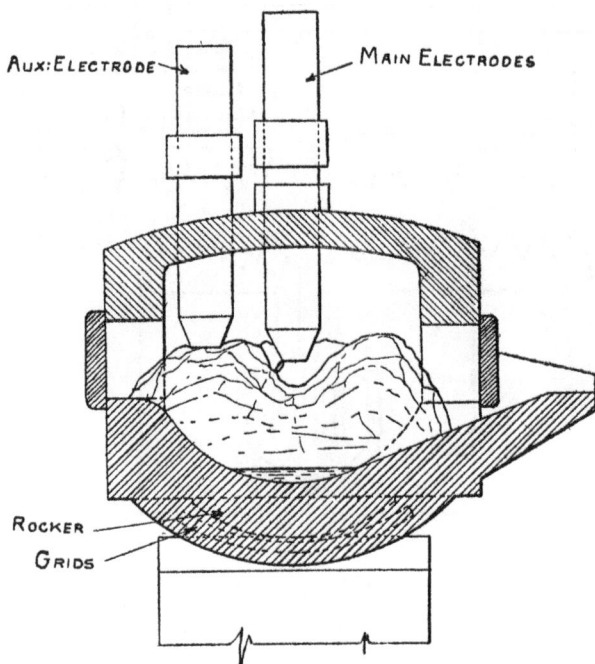

FIG. 140.

to press with its entire weight on top of the charge, and is connected in parallel with the conducting hearth. When enough metal has melted to form a pool on the bottom the hearth becomes conductive, and the auxiliary electrode is withdrawn from contact with the charge. A voltage of 125 to 150 is employed.

With the two-phase furnace which is operated by means of Scott-connexions from a three-phase supply, two sets of grids are embedded, which are insulated from each other and the two phases separated (cf. Fig. 142) and so related to the main electrodes that the current

PLATE XIV.—BOOTH-HALL FURNACE.

of the two phases crosses in the bath, thus causing circulation of the molten metal, and in conjunction with the bottom heating, promoting a mixing of the metal. The hearth is either acid or basic, as desired, and at least 24 inches in thickness, being sintered in place layer by layer. With a basic hearth, " dead-burned " dolomite is used, and with an acid hearth, ground ganister. The furnace is provided with a door on each side, and by mounting on trunnions, can be tilted forwards or backwards and the slagging operations handled from either side.

The furnace is particularly intended for melting down cold charges for steel-making foundries, and in continuous operation it is claimed will average a power consumption of 500 to 550 kw. hours per ton (2000 lb.) of product. A 4-ton furnace has been largely used for making low-phosphorus pig-iron and also special steels.

Mechanical Features of Booth-Hall Furnace.—The furnace is

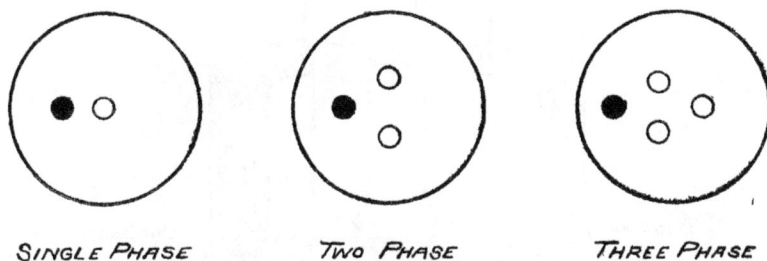

SINGLE PHASE   TWO PHASE   THREE PHASE

FIG. 141.

contained in an upright cylindrical shell provided with two main doors, slightly oval-shaped, at the front and rear of the furnace respectively. The furnace is supported on heavy cast-iron rockers mounted on steel rails, and provided with a tilting mechanism allowing of forward or backward displacement.

With the two-phase furnace the two main electrodes pass vertically downwards through the roof, while in some models the auxiliary electrode is admitted when required through an opening in the side walls, and in other models is admitted through the roof. The 3-ton furnace, which will, however, contain up to 9000 lb. of charge, has outer dimensions of 8 feet by 10 feet and will deliver an output of 32 tons per twenty-four hours from cold charges.

The largest furnace which is considered practical for working cold materials is a 12 to 15-ton unit with an output of 100 to 120 tons of steel per twenty-four hours.

The electrode consumption, including breakages, with amorphous carbon is 18 to 20 lb. per ton of steel. The roofs last on an average 150 heats.

Plate XIV gives a view of a Booth-Hall Furnace during teeming.

VOM BAUR FURNACE.[1]

The Vom Baur Furnace, as shown in Plate XV, is of oval form, with three vertically suspended electrodes of carbon or graphite arranged in a straight line.

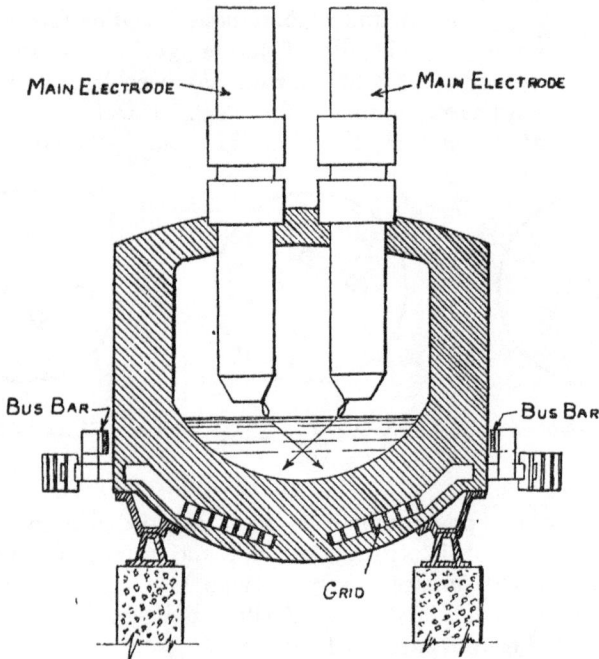

FIG. 142.

The contour of the inner lining is arranged so as to receive at all parts equal radiation from the carbon arcs. The design is made to be suitable for the use of two-phase current, when the central electrode serves as a common return and thus receives 41 per cent more energy than the outer electrodes, or else for a three-phase supply when the arcs are of equal intensity.

[1] "The Iron Age," 24 April, 1919; "Metall. and Chem. Engineering," 1919, **20,** 488; "Trans. Amer. Electrochem. Soc.," 1919, **33,** 237.

PLATE XV.—VOM BAUR FURNACE.

The doors are fitted into water-cooled frames and arranged to provide effective sealing against the entrance of air. The whole furnace is arranged to tilt by mounting on rockers. The roof is constructed of fireclay bricks. The furnace is made in sizes from ¼ ton up to 30 tons capacity.

The main installation of these furnaces is at Milwaukee, Wis., U.S.A., where they are applied for melting down steel scrap, cast-iron borings, and producing malleable iron.

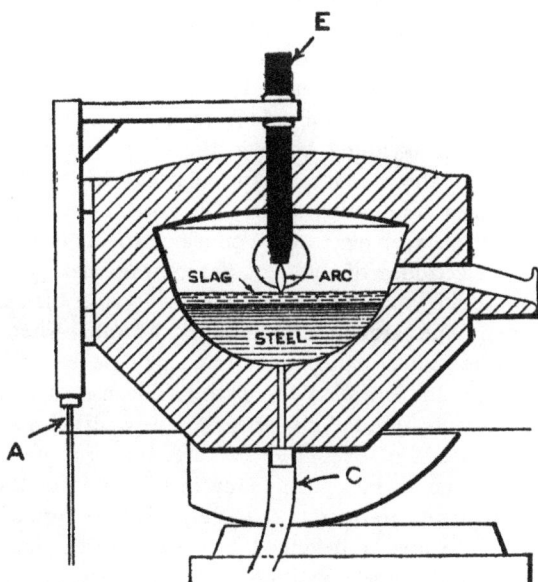

FIG. 143.

THE SNYDER FURNACE.[1]

The original type of Snyder furnace was designed for single-phase current and operated with a high voltage single arc. The furnace consists, as illustrated in Fig. 143, of a circular shell lined with refractories. The current enters at A at the bottom of the crane and passes into the furnace by an electrode E, arranged centrally. The heat is generated by the arc which forms between the electrode and the surface of the bath, and the current leaves the furnace at C through a water-cooled metal contact connecting with the bath.

[1] F. T. Snyder, "Trans. Amer. Electrochem. Soc.," 1915, 28, 221. (The writer is also indebted to the Industrial Electric Furnace Company, Chicago, for further data supplied on this subject.)

This particular design was selected on the assumption that the best economy would be obtained by the provision of the highest energy input in proportion to the radiation surface, and that to minimise the losses through electrode conduction and water coolers, it was considered that the higher the voltage and the fewer the electrodes, the greater would be the efficiency obtained. A lowering of the power factor is considered no detriment, as in any case with direct-arc furnaces, it is generally necessary to reduce the power factor to between o·7 and o·8 by the introduction, where necessary, of outside reactance in order to protect the source of power supply against too violent oscillations of the load.

Further advantages which are claimed for this type of furnace are that the wear of the roof, which is attributed mainly to the action of hot gases escaping between the electrodes and the roof, is less than with three electrode furnaces of the same capacity. The roof becomes conducting when hot, and allows a passage of current to occur between electrodes connected to a polyphase supply. With multiple electrodes of different phases, this leakage of current necessitates spacing the electrodes a considerable distance apart and thus brings the arcs near the furnace walls where erosion occurs. On the other hand, with the single-phase arc, the arc is maintained in the centre of the bath. The use of single-phase furnaces is, however, limited to sizes of a capacity of about 1 ton, as single-phase loads of over 400 kw. are not as a rule permitted by power supply companies. In many installations of the Snyder furnace, it is arranged to operate a battery of three furnaces simultaneously so as to give a balanced load on a three-phase supply. Further, with single-phase operation, it is necessary to conduct the whole of the current used through the electrode in the base of the furnace, and with larger units the difficulties in the construction of these electrodes increase.

For larger furnaces a system has consequently been adopted which makes use of three-phase current, two phases being connected to two top or movable carbon electrodes, while the third is connected to the metal electrode through the hearth.

Mechanical Features of the Snyder Furnace.—Special attention has been given in the Snyder furnace to the refractory lining, this being composed of a series of concentric layers of refractories with a view to heat insulation, so that the temperature of the outside of the furnace does not exceed 60° C. The use of a sliding door for feeding in the charge is obviated, and in place a plug type of door is employed. The loss of heat through this source, which, with a usual door of bricks 4½ inches in thickness, is said to amount to 100 kw. hours per ton of steel produced, is very much lowered, while access of air is more completely excluded than with furnaces

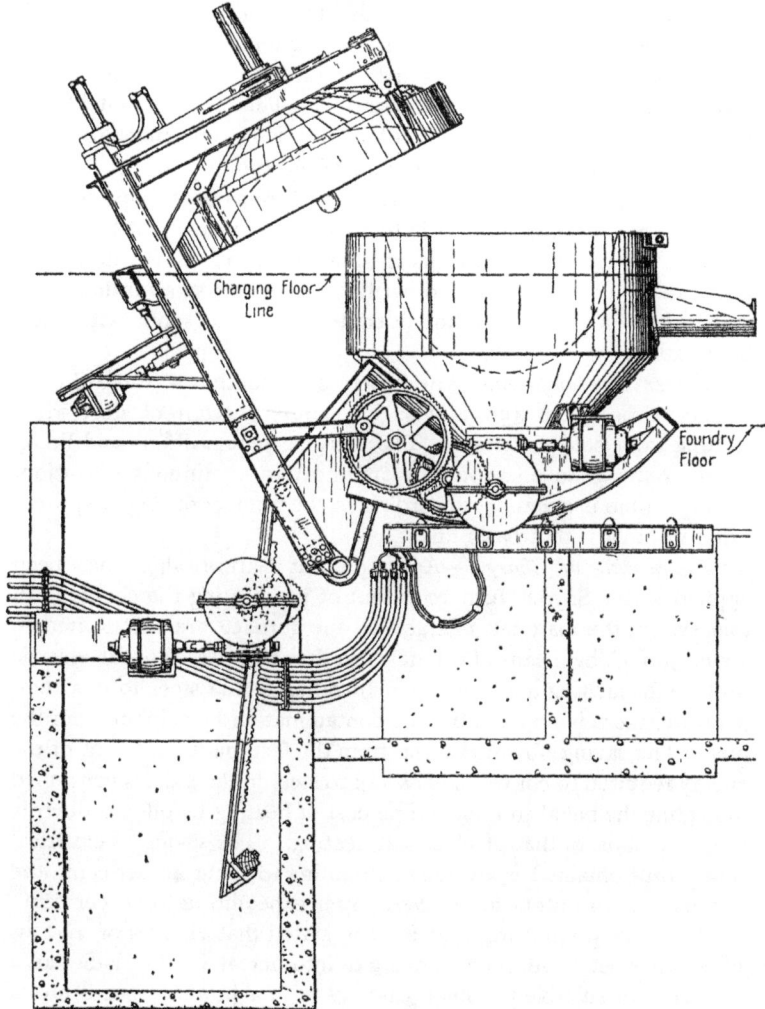

Fig. 144.

provided with doors. The charging of the furnace is accomplished by removing the roof by means of the mechanism illustrated in the diagram in Fig. 144. This device allows of mechanical charging

16 *

whereby a considerable saving in labour is obtained.   The furnace
is furnished with either an acid or basic lining, the total thickness
of the furnace walls in small units being about 14 inches.

Tilting Mechanism.—The tilting mechanism of the Snyder furnace
is arranged according to one of the three following types which
vary for the greatest convenience in pouring under varying conditions.

1. *Rocking Type.*—In this method the end of the spout moves
forward and down during the pouring.   This type is suitable, in
cases, where through provision of cranes, it is convenient to move
the ladle to correspond with the path of the spout.

2. *Heaving Type.*—In this the spout remains stationary, and
the body of the furnace is made to describe a circle about the spout.

3. *Dumping Type.*—In this method the spout moves down and
backward.   This system is ordinarily preferred for small installations
in which it is easy for the moulders to follow the spout with their
hand ladles.

Power Consumption.—The power consumption with the Snyder
furnace varies from 450 to 700 kw. hours per ton of steel when
starting from cold materials.   The working time with an acid lining
can be reduced to one hour for each heat in continuous operation.
A longer time is necessary according to the number of slags required,
when working with a basic lining.

Pre-heating of Charge.—An important feature which has been
applied in the Snyder furnace is that of pre-heating the charge.   In
this system the materials charged to the furnace are first heated to
about 900° C. by means of oil fuel, and then transferred to the electric
furnace for melting and refining.   In bringing the steel to this temperature, there is no oxidation or contamination by sulphur from the
fuel.   The saving obtained arises from the fact that 60 per cent of the
energy required to convert steel scrap to steel in the ladle is consumed
in heating the metal to 900°.   The cost of heating by oil is ordinarily
only a fraction of that of electrical heating.   The saving of electrical
power thus obtained is given as amounting to about 40 per cent, and
the increase in output for a given power expenditure to 50 per cent.

With this pre-heating system, it is stated that charges of 2½ tons
of scrap steel have been converted into metal in the ladle at a
temperature suitable for small castings with a power expenditure of
800 kw. in one hour five minutes, corresponding to 360 kw. hours
per ton of metal.   A further advantage gained is that the arcs are
steadier and consequently the power fluctuations are less when
working with pre-heated charges than with cold metal.   An advantage
claimed for the method of pre-heating to below the melting-point

in place of the duplex process, whereby the steel is first melted in the open-hearth furnace, is that greater economy results, and further,

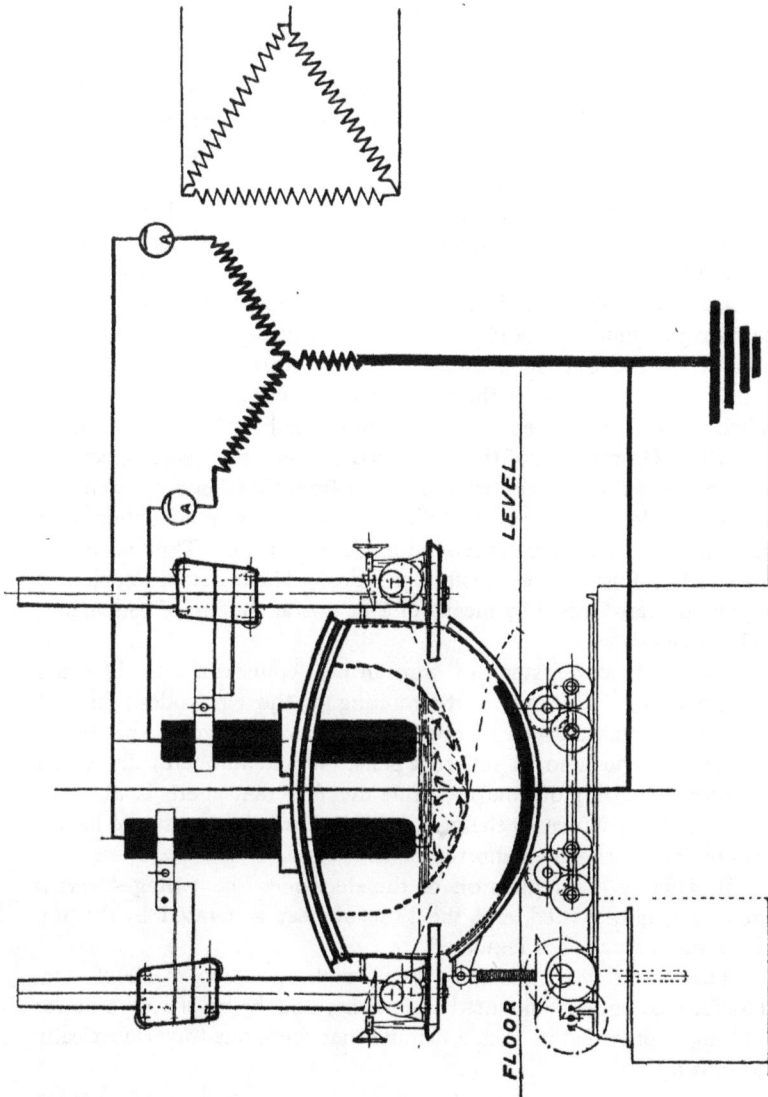

FIG. 145.

the cycle of operations of an open hearth does not at all synchronise with that of an electric furnace. It is claimed that a 4-ton electric furnace with a pre-heating system will have an output equal to that of a 20-ton open hearth, while the size of the ladles, cranes, etc., is very much in favour of the electric furnace.

THE GREAVES-ETCHELLS FURNACE.[1]

The Greaves-Etchells furnace is designed for operation from a three-phase supply, using two vertical carbon or graphite electrodes and a conducting hearth. Fig. 145 shows a diagram of a vertical section through the centre of the furnace, and illustrates the electrical connexions and convection currents set up in the molten metal in the bath. The hearth lining in the smallest furnaces is 20 inches thick, and is constructed mainly of dolomite and magnesite in such a manner that the electrical resistance is high at the inside of the bath in proximity to the charge and decreases rapidly to a negligible quantity at the outside. Two phases of a three-phase low-tension supply are connected to their respective upper electrodes while the third phase is connected to the bottom of the hearth and also joined to earth. On account of the resistance between the hearth electrode and the carbon electrode being different from that between each carbon electrode, it is necessary to adopt some means of obtaining a load which is equally balanced on all three phases. This balance is effected by adjusting the transformer ratios. The high-tension electric supply is transformed by means of a delta-star system of connexions to low tension.

The delta-star system of transformer connexions enables the equipment to withstand short-circuiting of the electrodes; in that the short-circuit current of one electrode must traverse two transformers in series and in different phase, which automatically lowers the power factor momentarily, while the fact that there is always a permanent resistance in the path of the current through the hearth also limits the effects of short-circuits.

By altering the regulation of the electrodes the voltages across the arcs can be varied, and the ratio of heat generated by the top and bottom electrodes regulated.

The power factor obtained in normal operation is about 0·9. The furnace body is mounted on rollers, and by means of a motor a tilting is provided in such a manner that the spout travels vertically downwards.

In furnaces of ½ ton capacity which are largely used for the

---

[1] From data kindly supplied to the writer by Messrs. T. H. Watson & Co., Sheffield. Cf. " Elec. Rev." (London), 1917, **80**, 395.

preparation of high-speed tool steel and alloy steels, the power applied is 260 k.v.a. When starting from scrap metal, eight heats are produced per twenty-four hours with this furnace, and the power consumption amounts to 900 kw. hours per ton of finished steel; or starting with molten open-hearth steel twenty heats are produced per twenty-four hours. With a 12-ton furnace, 2600 k.v. amperes are applied, and three to four heats produced per twenty-four hours from cold scrap and eight heats from molten charges of open-hearth metal.

The largest furnace of this type at present in operation is installed at Domnarfvet, Sweden, and is of 20 tons capacity. Furnaces of 3 tons and larger capacity are operated with four upper electrodes, each pair of which is connected to a separate transformer and controlled independently.

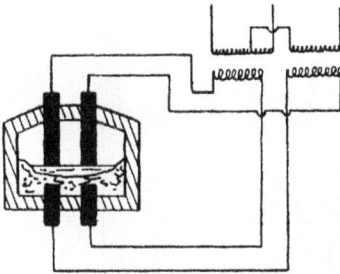
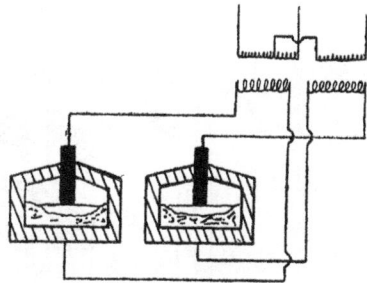

FIG. 146. FIG. 147.

THE STOBIE FURNACE.

The Stobie furnace is operated with the use of a conducting hearth, and is made according to one of two types :—

1. With one carbon upper electrode connected, together with the hearth, to a single-phase supply as shown in Fig. 147. In this type two units are usually worked together to give, by means of Scott-connexions, a balanced load on a three-phase supply.

2. With two upper carbon electrodes and two electrodes in the conducting hearth, making a four-phase system operated by means of Scott-connexions from a three-phase circuit. This type is illustrated in Fig. 146.

The Stobie furnace is in operation at Dunston-on-Tyne for the production of steel billets. The furnaces in use include one of 20 tons, two of 15 tons, and one of 5 tons capacity.

Nathusius Furnace.[1]

In the Nathusius furnace, an attempt is made to combine the surface heating of the Héroult system with the internal heating of the Girod furnace. As shown in Fig. 148, three vertical electrodes are introduced through the circular arched roof, while in the hearth of the furnace three other steel electrodes are symmetrically placed, and make contact with the metal bath through an intervening layer of the conducting refractory which forms the bed of the furnace. The three vertical electrodes are each connected to one end of the different windings of a three-phase supply, star-connected, while the other ends of the windings are connected to the different electrodes in the hearth. By means of these connexions, a current flows

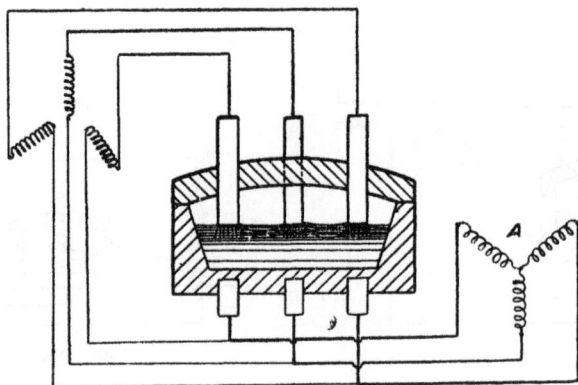

Fig. 148.

horizontally between the different electrodes, and leads to effective heating of the slag while, at the same time, a path of current is obtained vertically through the bath of metal. Through the arrangement of base electrodes, the whole of the furnace hearth becomes a heating surface. By means of the auxiliary transformer A, a strong current can be sent through the hearth electrodes, and this circuit can be applied either alone or together with the main heating circuit.

During the final refining period when the power applied to the furnace can be considerably reduced, the arc heating can be eliminated altogether by withdrawing the electrodes and connecting them together so as to form a neutral point for the three phases connected to the base electrodes. In this way base heating alone is applied.

[1] " Stahl u. Eisen," 1910, **30** (2), 1410.

In a 5-ton furnace the main transformer is of 550 k.v.a. capacity and the high-tension side is connected to a supply of 6000 volts. The low-tension voltage between each phase is 100 volts and the potential between upper and lower electrodes about 63 volts. The auxiliary transformer is built for 150 k.v.a., and connected to the high-tension supply. On account of the varying resistance of the hearth of the furnace, the auxiliary transformer has, on its low-tension side, two tapping points, and for each of these the connexions can be made either in star or delta form, so that the secondary volts can be adjusted to either 16·2, 19, 22, 28, 33, or 38.

A 5-ton unit of the Nathusius furnace is in operation at Friedenshütte, in Oberschlesien, Germany. The base electrodes are water cooled, of 8·6 inches diameter, and covered with a layer of dolomite 7·8 inches thick. The metal bath is 1 foot deep and 6½ feet in diameter. The furnace is used for refining the molten charge from an open-hearth furnace. Treatments with an oxidising slag to remove phosphorus and basic slag to remove sulphur are given in succession, the total time of treatment extending over three and a half to four hours. The power consumption amounts to 300 to 400 kw. hours per ton of metal.

Temperature measurements have shown a value of 1450° C. to 1470° C. for the metal admitted to the furnace, while the temperature of the slag rises to 1650° C. and the steel to 1500° to 1560° C. An important advantage claimed for this furnace is the circulation of the metal brought about through the distribution of the paths of current in the bath.

THE HÄRDEN FURNACE.[1]

The Härden furnace, also known as the " Paragon " furnace, is heated both from the surface of the slag, by means of an arc from vertically suspended carbon electrodes and also, at the same time, from the sides and beneath the bath, by means of side plates of second-class conductors, similar to those in use in the Röchling-Rodenhauser furnace.

Fig. 149 shows a sectional elevation through the plane of one of the vertical electrodes, and Fig. 150 a sectional plan of a furnace designed for operation by three-phase current. The electrodes C traverse the roof B of the furnace through the water-cooled collar as shown at C^1. The composite plate electrodes forming the end walls

---

[1] " Brit. Pat.," 26,251 of 1909; " Metall. and Chem. Engineering," 1911, **9**, 595; " Trans. Faraday Soc.," 1912, **7**, 183.

of the hearth D consist of cast-steel plates E, covered with a facing
of a refractory material F which becomes electrically conducting

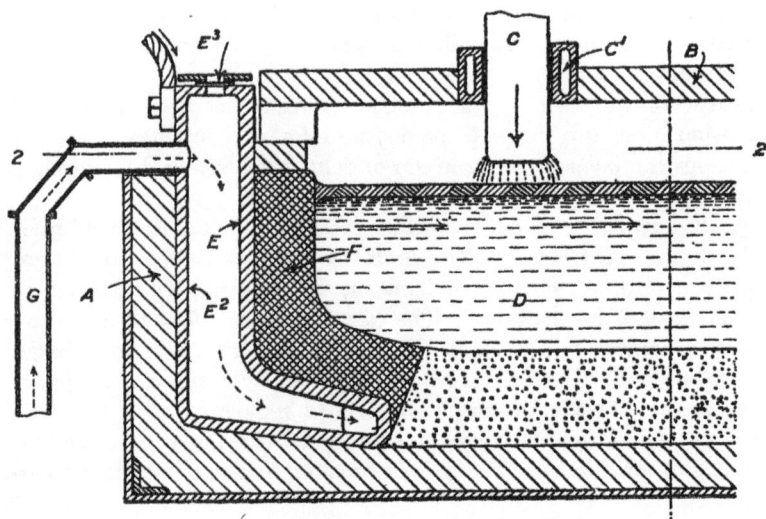

FIG. 149.

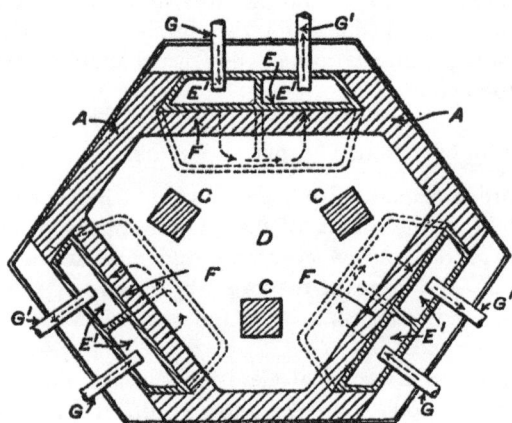

FIG. 150.

when hot and forms part of the lining of the furnace. The material
is arranged of graded conductivity so as to be of higher conductivity
near the metal plates.

The plate electrodes are provided with a cooling arrangement by the circulation of air which enters through the pipe G and flows out through the exit pipe G^1.

A mica window is placed at E^3 to permit inspection of the walls of the metal plate E.

The current is conducted along two independent paths, viz. be tween the respective carbon electrodes and between the composite plates. The control of either of these circuits may be arranged to take place independently of the other circuit. The advantages obtained with this furnace over types operated solely by direct arc heating are that (1) the electrode area for a given capacity of furnace can be made smaller, since only about half the power is required to pass through the carbon electrodes. In a furnace designed for 30 tons capacity, the electrodes have a cross section of 16 inches by 16 inches. (2) The destructive action on the roof is minimised, as there is less localised heating from the arc at the surface of the bath. (3) The furnace is suitable for the treatment of a cold charge as the current traverses the contents of the bath.

In a subsequent design of furnace,[1] the vertical carbon electrodes are eliminated and a gas-heating system is applied as in the open-hearth furnace, combined with supplementary electrical heating through the hearth plates.

THE STASSANO FURNACE.[2]

The Stassano furnace as originally introduced for the smelting of iron-ores (cf. p. 164) is a development of the system introduced by Siemens whereby a heating of the material under treatment is effected by means of the heat radiated from an arc which is formed between electrodes placed above the charge and is independent of the materials under treatment.

In the design of the Stassano furnace for the production of steel from iron-ore or iron, the main requirements which it was aimed to satisfy were the following :—

1. The vessel in which the energy is to be converted into heat should be hermetically closed so that a chemically neutral atmosphere is maintained.

2. The heat developed to be generated at the highest possible temperature.

3. No foreign substance to be brought into contact with the charge in order to avoid contamination.

[1] " Brit. Pat.," 3739 of 1910.
[2] " Trans. Amer. Electrochem. Soc.," 1909, 15, 63.

The original application of the Stassano furnace to the production of malleable iron or soft steel in one operation from high-grade ores (cf. p. 167) was later abandoned for the production of a high-grade steel from cast-iron or scrap metal. The type of furnace finally employed consists of an enclosure with a thick lining of highly-refractory material as shown in Fig. 107. Three electrodes for use with three-phase current are admitted through openings in the side walls equally spaced around the periphery of the furnace.

The electrodes where they pass through the furnace walls, are surrounded by double-walled metal cylindrical chambers. A circulation of water is provided in the space between the walls in order to keep low the temperature of the part of the electrodes which project outside the furnace. Over each cooling cylinder is a hydraulic regulating cylinder to aid the setting of the electrodes to any desired position. The piston-rod is connected at its outer end by means of a sliding guide rod with a rod which carries the electrode.

The electrodes are inclined slightly to the horizontal so as to point downwards towards the centre of the furnace.

Provision is made for bringing about a mechanical agitation of the charge by supporting the furnace on rollers, inclining to an angle of about 7° from the vertical axis, and providing for rotation by means of a geared axle fitted vertically under the furnace.

An important distinction possessed by the Stassano furnace from the direct-arc types is the higher current density employed in the electrodes. This varies from 130 to 250 amps. per square inch with a voltage of from 100 to 140.

The length of arc employed is on first heating about 4 inches from electrode to electrode, but during the run increases up to about 12 inches. Through the electro-magnetic action of the current, the arc is deflected downwards slightly towards the charge.

The lining of the furnace is made of magnesia brick, and its life when working continuously is from three weeks to one month when it is completely replaced. In the meantime, no repairs are made to the furnace.

The water-cooled casings used to protect the electrodes are inserted with air-tight joints in the cover, thus enabling the enclosure to be hermetically sealed. The different current phases are connected with fixed collector rings, and a sliding contact is made between these and brushes which lead the current to the respective electrodes.

The main advantages claimed for the Stassano furnace over other types are :—

1. A more steady consumption of power than in the types where the arc is formed directly on the surface of the charge.

2. The provision of a thick layer of slag in order to protect the metal from the carburising action of the arc is not required.

3. The furnace is more completely sealed against the entrance of air than in the direct-arc types, and hence a better control obtained over the furnace atmosphere.

4. In common with processes of the direct-arc type, an advantage obtained with a radiation-arc furnace over the induction furnace is the higher temperature brought about in the zone of reaction between slag and metal.

On the other hand the disadvantages of radiation-arc furnaces compared with the direct-arc type are :—

1. The lining and roof of the furnace are exposed to a higher temperature through direct radiation from the arc.

2. With the Stassano furnace, the rotating mechanism and the fittings for the water supply for electrode regulation and cooling are somewhat complicated.

3. Teeming of the charge is made by tapping, in place of being poured.

Production of Steel in Stassano Furnace.—At the works of the Forni Elettrici Stassano at Turin ordinary casting-steel is produced rather than high-grade metal. A typical method of operation for the production of a semi-hard steel (of 0·6 per cent carbon) in a furnace of 200 kw. capacity is to add a charge consisting of

> 450 kg. scrap,
> 250 kg. oxidised turnings,
> 50 kg. lime,

and, after heating for three hours, by which time the material is completely melted, the slag is removed and an addition made of the following mixture :—

> 50 kg. cast-iron.
> 1 kg. ferro-silicon (50 per cent Si).
> 6 kg. ferro-manganese (80 per cent Mn).
> 15 kg. calcium carbide (in granular form).

After thirty minutes' heating the slag is removed a second time, and an addition made of

> 1 kg. ferro-silicon.
> 4·5 kg. ferro-manganese.

After a final heating of about ten minutes the metal is cast. For the production of extra soft steel (0·08 carbon) the procedure is similar, except that a quantity of iron-ore is added before the last charge.

Consumption of Energy and Electrodes.—With furnaces of small capacity the power consumption amounts to 1000 kw. hours per ton of steel produced, and with units of 750 to 1500 kw. capacity the energy expenditure is reduced to from 700 to 800 kw. hours per ton.

The expenditure of electrodes, including the unused ends, amounts to 8 to 12 kg. per ton.

Power Factor.—The power factor obtained in the Stassano furnace is generally as high as 0·9 to 0·95.

Production of Steel from Iron-Ore.—In the smelting of iron-ore direct to steel (cf. p. 167) it is claimed that 1 metric ton of steel can be produced with a consumption of 3000 kw. hours of current and 15 to 20 kg. of electrodes.

RENNERFELD ARC FURNACE.[1]

This furnace resembles that of Stassano in applying a radiating arc formed between three electrodes, but by employing a different method of electrode connexions the arc is deflected downwards away from the roof and on to the charge in the furnace. In Fig. 151 the arrangement of the electrodes in the furnace when in operation is shown. The two side electrodes are arranged slightly inclined to the horizontal, while the third electrode is suspended vertically in the centre of the furnace in a position equidistant from the ends of the side electrodes. The electrodes are connected to a two-phase system, transformed by means of Scott-connexions from a three-phase supply (cf. p. 68). The middle or vertical electrode is arranged as the common return for the two phases, and thus carries about 40 per cent more current than the side electrodes. The electro-magnetic action of the current arranged in this manner causes a deflection of the arc downwards on to the surface of the bath. The operating distance between the ends of the side electrodes is usually from 18 to 22 inches, and if the electrodes are horizontal, the distance above the surface of the bath is about 15 inches. The present practice, however, is to incline the electrodes downwards so that during melting the tips of the electrodes are about 1 inch above the surface of the charge, and as this melts the electrodes are lowered and adjusted so that the ends of the side electrodes are maintained at $2\frac{1}{2}$ to 3 inches above the surface of the charge.

[1] C. H. Vom Baur, "Trans. Amer. Electrochem. Soc.," 1917, **31**, 87; "Electric Furnaces in the Iron and Steel Industry," Rodenhauser, Schoenawa, and Vom Baur, 1920. New York, J. Wiley & Sons.

The flame of the arc strikes the bath and spreads over the surface of the thinned slag.   Direct heating of the roof is thus avoided and its preservation favoured.   The furnace is supported on trunnions and arranged to tilt.   The roof is removable, and the hearth, being round, and having one or two doors, is readily inspected and repaired.

The electrodes, where they pierce the walls of the furnace, are surrounded by iron or copper water-cooled cylinders, so that the portion of the electrodes outside the furnace and the brickwork are

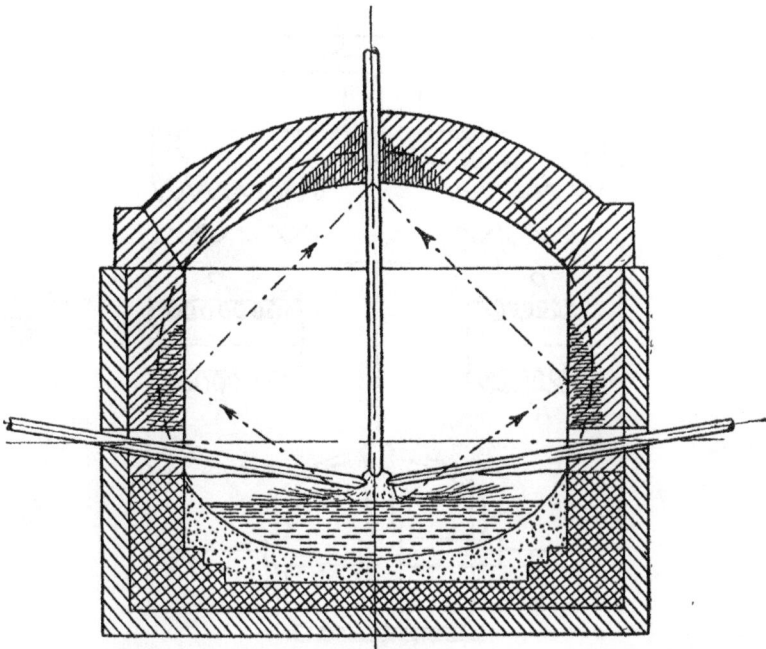

FIG. 151.

cooled.   Water-cooled copper clamps are used for connecting the conductors to the electrodes.   The electrodes are adjusted by worm-gearing and controlled either by hand or automatically.

The usual size of Rennerfeld furnace is 3 tons capacity, though 12-ton and larger units have been designed.

The current supply for a 3-ton furnace is usually as follows :—

3750 amperes × 100 volts × 2 = 750 k.v.a.

and with a power factor of 90 per cent

0·9 × 750 = 675 kw.

The electrodes are usually of graphite $5\frac{1}{4}$ and 6 inches diameter, which, with the above current, gives a density of 182 amps. per square inch for each side electrode and about 2 per cent greater density for the vertical electrode.

The electrode consumption is less than 6 lb. per ton of steel when operating continuously.

The roof of the furnace is composed of silica bricks or carborundum bricks covered with silica or, in some cases, with kieselguhr bricks.   The roofs last from 100 to 200 heats.

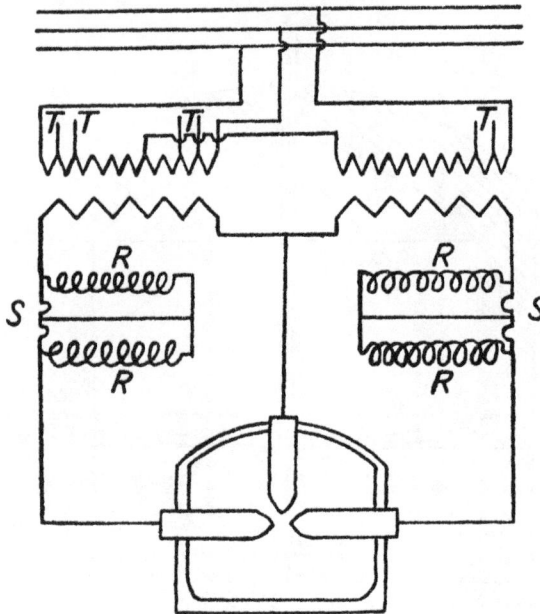

FIG. 152.

The shaded portions of the lining in the diagram in Fig. 151 show the places which are the most subject to wear on account of heat of the arc reflected from the surface of the bath.

Output and Power Consumption.—In furnaces of 300 kw. per ton of charge, heats are made on basic hearths producing good steel from cold scrap, and taking off one slag in two and a half hours with an expenditure of 670 to 720 kw. hours per ton (2000 lb.) when working continuously.   With acid hearths and 200 kw. power per ton, taking off one slag, steel suitable for castings is produced in three and a quarter hours with 635 kw. hours per ton (2000 lb.).

Periodicity and Power Factor.—The Rennerfeld furnaces operate with current of frequencies varying from 25 to 60. A power factor of 90 per cent or over is in all cases obtained.

It is claimed that an important advantage obtained by the Rennerfeld arc over that in the Stassano furnace is in the steadiness of the arc, the power fluctuations being stated to be much less in the Rennerfeld furnace than in the latter.

Fig. 152 (from "Electric Furnaces," General Electric Company, N.Y., U.S.A.) illustrates the transformer connexions used with the Rennerfeld furnace. The three-phase high-tension side is provided with the tappings TT for varying the voltage on the two-phase side. Reactances are included in the circuit at RR for use during the starting of the heating, and afterwards can be cut out by means of the short-circuiting switches SS.

Application to Non-ferrous Metals.—The Rennerfeld furnace has been largely applied for the preparation and melting of non-ferrous metals. A considerable advantage is stated to be obtained with the radiation arc as compared with direct-arc heating, on account of the less localised heating.

With low-powered Rennerfeld furnaces the power consumption per metric ton of product when treating various metals has been found to be as follows :—

Melting red brass	168	kw. hours.
„ pure copper	197	„
„ white iron	290	„
„ grey iron	325	„
„ 80 per cent ferro-manganese	441	„
„ steel scrap not ready to pour	455	„
„ and "killing" steel scrap on an acid hearth, about .	600	„
„ and refining steel scrap on a basic hearth, about .	700	„
„ 80 per cent ferro-manganese and holding, tapping, and charging	741	„
„ 6/ per cent ferro-tungsten, small scale . . .	5730	„

The present type of Rennerfeld furnace is of upright cylindrical shape, with side tilting electrodes passing through the sides and a dome-shaped roof.

INDUCTION FURNACES : FERRANTI FURNACE.[1]

The first example of an induction furnace was designed by Ferranti [2] and consisted in principle, as shown in Fig. 153, of a shallow,

[1] V. Engelhardt, "Elektrotech. Zeit.," 1907, **28**, 1051 et seq.; "Met. and Chem. Engineering," 1908, **6**, 143.

[2] Eng. Pat. 700 of 1887.

17

ring-shaped or oval crucible surrounded by a double-yoke magnetic core. Disc windings are provided above and below the metal bath and placed in close proximity to the metal.

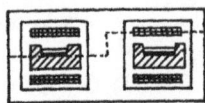  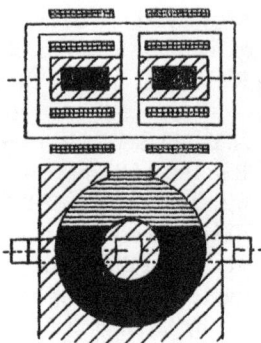

FIG. 153.—FERRANTI, 1887.

FIG. 154.—FRICK, 1904.

FIG. 155.—WALLIN, 1904.

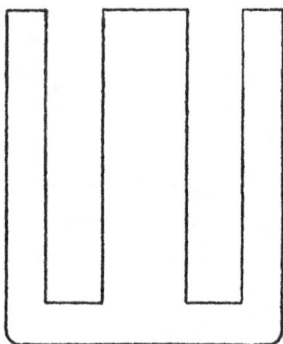 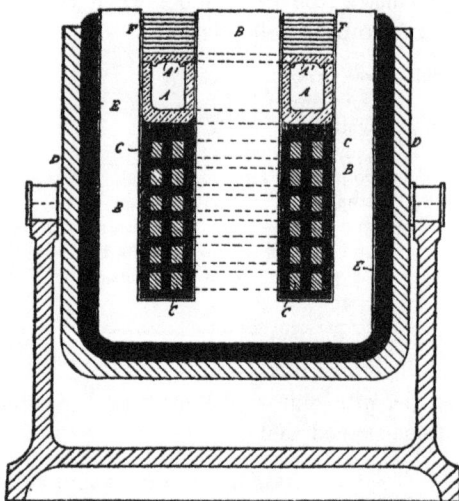

FIG. 156.

FIG. 157.

Fig. 156 shows the design of the magnetic core of transformer sheet-iron, and Fig. 157 a vertical cross section of a type of furnace in which the windings are made around the central arm of the core and immediately above and below the annular crucible A. In the type shown in Fig. 158 the windings and crucible are placed outside the magnetic core. By using a metal trough this furnace was also applied to the heating of water and other liquids.

This device attracted no serious interest at the time, and on account of the low state of development of electrical engineering in the generation of alternating current the furnace received no practical application.

It was later proposed by Dewey in America to apply the principle of induction heating to the treatment of wheel rims.

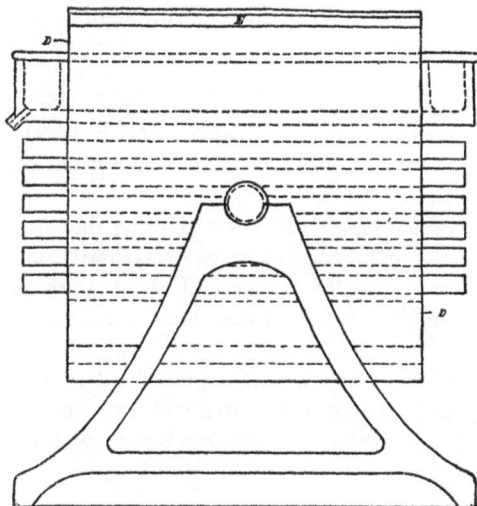

FIG. 158.

COLBY FURNACE.

In 1890 Colby, in America, developed a practicable type of induction furnace.[1] This contained a double-yoked magnetic core and a primary winding which, in an early type, was arranged outside the metal bath as shown in Fig. 159, but in a later development the primary winding was placed in the centre of the metal ring, as in the Kjellin furnace. A unit of 131 k.v.a. capacity was installed about 1906 at the works of H. Disston & Sons, near Philadelphia, for the manufacture of high-grade crucible and special steels. The furnace had a capacity of about 190 lb. of steel. The primary consisted of 28 turns of copper tube cooled by water circulation and

[1] U.S. Pats. 428,378 et seq.; "Trans. Amer. Electrochem. Soc.," 1907, 11, 411; "The Iron Age," 1906, 77, 1811.

arranged to utilise 541 amps. at 240 volts, while the secondary current induced in the metal ring had a voltage of about 8. The furnace was mounted on a pivoted support, and by means of a spout could be poured by tilting.

For melting 1 ton of steel a power expenditure of about 640 kw. hours was required.

KJELLIN FURNACE.

The Kjellin furnace was introduced in 1899, and as seen in Figs. 160 and 161 contained either a single or double-yoke core with one primary winding in the centre. The Kjellin furnace was erected at Gysinge in Sweden and applied to the production of steel in 1900 (cf. p. 264).

FRICK FURNACE.[1]

In 1904 a furnace was designed by Frick, in which the primary winding was of disc form and placed above the metal bath as shown in Figs. 154 and 162.

This design was adopted in order to bring the primary winding nearer to the metal bath composing the secondary than is possible when the refractory base of the crucible is interposed. With this arrangement a higher power factor and lower self-induction is obtained.

In the above types of furnaces the secondary current is conducted mainly through the mass of metal, and to a small degree through the slag, in proportion to the relative conductivities of the two media.

WALLIN FURNACE.

In a furnace designed by Wallin in 1904 the annular metal bath is arranged vertically as seen in Fig. 155. The amount of metal present is not sufficient to close the circuit at the top and the electrical circuit is completed by the superimposed layer of slag. The primary windings are arranged in the form of discs on either side of the metal bath. With this arrangement the metal and slag are joined in series in the electrical circuit and both carry the same current.

SCHNEIDER-CREUZOT FURNACE.

' In this design which was introduced in 1903 the bath consists, as seen in Fig. 163, of a larger container of rectangular horizontal section, and a narrower circular channel hollowed in the body of the refract-

---

[1] Eng. Pat. 4866 of 1904.

ory so as to form a circuit for the circulation of the secondary current. The increased resistance obtained by the narrowing of the path of current in this side channel enables the production locally of a high temperature at the junction of the channel and the metal bath and leads to an active circulation of the metal.

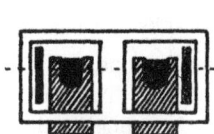

FIG. 159.—COLBY, 1890.

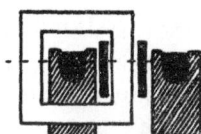

FIG. 160.—KJELLIN, 1900.

FIG. 161.—KJELLIN, 1905.

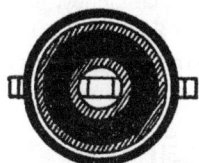

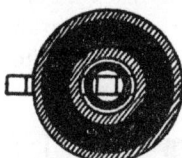

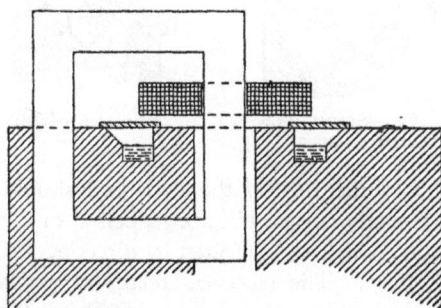

FIG. 162.

GIN FURNACE.

The Gin furnace as introduced in 1906 consists of two wide reservoirs of rectangular horizontal section, which are connected to each other by four inclined channels, each pair of which is arranged in X-form. The primary winding and the metal bath are arranged concentrically with each other (cf. Fig. 164).

RÖCHLING-RODENHAUSER FURNACE.

The Röchling-Rodenhauser Furnace, which was brought into operation in 1905, combines the method of induction heating together with that of direct heating by the passage through the bath of current from a circuit introduced by electrodes or pole plates of refractory earths (*vide* p. 268).

GENERAL FEATURES OF INDUCTION FURNACES.

To obtain the maximum efficiency, the primary and secondary circuits should be arranged as close together as possible. The transformer windings may consist of tubes which permit of cooling by the circulation of water, or of discs. For the highest efficiency, the primary winding should be arranged in the interior of the metal bath. With disc windings, the primary coil should be arranged as far as

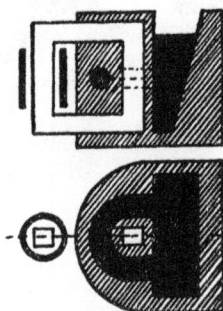

FIG. 163.—SCHNEIDER-CREUZOT, 1903.

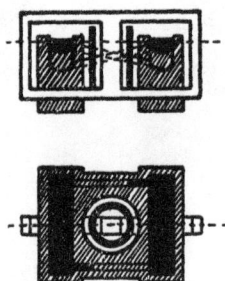

FIG. 164.—GIN, 1906.

possible parallel to the ring of the bath and should be wide and shallow. The primary coil may be above, below, or in the same plane as the metal bath. Cooling is effected by directing a rapid stream of air over the windings. The electrical relations vary largely with the size of the furnace. With increasing size of the bath the ohmic resistance falls, and the inductance increases, causing a lowering of the power factor. In the case of the Kjellin furnace, this tendency for a fall in the power factor is counteracted by operating the furnace with a lower frequency of current with increasing size of furnace. In a 750 kw. furnace, the frequency is taken as low as 5. It is desirable, however, that a furnace should be adapted for operation from a current supply of normal periodicity (or 25 to 50 cycles). With the Kjellin furnace, this is only possible with the smaller units (*vide* p. 264).

The Röchling-Rodenhauser furnace (cf. p. 268) was designed

mainly with a view to utilising normal power supplies for furnaces of large capacity. In this case, on account of the ∞ -shape of the metal bath, a narrowing of the bath cross section in the circular channels accompanied by an increased electrical resistance can be effected. With increasing capacity of furnace the cross-section area of this portion of the furnace, which forms the induced secondary circuit, is not increased in the same ratio as the increased content of metal. The power factor is further raised by the employment of the second direct resistance circuit.

Circulation of Metal in the Induction Furnace.—In the circular channel of an induction furnace a circulation is brought about through the influence of the magnetic lines of force accompanying the current circuits.

As the parallel lines of force repel each other, the iron ring will tend to extend as with a centrifugal force. The metal is conse-

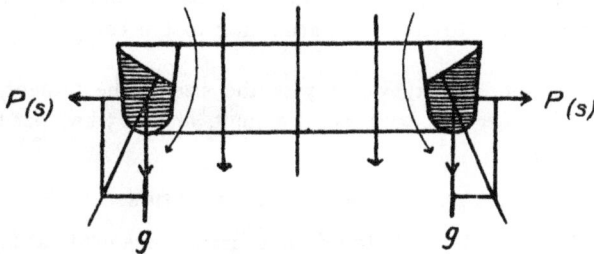

FIG. 165.

quently under the influence of two forces at right angles to each other as seen in Fig. 165, viz. the gravitational force g, and the magnetic effect P_s. The surface of the metal will accordingly assume an angle of slope which is normal to the resultant of these forces.

The effect of this condition on the liquid metal is to cause a rotation of the whole charge in a plane at right angles to the ring of metal, and in an inward direction as with a vortex ring. In this way continued renewal of the metal with the surface of the slag is ensured.

Applications of Induction Furnaces.—The main application which has been made of induction furnaces is for the preparation of steel from pig-iron or scrap steel starting with a cold charge or with a previously melted charge.

To provide the conducting circuit in the crucible when starting with cold materials, sufficient metal to complete the ring is, as a rule, left in the crucible after teeming, for the operation of the following charge.

Relation between Capacity and Periodicity of Current in Induction Furnaces.—In the case of the Kjellin furnace, the capacities of a number of installations together with the periodicity of the current which has been adopted is given in the following table :—

Place of Installation.	Capacity of Furnace. Tons.	Power Consumption of Furnace. Kw.	Periodicity.
Völklingen, Essen . . .	8·5	750	5
Kladno.	4	440	7
Gurtnellen	3·8	330	10·3
Gysinge	1·5	215	15
Vöcklabruck.	0·4	65	24
Laboratory demonstration furnaces	4 kg.	6	150

Temperature.—The temperature of the bath in induction furnaces varies from 1600° to 1850° C., according to the composition of the metal.

The power capacity varies with the size of the furnace, being about 90 kw. per ton for a 750 kw. unit, and 165 kw. per ton for a 65 kw. unit.

KJELLIN FURNACE INSTALLATIONS.[1]

The first installation of the Kjellin furnace was made at Gysinge, in Sweden, in 1899, with a 78 kw. unit. This furnace, containing a single-yoke core, is illustrated in Fig. 166 in vertical section, and in Fig. 167 in horizontal plan through the centre of the annular crucible. When using an effective power of 58 kw., six charges of about 100 kg. each were treated per twenty-four hours, corresponding to 2300 kw. hours per ton. In 1902 a furnace of 170 kw., with a capacity of 1 to 1·5 ton was installed. This furnace treated daily five charges of 1 ton each from cold materials, and the process was devoted to the preparation of high-class tool steels from the Dannemora iron produced at Gysinge.

The Gysinge furnace was supplied with single-phase current at fifteen cycles and 3000 volts. With a charge of 3000 lb., the power factor obtained was 80 per cent., and with 4000 lb. 68 per cent. The fall of the power factor with increasing weight of charge is due to the decreasing resistance, and consequently the greater the proportionate influence of the induction factor (cf. p. 54).

[1] " Trans. Amer. Electrochem. Soc.," 1909, 15, 173.

The primary coil of this furnace was wound with 295 turns, thus giving a secondary voltage corresponding to $\frac{3000}{295}$ or about 10 volts.

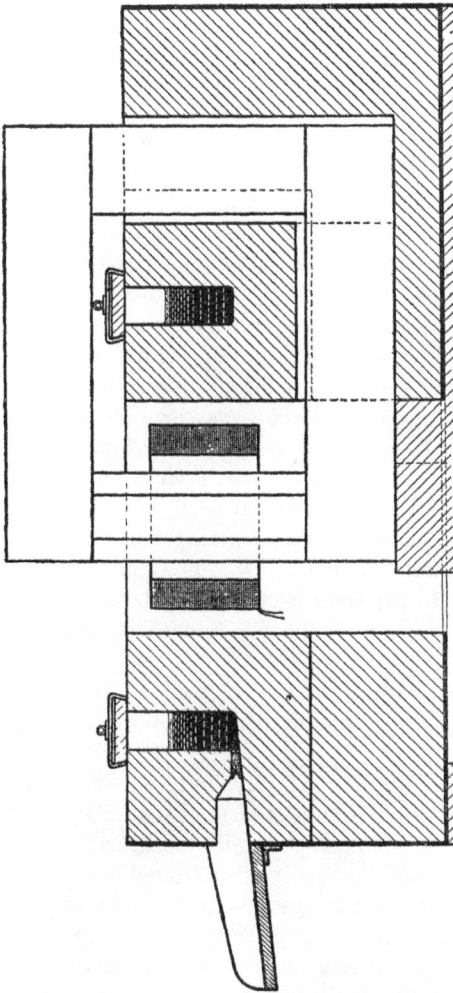

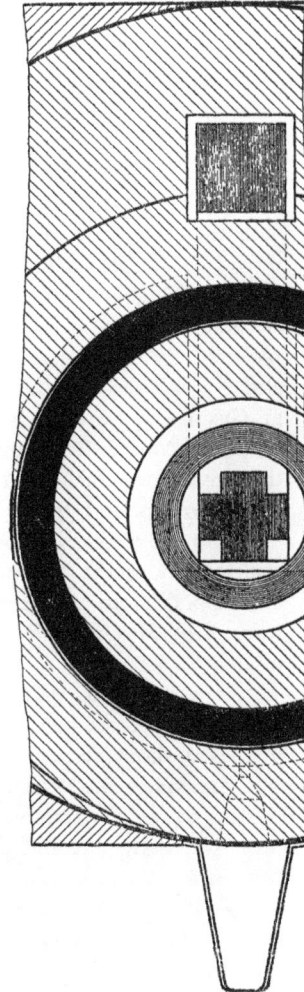

FIG. 166.

FIG. 167.

.A magnesite lining was used in this later furnace at Gysinge in place of silica bricks which were employed with earlier units.

A Kjellin furnace was later installed at the Röchling Steel Works at Völklingen, Essen, Germany, with a capacity of 750 kw.

and operated with single-phase current of five periods at 4500 to 4900 volts. The circular crucible of this furnace which had a capacity of 8·5 tons, was built up from masonry and lined with a suitable basic material as used in the Bessemer converter or Siemens-Martin furnace. During operation the annular crucible is roofed over by covering with segmental iron plates. The furnace is arranged to tilt, and provided with a pouring spout for emptying.

Electrical and Thermal Efficiency of Kjellin Furnace.—The induction furnace has the highest electrical efficiency of any type since all electrode losses are obviated. The only losses are those due to the transformer which are also present in the transformers used with arc furnaces. The thermal efficiency is good and increases with increasing size of furnace and speed of operation. In spite of the radiation losses the total efficiency of the induction furnace is said to be higher than that of arc furnaces.

According to Engelhardt[1] the theoretical power consumption required for the preparation of steel from a mixture of 1/3 pig-iron and 2/3 scrap, is 489 kw. hours per ton. With the 1·5 ton furnace at Gysinge, a power expenditure of 966 kw. hours was required to prepare a ton of steel when the melting was extended over six hours, thus giving an efficiency of 50 per cent, and 800 kw. hours during a four-hour period, thus giving an efficiency of 60 per cent.

With the 750 kw. furnace at Völklingen, when working with cold pig-iron and scrap a power consumption of 590 kw. hours per ton, corresponding to an efficiency of 80 per cent, is said to be obtained, while the efficiency with the Ròchling-Rodenhauser furnace is still higher.

At Volklingen, when working with a mixture of cold pig-iron and iron-ore briquettes, the power consumption per tòn of steel is increased about 50 per cent.

Advantages of Kjellin Induction Furnace.—The Kjellin furnace has proved to be very suitable for making the highest class of steel from pure raw materials, and has been able to compete with the crucible process in this connexion, even when the power is generated from coal. The steel can be still better protected against injurious effects than in the crucible process. Tappings are obtained which are very large compared with the contents of a crucible, and the steel produced is stated to be of a better quality than that given in crucibles from the same raw materials. As described on page 263, the alternating current, through its electro-magnetic action, causes an efficient cir-

---

[1] " Stahl und Eisen," 1905, 205.

culation of the metal. Compared with other types of electric furnace the induction furnace is characterised by the absence of any sudden fluctuation of the load, and gives a very steady power consumption. The plant can consequently be operated directly on any system of power supply. When worked from special generators, these need only be constructed to give the steady maximum consumption demanded, whereas, with direct-arc furnaces, the generators must be of larger capacity to cope with the heavy fluctuations in the load.

Disadvantages of Kjellin Furnace.—The main disadvantages of the Kjellin furnace are (1) the inconvenient shape and narrow width of the annular crucible which prevents the hearth from being readily

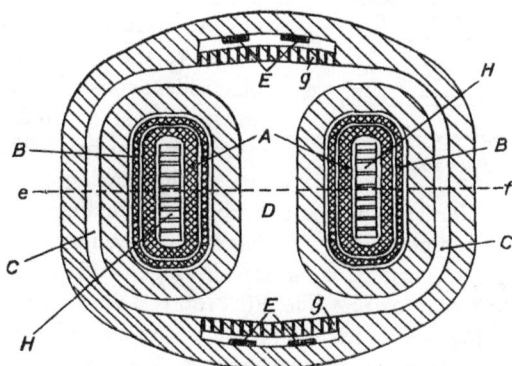

Fig. 168.

accessible and easily surveyed; (2) the limitation in the temperature, since the refractory lining is heated to the same degree as the metal. This limitation does not enable the reaction with the slag to take place at the same high temperature as with direct-arc furnaces, and in consequence no considerable refining can be effected; (3) the specially low frequency of the alternating current required in the larger furnaces, which involves the construction of special expensive generators.

The above disadvantages have for the most part been overcome in the modified type of induction furnace devised by Röchling and Rodenhauser, and this has now generally replaced the earlier type of Kjellin furnace.

RÖCHLING-RODENHAUSER ¡FURNACE.[1]

The Röchling-Rodenhauser furnace was first brought into operation at Völklingen on the Saar, Germany, in 1906. The chief features
which distinguish it from the Colby-Kjellin type is that the crucible
instead of being of the form of a plain ring, is arranged in the shape
of a figure 8, with a wide central space. Heating is further effected
by a combination of two systems.
Firstly, by induction currents,
whereby the furnace acts like two
combined ordinary induction furnaces, and secondly, by an auxiliary
circuit in which by means of a separate winding around the primary
coil an induced current is introduced to the metal bath through
electrodes consisting of plates of
soft cast-steel covered by grids of
refractory earths. These earths
are of the same nature as the fila-
ment of the Nernst lamp, and become conducting by electrolytic
action at high temperatures. A diagram of the furnace as designed
for single-phase current is shown in Figs. 168 and 169. Fig.
168 is a horizontal plan, while Fig. 169 is a vertical section along
the line *ef* of Fig. 168. Like the ordinary induction furnace, this furnace is essentially a transformer with a single primary winding A
around both iron cores H of the transformer. The secondaries are
two in number, one is the molten bath C in form of a figure 8, the
channel D between the two cores being comparatively broad, and
the other consists of extra windings B surrounding the primary
coils and conducting currents to the metallic plates E which are
inserted in the furnace walls, and are raised to a high temperature
by the passage of the current to the metal bath.

In order to protect the windings against the effect of high temperatures, and also to diminish the stray fields of magnetic flux,
thin-walled copper cylinders MM (Fig. 169) are provided, through
which is conducted an air blast passing out of the tubes N_1N_2. The
transformer iron contains ventilation slits H (Fig. 168). The charging
door is at one end of the furnace and the tapping door at the other

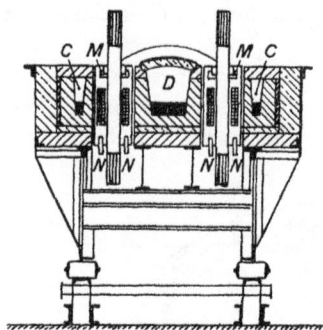

FIG. 169.

[1] "Electrochem. and Metall. Ind.," 1908, **6**, 10, 143, 458; "Stahl und
Eisen," 1908, **27**, 1605; "Trans. Faraday Soc.," 1908, **4**, (2), 120; F. A.
Kjellin, "Trans. Amer. Electrochem. Soc.," 1909, **15**, 183.

end. During operation the molten charge is kept in good circulation by the electro-magnetic influence of the current. The whole furnace is built as a tilting-furnace by mounting on rollers, and in its general design is similar to the Siemens-Martin open-hearth furnace. The hearth and walls of the furnace are lined with magnesia containing 10 to 12 per cent of tar, and the arch is composed of firebrick. The Röchling-Rodenhauser furnace was introduced at Völklingen for the refining of fluid steel from the Bessemer converter, in order to produce a higher class of steel, mainly for use as rails.

When applied for the treatment of a molten charge, the mode of operation on first starting up is to place rings of soft steel in the furnace, and heating to about 900° C. by induction currents, until the hydrocarbons of the tar are decomposed, when fluid pig-iron is introduced from the blast-furnace. The electric circuit is then closed and the temperature rises slowly, until after eighteen hours the furnace is up to full heat and there is no further smoke. For the treatment of steel which has undergone a preliminary refining in the Bessemer converter, the bulk of the charge used for the heating of the furnace is removed and finished steel from the basic converter introduced. Burnt lime containing some magnesia is placed on the bath, and about 8 kg. of fluorspar is added to produce a sufficient degree of fluidity and the refining begins. If the slag becomes too fluid, lime is added, if too thick, fluorspar. The operation is complete when bubbles no longer rise up from the liquid bath, and when samples taken from the bath give satisfactory tests. The slag, which contains generally 25 per cent of iron in the form of oxides, is removed, and a pure lime slag is formed from fresh burnt lime and fluorspar, and deoxidation is brought about by the addition of ferrosilicon.

For the preparation of high-carbon steel, powdered coke is added to the furnace, which quickly dissolves in the bath, otherwise spiegeleisen is added as usual. Each heat lasts in general for two to three hours.

These furnaces can be operated at fifty periods, and even for the largest sizes it is not necessary to go below twenty-five periods. In this way electrical power can be applied in the form usually generated and transmitted. With increasing size of furnace it is chiefly the middle channel which is increased, so that the cross section of the other secondary channels, which are heated by induction only, are not increased in the same proportion, and the power factor is thereby improved.

With a 3 to 3½-ton furnace taking 330 kw., a power factor of 0·87 is obtained.

With the Röchling-Rodenhauser furnace the disadvantage attending the generation of single-phase current has been overcome by a three-phase induction furnace, which in furnaces of from 3 to 15 tons capacity are operated with a frequency of fifty periods, and thus enable the use of standard three-phase generators. The design of a 1·5-ton furnace of this type is shown in Figs. 170 and 171.

A is the hearth, which in the 1·5-ton furnaces is 1 ft. 7 ins. wide and 4 ft. 9 in. long. The three transformer cores are surrounded by the heating channels R. At the places where two such channels R enter into the main hearth A, the special electrode plates are arranged as shown in Fig. 170 in the form of rectangular black slots built in the furnace walls. These electrodes are embedded in the

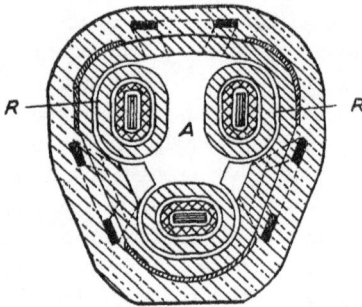

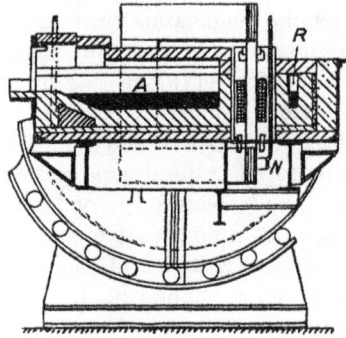

FIG. 170.   FIG. 171.

furnace wall and separated from the fused charge by a refractory wall which becomes conducting when heated. Each of the three transformer cores is provided with a primary winding, and above each primary winding a secondary winding is also arranged. While one end of the latter is connected to the bus-bar N, the other ends of the three windings are connected to the three electrodes.

The furnace is built as a tilting furnace, and as far as possible made to resemble the open-hearth furnace.

In the three-phase furnace a characteristic rotation of the charge is brought about due to the presence of a rotary field as in an induction motor, and this leads to effective automatic circulation.

For the manufacture of steel rails the electrically-produced steel is superior to open-hearth metal in respect to greater density and homogeneity.

At the Ròchling Steel Works in Germany in the method most generally in use, the metal is submitted to a "triplex" treatment, the steel being first blown in the converter, further purified in a basic open-hearth, and then subjected to a final refining treatment in the electric furnace, where the steel is allowed to stand and give off its gases.

The main advantages of the Röchling-Rodenhauser furnace are that the metallurgical process takes place in a large container, as in the open-hearth method, in place of the narrow channel of other types of induction furnaces. The furnace has a high power-factor and electrical efficiency and is particularly suited for the refining of impure raw materials. Large units can be efficiently worked with current of the periodicity of usual large power supplies (50 cycles). The furnace is applicable for using directly current from a three-phase supply, and is suitable for extensive refining.

Efficiency of Röchling-Rodenhauser Furnace.—With a unit of 8 tons capacity, the power consumption required to melt 1 ton of common scrap is stated to amount to 580 kw. hours. On the basis of 489 kw. hours being required by theory for this operation, the efficiency corresponds to a value of 85 per cent.

Data on Röchling-Rodenhauser Three-phase Electric Furnace.[1]

Contents, Lb.	Electrical Data.				Tapping, Lb.		Output per 24 Hours.	
	Kw.	Voltage.	Cycles.	Cos φ.	Cold Charging.	Hot Charging.	Cold Charging. Lb.	Hot Charging. Lb.
2,200	175	500	50	0·8	1,540	2,200	9,240	35,200
6,600	350	3000	50	0·6	4,620	6,600	27,720	106,600
11,000	550	3000	25	0·65	7,700	11,000	46,200	176,000
15,400	750	3000	25	0·6	10,780	15,400	64,680	246,400

The weights given are calculated on the basis of six charges per twenty-four hours for cold melting and sixteen charges for hot melting, whereas the actual times are three and a half hours per cold charge and one hour twenty minutes per hot charge.

Analysis of Product.	P.	S.	Mn.	C.
Basic Bessemer steel charged into furnace having	0·08	0·08	0·5	0·1
Can be finished for rails, with 100-125 kw. hours per ton to	0·05	0·04	0·8	0·5
Or for high-grade steel with 300 kw. hours per ton to	trace	trace	0·25	0·05

[1] F. A. Kjellin, loc. cit., p. 192.

Open-hearth Steel.—Martin steel which has been previously refined, dephosphorised, and desulphurised, containing about 1·22 carbon, 0·38 manganese, and 0·209 silicon, has been transferred to the electric furnace, and allowed to remain there to remove gases, regulate the carbon content, and adjust the alloys, with an energy expenditure of 200 to 250 kw. hours per ton.

DEVELOPMENT OF ELECTRIC STEEL PROCESSES IN GREAT BRITAIN UNITED STATES AMERICA, AND FRANCE.

The production of steel in electric furnaces during 1918 in Great Britain is estimated at 110,000 tons. In 1919 the following furnaces were installed or under construction:—

Type.	Number Installed. Total Capacity. Tons.	Total Capacity. Tons.
Héroult	49	195
Electro-metals	34	79
Greaves-Etchells	32	78
Stobie	8	88
Snyder	8	—
Rennerfeld	7	10
Stassano	4	—
Girod	1	—
Special	1	—
Total	144	.

The United States in January, 1919, had in operation 287 electric steel furnaces, including three of the induction type, and Canada, forty-three.

ELECTRIC STEEL FURNACES IN FRANCE.

The main electric steel works in France are those of the Société des Aciéries électrique Paul Girod, at Ugine, Savoy, where there are nine Girod furnaces in operation of a total capacity of 96 tons. Power for the operation of electric steel furnaces is derived from two sources : (1) hydro-electric power, and (2) by gas engines operated by blast furnace gas. For the operation of larger scale plants the use of power from blast furnace gas is being extended more than that from water power. The electric furnace is taking a larger place in the reconstruction of steel works which were devastated during the war. In 1919 the production of steel in the electric furnace in France amounted, however, only to 40,000 tons of ingots and 10,000 tons of castings.[1]

The following installations are now in operation in France :—

Type of Furnace.	Total Capacity. Number Installed.	Total Capacity. Tons.
Girod	9	96
Keller	9	22
Héroult	6	25
Chaplet	7 and others	—
Greaves-Etchells	2	1

ESTIMATED COST OF PRODUCTION OF ELECTRIC STEEL.

An estimate of the cost of production of electric steel from cold scrap in the United States, given in 1918, by the General Electric Company, Schenectady, is as follows :—

1. *Capital Cost of Furnace.*—In the case of the most popular type of furnace, that of the direct-arc type of 6 tons capacity, having three carbon electrodes of 17 ins. diameter, and a transformer bank of 1200 to 1500 k.v.a. capacity. Such an equipment, taking power from three-phase 11,000 voltage circuit would have, in normal times, a selling price of about £4200. (£1 = $4·80.)

2. *Working Costs.*

Materials.	Cost per ton, £1=$4·80.
Energy, 600 kw. hr. at 0·5d.	£1·25
Electrodes, 30 lb. at 2·75d.	0·34
Refractories (roof, walls, and hearth)	0·16
Slag	0·10
Alloy additions	0·16
Scrap (£4·6 per ton)	4·60
Labour	0·35
Overhead charges	0·25
Total per ton of molten steel in the ladle.	£7·21

In making castings, approximately 60 per cent of the metal will be recovered, which gives a figure of £12·0 per ton, and adding a royalty charge of £0·1 and customary moulding charges will probably bring the cost well over £25 or 3d. per pound.

[1] " J. d. Four élec.," 1920, **29**, 19.

SECTION XII.

THE ELECTRICAL PRODUCTION OF FERRO-ALLOYS.

THE growth of the ferro-alloy industry is contemporary with that of steel but considerable advances were made after the year 1900 by the application of the electric furnace to this manufacture. Ferro-alloys were formerly prepared in the blast furnace and crucible furnace, and to a smaller extent in the open-hearth furnace. Ferro-manganese, spiegeleisen, and low-grade ferro-silicon are still produced mainly in the blast furnace. However, the limitations imposed by this system are that the temperatures attainable are not sufficiently high for the reduction of certain oxides, and in any case, the large quantity of carbon used in the charge does not permit low-carbon alloys to be prepared. With the crucible process the manufacture could only be carried out on a comparatively small scale and under disadvantages, and this method has now been abandoned.

Ferro-alloys are used in the manufacture of steel and serve firstly, by combining with different elements, to bring about the purification and deoxidation of the steel, during which process the alloyed metal passes off into the slag, and secondly, for introducing into the steel a certain proportion of the alloyed metal and thus conferring certain physical properties according to the alloy admitted. The alloys in general use for the first reaction are ferro-silicon and ferro-manganese, and for the second the alloys of iron with manganese, chromium, tungsten, molybdenum, vanadium, titanium, zirconium, and uranium.[1]

The preparation of ferro-alloys can be conducted without the use of any special type of electric furnace. The manufacture was first commenced industrially by adapting furnaces which had been constructed for calcium carbide at the time the boom in this industry ended. The furnaces subsequently constructed for ferro-alloys are of the same general type as those used for the smelting of iron-ores. On account of the property possessed by all iron compounds at high

---

[1] " Met. and Chem. Engineering," 1910, **8**, 133 ; C. B. Gibson, " Trans. Amer. Electrochem. Soc.," 1920, 265 ; R. J. Anderson, *ibid.*, 1920, 177.

temperatures of absorbing carbon, it was, in both cases, aimed to arrange that the product from the reaction is withdrawn from contact with the carbon electrodes as quickly as possible or else separated by means of a slag during the heating. In this way contamination by carbon, and the consumption of the electrodes, is reduced to a minimum. The temperature necessary for the formation of ferro-alloys varies within very wide limits, and in general can be assigned to between 1200° C. and 1800° C. The material in this class of earliest development, and which has received the greatest application, is ferro-silicon. Another alloy which has attained the greatest importance in the iron and steel industry is ferro-manganese, while ferro-chrome and ferro-tungsten are largely used for the preparation of special hard steels, and to a minor degree ferro-molybdenum, ferro-titanium, and ferro-vanadium.

The consumption of electric power per unit value of product is high in the case of most of the ferro-alloys (cf. p. 404), so that the industry can usually only be economically undertaken with the use of low-price power. The conditions for this industry should be favourable in many of the British Colonies where water power is abundant. The minerals providing the raw materials are very widely distributed throughout the Empire, and the consumption of ferro-alloys in the manufacture of steel is large.

France has, however, been the pioneer and leader in ferro-alloy metallurgy, and particularly with ferro-alloys made in the electric furnace. In France three of the principal works engaging in the manufacture of ferro-alloys are the following :—

1. *Keller and Leleux, at Livet (Isère).*—In the works at Livet, there is now (1920) in operation or course of installation a total power of 35,000 h.p. The power actually in operation in 1914 amounted to 20,000 h.p. The main products now prepared at Livet consist of ferro-chromium with 1 to 10 per cent carbon ; ferro-silicon with 1 to 10 per cent carbon and 12 to 85 per cent silicon ; silico-spiegels of 23 to 25 per cent silicon, and 40 to 70 per cent manganese ; ferro-tungsten ; various special steels ; and calcium carbide. The Keller furnaces for steel and ferro-alloy manufacture are also being worked under licence at The Holtzer Company, at Unieux (Loire), Bohler Company at Kappenberg, Styria, at Jeumont (Nord), at Darfo and Mazzano in Italy, and at Burbach and Neunkirchen in Germany.

2. *Société Anonyme Electro-Metallurgique Procédés—P. Girod, at Ugine, Savoy.*—The chief ferro-alloys now being made at these works include :—

18 *

Ferro-silicon of all grades, including an alloy with 95 per cent Si for the preparation of hydrogen (1 kg. giving 1500 litres).

Ferro-chromium, 70 per cent chromium, and 1 to 10 per cent carbon.

Ferro-titanium with 16 to 60 per cent titanium.

Ferro-tungsten with 81 per cent tungsten, and 0·3 per cent carbon. The tungsten ore is mined from a deposit at Puy-les-Vignes.

Ferro-vanadium with 40 per cent vanadium, and 4 per cent carbon.

Silico-titanium, silico-aluminium, silico-calcium, copper-vanadium, etc.

3. *La Néo-Metallurgie, Société Electro-Chimique du Giffre, at Saint-Jeoire (Haute-Savoie)*.—Ferro-alloys are made at these works by the Chaplet electric furnace of a resistance type, the charge forming the conducting medium between a suspended upper electrode in the centre of the furnace, and one in the base (cf. p. 189).

The products manufactured include : ferro-silicon, ferro-chromium, ferro-tungsten, ferro-nickel, copper-manganese (30 per cent manganese), iron-nickel-chromium (30 per cent Fe, 20 per cent Ni, 50 per cent Cr), nickel-molybdenum (33 per cent Ni, and 66 per cent Mo), manganese-silicon (60 per cent Mn, 25 per cent Si), manganese-silicon-aluminium in various proportions, nickel-boron (85 per cent Ni, and 15 per cent B), ferro-boron (14 per cent boron), copper-silicon, and tungsten metal (98·6 per cent W).

Ferro-Alloy Manufacture in the United States.—The manufacture of ferro-alloys in America was first undertaken at the Willson Aluminium Company at Kanawha Falls and at Holcomb Rock (W. Va.). In 1904 there were in operation at the former place three dynamos each of 800 kw., and at the latter one-third of that power.

Ferro-chromium was made from Cuban and Turkish chrome-iron ore and from ore from Canada and New Caledonia. The product obtained had the percentage composition

Cr 70·9, Fe 23·2, Si 0·5, C 5·2, P 0·008.

The consumption of energy amounted to 7800 kw. hours per metric ton of product. At Kanawha Falls, alternating current of 110 volts was used, and the current of 22,000 amps. divided on seven furnaces. Each unit consisted of an iron box with a thick lining of pieces of anode carbon together with tar as a binding material, and provided with a tapping hole. In some furnaces the lining constituted one electrode, the other being formed by a bar of carbon suspended vertically, while other furnaces had two vertical adjacent electrodes.

The more recent position of ferro-alloy manufacture in the States is that until 1916 there was only one ferro-alloy works of any considerable size, viz. the Electro-Metallurgical Company at Niagara Falls. Subsequently there have been brought into operation four large and several small works. These have been mostly established in centres where electric power is relatively cheap, as at Niagara Falls, at Keokuk, Iowa, and the Pacific Coast, though to meet the increased demand during the war, plants were established in some districts where power is relatively dear. Since the termination of the war many of these plants have suspended operation.

Ferro-manganese was produced by the Anaconda Copper Mining Company in 1918. At the end of 1918, about 15 per cent of the total of this alloy made in the United States was produced in the electric furnace.

With regard to ferro-tungsten, under the stimulation of war requirements, the United States now leads both in the production and consumption of tungsten alloys which are applied to the preparation of alloy steels for high-speed cutting tools.

Several hundred tons of ferro-molybdenum were produced in the United States in 1918 mainly for export to Europe for the preparation of specially hard steels used in connexion with munitions.

FERRO-SILICON.

Ferro-silicon and ferro-manganese received an extensive application in the manufacture of steel after the introduction of the Bessemer process.

Low-grade ferro-silicon alloys are still mainly manufactured in the blast furnace, whereby alloys are obtained containing from 10 to 12 per cent silicon. Alloys containing from 50 to 90 per cent silicon are made exclusively in the electric furnace, while this method has also been extended to the regular production of low-grade alloys.

The chief use of this alloy is as a deoxidiser in steel manufacture, ferro-silicon having a greater affinity for dissolved oxygen than ferro-manganese. The alloy is also used in large amounts for the preparation of silicon steels, which on account of the special properties they possess of low hysteresis and high permeability, are used almost entirely in the construction of transformer cores.

Two systems are in use for preparing ferro-silicon in an electric furnace.

1. Reducing silica by carbon in a type of furnace similar to that used for carborundum, collecting the fused silicon and absorbing it with metallic iron.

2. Reducing in an electric furnace iron-ores of high silicon content, or after admixture with a silicious mineral, or by smelting the slag obtained with the Bessemer or Siemens-Martin furnace. For this purpose a strongly acid slag must be employed during the smelting, and a product can be obtained containing up to 80 per cent silicon. The composition of typical products obtained by the blast and electric furnace respectively is shown in the following table :—

	Si.	Fe.	C.	S.
Blast-furnace ferro-silicon . .	10·5	83·1	2·3	0·03
Electric-furnace . . .	51·8	46·1	0·1	0·003

Furnace Construction.—Resistance furnaces are used throughout for the manufacture of ferro-silicon. The main differences in different types are in the external shape, in the size and shape of the upper electrode, and in the mechanical means adopted for raising and lowering, and usually for water-cooling this electrode. The furnaces are built up of firebrick, sometimes lined with a carbon composition which is pasted in, and the exterior of the furnace is usually braced with iron " stays " or completely enclosed in a circular or square iron casing made up of sections bolted together. Fig. 172 illustrates the principle of a single electrode type of furnace, with the walls at A, the electrodes at B and C, and the tapping hole at D. The electrode C is placed beneath the floor of the furnace, its upper surface being level with the furnace hearth. At the Giffre works, for instance, this lower electrode is formed *in situ* by ramming into an open space, in the floor of the furnace, a mixture formed by heating together powdered retort graphite and coal-tar.

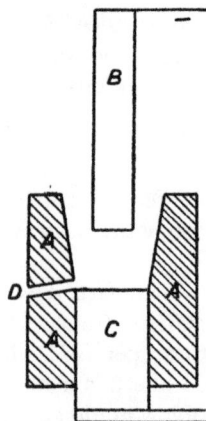

FIG. 172.

The Keller furnaces used at Livet are of the type shown in Fig. 109. This method obviates the use of the furnace hearth as a conductor, and enables the application of a higher voltage. The voltage in different cases ranges from 40 to 75, and the current from 10,500 to 15,000 amps. To ensure continuity of working, several electrodes are placed in parallel in the system, any one of which is renewable without the necessity of suspending or varying the working of the furnace as a whole. When once charged and started, the furnaces are run continuously for an average period of two years. While running, the furnaces are tapped at intervals of one to two hours, and fresh amounts of charge are added by shovels to the top of the furnace, around the upper electrode, as often as is rendered necessary by the melting and withdrawal of the charge below.

Evolution of Gases from Ferro-silicon.—Phosphorus and arsenic, if present in the raw materials used in the preparation of ferro-silicon, combine to form compounds in the final product, and by reaction with water vapour these generate hydrogen phosphide and arsenide which possess powerful toxic effects. Serious explosions have occasionally occurred during the transport of ferro-silicon through the bursting of the iron containing drums. These explosions with ferro-silicon have been considered to be due to the evolution of hydrogen silicide from included calcium silicide.

Samples of ferro-silicon containing from 42 to 60 per cent silicon show a distinct tendency to spontaneous disintegration, which is accompanied by the evolution of poisonous gases. It has consequently been agreed to restrict the manufacture of ferro-silicon to grades containing below 30 per cent or above 70 per cent silicon. The raw materials used are also more rigorously selected.

Originally the charge of ferro-silicon furnaces was composed of a mixture of (1) iron-pyrites or other form of iron-ore; (2) silicious material in the form of quartzite or sand; (3) carbon in the form of charcoal, coal, or coke, together with lime as flux. Owing to the impurities (especially sulphur and phosphorus) of earlier samples of electric furnace ferro-silicon, to which a number of explosions during transport have been attributed, scrap iron and steel shavings are now preferred to iron-ore. Quartzite is used in preference to sand, as being less productive of obstructions in the furnace. Further, the purer the starting materials the smaller the amount of slag which will be formed.

RAW MATERIALS USED IN THE MANUFACTURE OF FERRO-SILICON
IN FRANCE.

At the Bozel works of *La Compagnie Generale d'Electrochemie de Bozel* ferro-silicon is made from quartzite (obtained locally), steel turnings, and anthracite coal (obtained locally). The different grades of ferro-silicon produced contain percentages of silicon of 25, 50, and 80 respectively. In the manufacture of the 80 per cent grade, however, charcoal is used instead of anthracite on account of its greater freedom from impurities.

At the Giffre works, St. Jeoire (Haute-Savoie), of the Société Electro-Chimique de Giffre ferro-silicon is manufactured from materials similar to those used at Livet.

At the Ugine (Savoie) works of the Société Anonyme Electro-metallurgique Procédés Paul Girod—the constituents of the charge are steel and iron scrap, quartzite, and good quality coal obtained from St. Etienne, with an average content of 8 per cent ash, and not more than 0·005 per cent of S. or P.

At the St. Marcel (Savoie) works of the Société d'Industrie Electro-Chimique La Volta—the charge constituents are: quartzite obtained from the neighbourhood of the works; gas-coke for the production of the 50 per cent grade of ferro-silicon, while in the manufacture of the 25 per cent alloy, or other comparatively low grades, this is replaced by anthracite coal, which is cheaper but not so free from impurities; iron-ore (for 50 per cent ferro-silicon) or steel shavings.

At the Keller-Leleux Works at Livet (Isère), ferro-silicon is made from iron and steel shavings, quartzite, and anthracite coal. Furnaces of 1200 h.p. have an output of 500 kg. of ferro-alloy every two hours. Thirty per cent ferro-silicon is produced with an energy expenditure of 3500 kw. hours per ton.

FERRO-MANGANESE.

Ferro-manganese is used extensively in steel manufacture for two distinct purposes:—

1. It raises the temperature of the steel by the chemical actions brought about, at the same time removing dissolved oxygen and iron oxide.

2. It enhances the forging and malleable properties of the finished product, and confers hardness on the steel, the quantity of manganese present in the final metal being generally from 0·25 to 2 per cent. For certain special purposes 12 to 14 per cent manganese is introduced whereby a steel is obtained which is specially hard and tough and has been largely applied for the manufacture of structural parts which are exposed to much wear, such as points and curves of tramway rails, and in mining and grinding machinery.

Attempts were made as early as 1830 to prepare pure manganese for the manufacture of *spiegeleisen*, which contains 8 to 10 per cent manganese. About 1860 a ferro-manganese alloy containing 80 per cent manganese was prepared in a crucible furnace. This preparation of ferro-manganese was later conducted in a Siemens-Martin furnace, and then in a blast furnace. This procedure has now been largely replaced by the electric furnace, whereby a product is obtained containing a considerably higher content of manganese than that ordinarily obtainable by fuel-heated furnaces.

The carbon content is at the most from 4 to 7·5 per cent while in some processes carbon-free alloys are obtained. The standard content of manganese in spiegeleisen is 18 to 22 per cent, and ferro-manganese, 70 to 82 per cent, while silico-manganese and silico-spiegel have the following typical percentage composition :—

	Mn.	Fe.	Si.	C.
Silico-manganese .	55-70	20-25	25	0·35
Silico-spiegel . .	20-50	67-43	4-10	1·3-3·5

In the preparation of ferro-manganese the loss of manganese in the slag through oxidation is considerably less in the electric furnace than in the blast furnace.

In the manufacture of silico-manganese in the electric furnace the recovery of manganese is generally as high as 95 per cent.

The mineral most generally used for the manufacture of ferro-manganese is pyrolusite (MnO_2), which occurs mainly in Russia and the United States.

Manganese silicide of the composition $SiMn_2$ is also prepared on a large scale by heating manganese oxide and silica together with carbon in an electric furnace. A silico-spiegel, consisting of an iron-manganese-silicon alloy, is prepared by smelting manganese-containing slags.

The difficulties to be guarded against in the electrothermal production of manganese and its alloys, are the volatilisation of the metal which is apt to occur at the high temperatures applied and its combination with carbon, for which element it has a powerful affinity. In a furnace method devised by Héroult for the production of manganese and its alloys, the metal is protected from contamination by the electrodes, by maintaining a layer of slag on the surface.

In applying spiegeleisen or ferro-manganese to the manufacture of steel, the best procedure is to melt in a small electric furnace and tap from this into the bath of steel. The melted alloy mixes more quickly and reacts more actively, while less is oxidised by the furnace gases (cf. p. 197).

FERRO-CHROMIUM.

This alloy was formerly prepared in a blast furnace, and to a smaller extent in a crucible furnace, while a carbon-free product was also prepared by the thermite process. In the blast furnace an alloy containing only 30 to 40 per cent chromium is obtained. Even with 40 per cent chromium prepared by this method, high blast pressures and excessive coke consumption are required. For many purposes a product containing over 60 per cent chromium is required. The crucible process gives a high-grade product, but has only a small output.

The alloy is now prepared almost exclusively in the electric furnace. A standard grade is an alloy containing 6 to 8 per cent of carbon and 60 to 70 per cent of chromium and is generally prepared by smelting a mixture of chrome-iron ore and carbon in an electric furnace, various types of which are in use, from the model shown in Fig. 172 to the steel furnaces of Stassano and Héroult. The average power expenditure for an alloy containing 60 to 65 per cent chromium is 7000 to 10,000 kw. hours per ton.

When added to steel, chromium imparts an extreme hardness, and the steel is used for armour-plate, gears, and cutting tools. The proportion of chromium added is from 2 to 4 per cent. In armour-plate manufacture, the metal is largely used in conjunction with nickel and with vanadium. Steel with 10 to 12 per cent chromium is non-tarnishable and is used for the manufacture of cutlery and various articles under the name of "stainless steel".

Ferro-tungsten.—This alloy is most conveniently prepared in the electric furnace from wolframite, a black oxide of iron and tungsten, or scheelite, a white oxide of calcium and tungsten, or by smelting iron-ore containing tungsten with carbon. The alloy does not so readily oxidise or form carbides as chromium and uranium. Only the lower grade alloys can be tapped, and the usual practice in all cases is to prepare in block form and dismantle the furnace after each run.

The introduction of tungsten imparts to steel an extreme hardness, which is maintained at high temperatures. Tungsten steels are consequently largely used in the manufacture of machine tools which can be used for rapid machining and under conditions which raise their temperature to a red heat. The proportion of tungsten added to the steel for this purpose is from 15 to 25 per cent. Alloys can be obtained with a tungsten content up to 85 per cent, and an amount of carbon generally as low as 0·5 per cent. The quality of the alloy is higher the lower the amount of carbon.

Ferro-molybdenum.—The raw materials used for this alloy are molybdenite (MoS_2) concentrates, carbon, iron scrap or turnings and an excess of lime which are smelted in the electric furnace. Alloys containing above 60 per cent molybdenum cannot be tapped and are prepared in block form. Silicon or ferro-silicon can also be used in place of carbon as the reducing agent. Molybdenum added to steel imparts characteristics similar to tungsten, and has proved of value as an addition to binary and ternary alloy steels. The proportion of molybdenum added is from 6 to 10 per cent. Molybdenum steels have been applied to the linings of heavy ordnance.

Ferro-titanium.—This alloy is prepared by the thermite process as a carbon-free product with about 25 per cent titanium, and in the electric furnace with a composition of 15 to 18 per cent titanium and about 6 per cent carbon. The ore used in the latter case consists of the double oxide of iron and titanium or titaniferous iron ore.

Ferro-titanium is used largely in steel manufacture to remove oxygen and nitrogen from the metal and also partially to substitute ferro-manganese.

Ferro-vanadium.—Ferro-vanadium is mostly prepared by the thermite process, though increasing quantities are now being manufactured in the electric furnace. Vanadium oxide or vanadate of iron is, for this purpose, reduced by smelting with silicon or high-grade ferro-silicon together with iron, lime, and fluorspar; the use of carbon was originally abandoned on account of the difficulty of keeping this element low in the final product. A large proportion of the commercial alloy is now, however, produced by smelting the ore with carbon. A furnace in use for this purpose is a three-phase rectangular type with water-cooled cover, provided with water-cooled bushings for three graphite electrodes 12 inches in diameter.

A charge of ore, coke and fluxes in a finely-divided state is thoroughly mixed and by means of continuous automatic feeders fed into the furnace through water-cooled bushings in the top of the cover. The furnace is tapped for metal and slag at intervals of 6 hours. An alloy with 33 per cent vanadium has proved to be most suitable for use in the metallurgy of steel and the silicon content can be reduced to below 1 per cent.

Vanadium imparts to steel the property to resist shock and vibration, giving a metal which is suitable for axles, cranks, connecting rods, and similar parts.   Vanadium is also introduced, together with tungsten, for the preparation of high-speed tools.   The proportion of vanadium added to steel is from 0·1 to 0·5 per cent.

Ferro-uranium.—This is one of the more recent ferro-alloys to be used in the metallurgy of steel.   The method which has been generally applied for the manufacture of the alloy is by the reduction of uranium oxide with carbon in a tilting electric furnace.   An alloy of from 40 to 70 per cent uranium and about 2 per cent carbon can be made without a second refining operation.   Refining for lowering the carbon is undesirable because of the tendency to loss of uranium by oxidation into the slag.   Uranium added to steel imparts strength and resistance against shock.

Ferro-phosphorus.—Ferro-phosphorus is produced both in the blast furnace and electric furnace by the reduction of apatite and phosphate rock.   The material is largely used in the manufacture of sheet steel.

SECTION XIII.

THE APPLICATION OF ELECTRIC FURNACES TO THE MELTING AND PREPARATION OF ALLOYS AND NON-FERROUS METALS.

A LARGE development of the use of electrically heated furnaces has been made for the preparation and melting of alloys and non-ferrous metals whereby the following general advantages are obtained.[1]

1. *Saving of Metal.*—Through being able to maintain a neutral or reducing atmosphere, loss by oxidation is avoided.

2. *Improved Quality.*—It has been found in most cases that a more uniform quality of metal can be produced in an electric furnace than in one fired by fuel, operating under otherwise similar conditions, and that it is easier to produce an alloy of closely specified composition. This is due to the greatly reduced loss of volatile metal and the elimination of contaminating combustion gases.

3. *Exact Temperature Control.*—In the electric furnace temperature control can be brought about more readily than in fuel-heated furnaces. This is of importance as the thermal treatment and pouring temperature determines the physical property of many metals and alloys.

4. *Increased Production.*—In general, the speed of melting is greater in electric than in fuel-heated furnaces on account of their higher operating temperature and greater efficiency. The electric furnace can also be used in larger units than is commonly the case with fuel-heated furnaces.

5. *Elimination of Crucible Cost.*—Electric furnaces which use no crucible eliminate this considerable item of expense. Large fuel-fired furnaces effect the same saving but, from a metallurgical point of view, are seldom as satisfactory as fuel-fired crucible furnaces.

6. *Incidental Savings.*—The operation of large units results in an economy of floor space and of labour dependent upon the use of

---

[1] H. M. St. John, " Met. and Chem. Engineering," 1918, **19**, 321.

285

large electric furnaces where large fuel-fired furnaces are not practicable.

7. *Improved Working Conditions.*—More favourable conditions for the workmen as regards coolness of the furnace room and with the well-regulated electric furnace, incomparably greater cleanliness.

It is estimated[1] that in March, 1920, there were in operation 261 units installed of electric furnaces for melting non-ferrous metals and alloys, with a total capacity of 23,000 k.v.a., one-third of this power being taken by resistance furnaces and the remainder by induction furnaces of small capacity.

FURNACES FOR THE MELTING AND PREPARATION OF BRASS, ALUMINIUM, AND SIMILAR METALS: THE WILE FURNACE FOR MELTING OF FERRO-ALLOYS.[2]

A furnace which has been devised by R. S. Wile for the melting of ferro-manganese and other alloys is operated by a three-phase supply and provided with one base and two top electrodes, spaced so as to compose, as nearly as possible, an equilateral triangle. To start, the two top electrodes are lowered to the floor of the furnace, some granular graphite or plumbago is sprinkled in, and when this glows the slag-making materials are added. Window-glass is used for most purposes, as it is neutral to most of the metals to be melted. When the slag is molten the electrodes are raised to the necessary degree, and the alloy is charged into the furnace and melted beneath the slag. The alloy is then either poured into an auxiliary ladle, or the entire electric furnace is raised by a crane and the contents poured into the steel ladle, while the slag remains in the furnace to be ready for the next addition of alloy. With this method no appreciable oxidation or volatilisation of the manganese occurs, and the value of metal thus saved exceeds considerably the total cost of operating the furnace.

With ferro-manganese containing 80 per cent Mn, the average power consumption amounts to 800 kw. hours per ton.

[1] "Jour. Soc. Chem. Ind.," 1921, **40**, 24.
[2] "Trans. Amer. Electrochem. Soc.," 1915, **28**, 239.

Fig. 173.

THE BAILY FURNACE.[1]

The Baily furnace is designed mainly for the melting of non-ferrous metals and the production and heat treatment of alloys. As shown in Fig. 173, the furnace chamber consists of an enclosure with a bowl-shaped hearth contained in a thick lining of refractory bricks and the whole is supported by a steel case. The heating is effected electrically by means of an annular refractory trough composed of carborundum fire-sand, and containing in a recessed space a ring of pieces of carbon. The trough is built in segments to allow of expansion and contraction through temperature changes and is supported by radial firebrick piers. The current is introduced by two graphite electrodes placed diametrically opposite to each other and arranged to make contact with the resister material. The electrodes are protected against oxidation from air by being embedded in the resister material. The heat is radiated downwards from the resister and dome-shaped roof on to charge in the bath.

A charging door is placed below the resister trough and above the level of the metal, while provision is made for pouring the metal without opening the door by means of a tapping hole which can be closed by a plug. Access of air to the furnace is excluded during the heating, and a reducing atmosphere can be thus maintained. The roof is arranged to be removable.

In continuous operation the resister material requires renewal after about a fortnight, while the trough can generally be used for four months without replacement.

The furnace can be mounted on trunnions with suitable supporting brackets and tilting gear which is operated by a hand-wheel or motor.

The furnace is made in sizes varying from 50 to 1000 kw. and, in addition to the cylindrical shape, a non-tilting pattern is also made of rectangular form with a rectangular-shaped hearth and resistance trough.

For the melting of phosphor-bronze a furnace of 105 kw. is operated at a voltage of 93 to 100. The daily output is 6000 lb. with a total loss of metal of 17 lb. (or 0·28 per cent). The power consumption amounts to 330 kw. hours per ton (of 2240 lb.) of metal. The temperature reached by the roof is 1260° C. and that of the bath 1080° C. With furnaces for melting brass and copper, the average of a series of results showed a power expenditure of 408 kw. hours per ton (2240 lb.) of metal with a loss of 1·6 per cent. These furnaces

[1] From data kindly supplied by Mr. Verdon Cutts, Sheffield.

have also been applied on a large scale for the melting of cathode zinc, notably at the Anaconda Mining Company at Montana (cf. p. 313), where a unit of 1000 kw. is in operation. This furnace has a capacity of 200 tons of zinc *per diem*, the power consumption amounting to from 70 to 80 kw. hours per ton of metal melted, while the loss of metal through oxidation is as low as 0·024 per cent.

THE GILLETT AND RHOADS FURNACE FOR THE PREPARATION OF BRASS.[1]

This is a rotating type of furnace which has been developed at the United States Bureau of Mines for the preparation or melting of brass, and provides for heating by means of radiation from an arc in a cylindrical enclosure. The furnace rotates around its cylindrical axis which is arranged to form a slight angle with the horizontal position. The rotation is brought about by means of cogged wheels on which the furnace body is supported. The door openings are on the periphery of the drum, and the furnace is rotated backwards and forwards so that the molten charge just fails to reach the door at either end of its rocking angle. The electrodes enter through the ends of the drum enabling an arc to be maintained in the centre. The rotation, besides ensuring thorough mixing, enables the heat of the walls to be largely taken up by the metal instead of being conducted to the outside and prevents the temperature of the walls from rising but little above that of the metal, thus ensuring a good life of the lining. One of these furnaces was in 1918 in commercial operation at the Michigan Smelting and Refining Company, Detroit, and four additional units in course of construction, for the manufacture to special specifications of brass ingots.

The furnace is 5 feet diameter by 5 feet long with a lining 1 foot thick of a special heat-insulating brick covered by "corundite" bricks which consist of a refractory material high in alumina. The hearth, which is 3 feet long by 3 feet diameter, takes a charge of 1300 lb. and upwards. The electrodes are of graphite 4 inches diameter. Single phase, 60 cycle current stepped down to 120 to 130 volts is used. The power input can be varied by altering the length of the arc, and varies from 100 to 200 kw. averaging about 165 kw. The results show that with alloys high in zinc the zinc content of the product is higher than with that from the same charge melted in coke-fired furnaces. The average expenditure of energy measured on the primary side of the transformer including that re-

[1] "Met. and Chem. Engineering," 1918, **18**, 583.

quired for the rocking mechanism has been found to amount to 262 kw. hours, and the electrode consumption to 1½ lb., per ton of charge for red brass.

Compared with the crucible process, it is claimed that in the electric furnace the total working cost, including electric power, interest, depreciation, replacements, etc., is, under present conditions, about one-half the cost per ton of charge of the single item of crucibles.

Electric Furnace for Melting of Aluminium.[1]

An electric furnace has been adopted for the remelting of aluminium pig, preparation of aluminium alloys, and annealing of alloys, by the United States Aluminium Company, Massena, New York State. Pig aluminium as it is obtained from the reduction cells is remelted for the purpose of recasting into ingots, rolling billets, or for the manufacture of alloys.

The furnace selected for remelting is a stationary resistance type with a heating core of granulated carbon, and has a bowl-shaped hearth capable of holding 3 to 4 tons of molten metal which is discharged by tapping. The furnace is of 500 kw. capacity and designed for single-phase 25-cycle current. The enclosure is rectangular and made of sheet metal suitably insulated with high-grade fireclay bricks, with an intervening layer of heat-insulating material in the space between the bricks and the metal shell. The doors provided at either end of the furnace are heat-insulated in the same manner as the body of the furnace. Running lengthwise with the furnace along either side are resister troughs, rectangular in section and open at the top. These troughs are made of highly refractory carbide with a binder so that they can be easily moulded, and are filled with a resister material consisting of finely broken carbon or graphite and supported on separate brick piers which serve to dissipate the heat generated in the troughs and diffuse it uniformly. Current enters the troughs through large copper terminals attached at the ends. The heat generated from the incandescent resister material is radiated mainly to the furnace roof and from there reflected down on the hearth and the material being treated. When running continuously the furnace can melt 1 ton per hour, and the product is tapped into graphite crucibles holding about 125 lb. The daily output of the furnace is 20 tons. Current is supplied from a transformer at voltages varying from 500 down to 240, according to the fall of resistance with rising temperature.

[1] D. W. Miller, " Met. and Chem. Engineering," 1918, **19**, 251.

THE THOMSON-FITZGERALD RESISTANCE FURNACE.[1]

A type of furnace has been designed by Thomson and Fitzgerald in which the contents of the furnace are heated by radiation from a downward resistance core of carbon plates arranged longitudinally under the roof of the furnace.

This design of furnace serves the same purpose as the radiation-

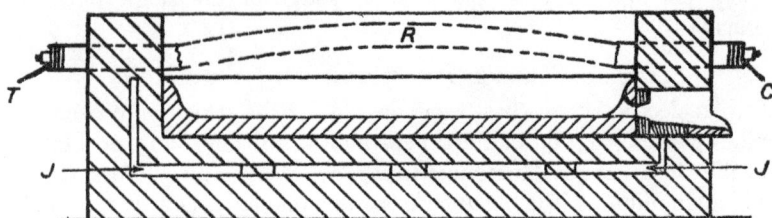

FIG. 174.

arc type, but a number of advantages are claimed to result from the use of this resistance core which operates at a lower temperature than the arc, and offers a steadier resistance to the passage of the current.

The furnace was developed mainly in connexion with the Imbert zinc melting process, but has been found applicable to the treatment of other materials.

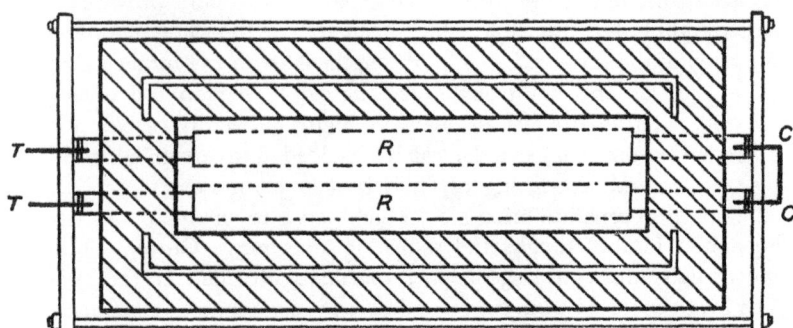

FIG. 175.

The arrangement of furnace which has been finally adopted consists of two resistance cores placed close to each other and connected in series by a common terminal connecting the two cores at one end of the furnace, while at the other end of the furnace the

[1] "Met. and Chem. Engineering," 1910, **8**, 289, 317; "Trans. Amer. Electrochem. Soc.," 1911, **19**, 273.

córes are connected with the respective current terminals. By this arrangement, with a given resistance, a reaction space more approaching a square in section is obtained than with a single core, and radiation losses are in consequence minimised. The temperature of the bath can be maintained at about 1450° C., while that of the resister is estimated at about 1900° C.

In its application to the Imbert process for zinc the atmosphere is reducing, and there is no need to protect the carbon resister against

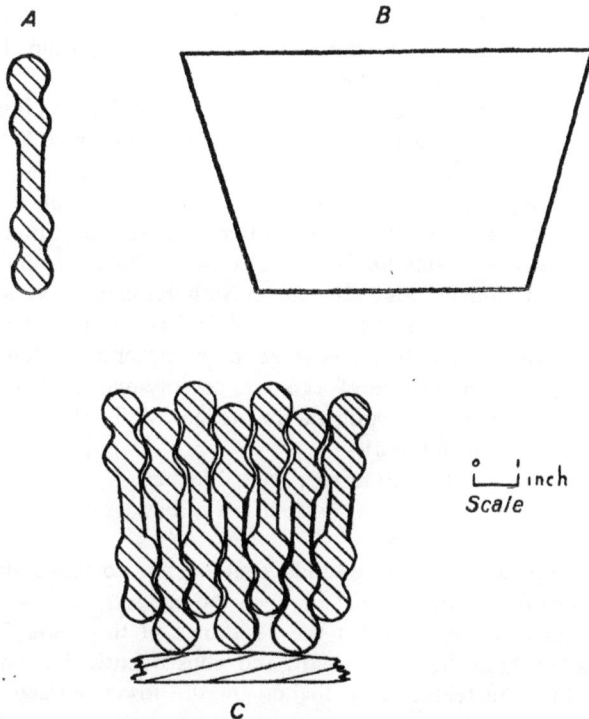

FIG. 176.

oxidation. For other purposes such as the melting of metals and alloys, it cannot, however, be arranged to exclude air from the furnace, and it is found desirable to enclose the resister in a sealed chamber. This is arranged by providing a partition of tiles of recrystallised silicon carbide, which serves as a floor to the resister chamber and as roof to the reaction chamber, the heat being transmitted through. To prevent all possibility of oxidation, it has been found advantageous to admit continuously to the resister

19 *

chamber a supply of oil such as kerosene and thus maintain a reducing atmosphere slightly above external pressure.

A longitudinal section of the furnace with cover removed is illustrated in Fig. 174 and a plan in Fig. 175. The walls and bottom of the furnace are cellular as well as the cover, so that the furnace chamber is almost completely surrounded with an air jacket as shown at J. Gases resulting from the reaction, or admitted from the outside, can be caused to burn in this space and thus enable auxiliary heating of the charge. TT and CC are carbon plugs which pass through stuffing boxes in the end walls of the furnace. TT form the furnace terminals and are connected to the cables while the ends CC are connected together by graphite conductors.

The resister RR is constructed of corrugated plates as illustrated in Fig. 176, where the end section of each plate is shown at A, the side section at B, and the method of packing together the plates at C. An arched support is provided during the filling in of the resistance plates, but the support is afterwards removed, and the interlocking of the corrugated plates holds up the resister, which is supported by the carbon terminals, and all contact with refractories is avoided. The arch form of the resister is ensured by having the plates shown in Fig. 176 of tapering thickness so as to be thicker at the top than at the bottom. In a furnace of 150 kw. seventy-one of these plates are contained in each section of the resister, while the dimensions of the plates are as follows :—

> Length at top, 16 inches.
> Length at bottom, 10 inches.
> Width, 6·5 inches.

When in operation about 98 per cent of the total resistance is found to be due to the contact between the plates.

With this design of plates the resistance of the lower layer of plates is less than the higher part, and consequently the current is greater and the temperature higher on the lower surface of the resister than on the higher surface, whereby radiation of heat to the roof is lessened. Measurements have shown that when the temperature of the bath was 1500° C., that in the space above the resister was only 1250° C. The arch form of resister also allows for expansion of the resister during heating.

When using the separating partition (Fig. 174) of silicon carbide or graphite coated with silicon carbide, it is found that with a temperature of 1400° C. in the resister chamber, that in the melting chamber amounted to 1290° C.

SECTION XIV.

THE ELECTRO-METALLURGY OF ZINC: THE HYDRO-METALLURGY OF COPPER.

THE production of zinc in the usual type of fuel-heated furnaces is a process of low efficiency and many disadvantageous features. Heat is applied uneconomically to the outside of retorts which are necessarily small and involve a large item of expenditure for renewals, and a high labour cost in handling the ore and conducting the distillation. The amount of fuel consumed amounts to from $2\frac{1}{4}$ to $3\frac{1}{2}$ tons of coal per ton of ore treated.

In distinction from this, electric furnace methods provide internal heating, permit of continuous operation, and in some processes dispense with preliminary roasting of the sulphide ores, which, in the retort method, it is necessary to carry out to a thorough degree. The direction in which the electrical production of zinc is finding its main application is in the treatment of complex lead-zinc ores which cannot be successfully treated in lead-smelting furnaces on account of the infusible slags produced by the high proportion of zinc, nor in zinc furnace retorts on account of the corrosive action of the lead on the clay retorts.

The amount of power required for the electrical smelting of zinc-ores amounts, according to an estimate of F. Peters[1] in kw. hour per ton of ore, to $650 + 5$ times the percentage of zinc, in the case of a furnace operating with 88 per cent thermal efficiency.

A notable defect in electric furnace processes for zinc production, however, is due to the tendency of the zinc to condense in the form of a grey powder which can only be melted to liquid metal with considerable loss, or the so-called "blue powder," which cannot be melted with any efficiency. The formation of this "blue powder" is caused mainly by the presence of carbon dioxide which leads to a coating of oxide around the particles of metal. The formation of the grey powder is due to rapid cooling, whereby the metal passes

[1] "Eng. and Min. Journ.," 1910, **89**, 1017.

direct from the vapour to the solid condition, the boiling-point of the metal being only slightly removed from the melting-point.

A feature of the retort process which has not yet been entirely overcome in the electric furnace is the maintaining of the condenser at a temperature which allows the metal to condense in liquid form and the gradual heating of the whole charge whereby water vapour and other gases are expelled, and the reduction of the zinc oxide well advanced before any distillation occurs.

The different methods which have been applied for the electrical treatment of zinc ores may be classified in the following three categories :—

(a) Electric smelting or electrothermal treatment of the ore.

(b) Fused process or electrolytic separation of zinc from a fused electrolyte.

(c) Wet process or hydro-metallurgical treatment of the ore to bring the zinc in aqueous solution, followed by electrolytic deposition.

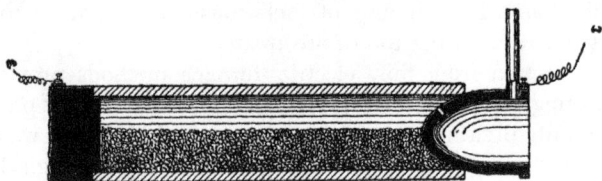

FIG. 177.

The earliest work, dating from 1884, on the use of the electric current in the winning of zinc was directed to the preparation of aqueous solutions of zinc salts by leaching the roasted ore in acid, neutralising and subjecting to electrolysis, using insoluble anodes. With the exception of one or two processes detailed below (p. 311), success in this direction was in the early days only obtained in the allied field of " electro-galvanising," where the solutions were obtained, however, not from the leaching of ore but by solution of zinc dust.

More considerable progress was made in electrothermal smelting methods, which found a successful application in the processes devised by de Laval, Snyder, Salgues, Côte and Pierron and Johnson.

Within the last few years, however, development has taken place more in the treatment of aqueous solutions, several processes having recently commenced successful operation.

A. ELECTRIC SMELTING OF ZINC-ORES.

The first electric furnace adopted in this connexion was designed by the Cowles brothers in 1885, and consisted of a fireclay retort (cf.

Fig. 177) of similar shape to the type used in the fuel furnaces. One end was closed by a carbon plug which served as one of the poles for the introduction of the current, and the other end by a carbon crucible of a form similar to the condenser. The crucible served as the second terminal for leading in the current, and also formed a receptacle for the condensation of the distilled zinc. An outlet was provided at the top of this carbon vessel for the escape of the gases produced in the reduction. The charge of roasted ore and carbon formed a resistance core, through which, by the passage of the current, the temperature was raised to the required degree.

THE DE LAVAL PROCESS.

In this method the finely divided zinc-ore is mixed with fluxing materials and the necessary amount of carbon, and admitted through

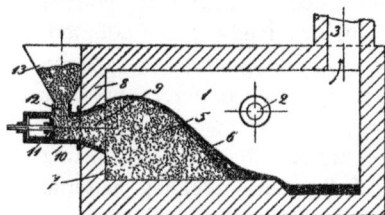
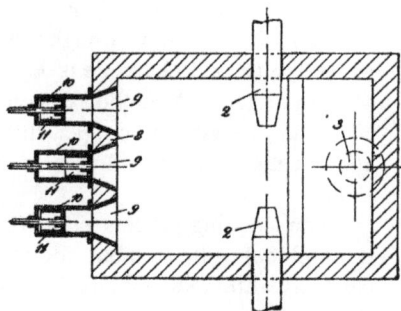

FIG. 178.   FIG. 179.

a hopper (13, Fig. 178) into a rectangular enclosure built up of refractory bricks, and containing in the centre at 2 (Fig. 179) an electric arc formed between carbon electrodes entering through the centre of the side walls in a plane at right angles to the section in Fig. 178. By means of the heat radiated from the arc, reduction and volatilisation of the zinc are brought about, and the zinc vapour led through the flue 3 into a suitably arranged condenser. Lead is to some extent volatilised together with the zinc, while the remainder together with silver, gold, etc., flows to the hearth of the furnace from whence it can be tapped. Tapping of the slag is also provided for.

By mixing iron with the charge of sulphide ore the preliminary roasting can be dispensed with, the sulphur entering into combination with the iron, giving iron sulphide in a fused condition which can be tapped.

The de Laval furnace has also been applied to the distillation of zinc from crude spelter, a product of 99·9 per cent Zn being obtained.

The de Laval process was operated in the first place at Trollhättan in Sweden, and afterwards with a total power of 2400 h.p. at Hafslund in Norway. The patent rights were also acquired by a Belgian Compant, the *Société Anonyme Metallurgique de Laval* of Brussels.

The process was afterwards extended at Trollhättan and installed at Sarpsborg in Sweden with a combined equipment for 18,000 h.p. and an annual production of 6000 tons of refined zinc.

According to a report by F. W. Harboard[1] success in obtaining solid zinc by the de Laval process had up to 1911 been very partial, most of the spelter which was produced being obtained by smelting dross, scrap, and other secondary products rather than ore. The arc furnace was replaced by a resistance type, in which one large electrode passes down through the roof and on to the furnace charge, while the second electrode is a carbon block bedded in the bottom of the furnace. A method of continuous side feed is being adopted. The energy-consumption per furnace is 350 h.p. and the capacity about 3 metric tons, 2·8 tons of ore being smelted per twenty-four hours.

The smelting process consists in charging the ore (after roasting if sulphide) with suitable additions of flux and reducing material (such as anthracite or coke) into the furnace, where most of the zinc and some of the lead are volatilised and condensed. The zinc condenses partly as a metal and partly as blue powder and oxide, containing about 54 per cent of zinc and 20 per cent of lead. This powder is then mixed with fresh ore and recharged, when a much larger percentage of the metal volatilised is recovered as metal. The first operation of smelting ore alone may be regarded mainly as a concentrating process for the production of a rich oxide, which is reduced to the metallic state by subsequent treatment. The other portion of the lead, carrying a considerable proportion of the silver, is mainly reduced to metal in the smelting hearth and is tapped out with the slag. The crude metal recovered from the distillate in this smelting operation has an average composition of about 79 per cent zinc, 20 lead, and 0·6 iron. This product is redistilled in an electric furnace to give a spelter of 99·9 per cent zinc.

In a test extending over twenty-seven days, when a total of

[1] "Eng. and Mining Jour.," 1912, **93**, 314.

some 550 tons of ore, consisting mainly of roasted Broken Hill slimes, were treated, it was found that the consumption of electric power averaged 2078 kw. hours per ton of ore smelted. The average consumption of electrodes was 31·5 kg. per metric ton of ore at Trollhättan, and 40·5 kg. per metric ton of ore at Sarpsborg.

The percentage recovery of the metals obtained, including metal in powder form, was 73·4 zinc, 79·3 lead, and 49·5 silver, or excluding the powder, 64, 73·9, and 45·9 per cent respectively.

The Trollhättan works has in operation eleven smelting furnaces of the resistance type, together with a number of the arc type, which, however, are being transformed into the former. The arc type of furnace was employed at Sarpsborg in 1911, and as far as the process of smelting is concerned was found to give as good a product as the resistance furnace, but the consumption of energy was over 70 per cent more, and the consumption of electrodes was also materially higher than in the case of the resistance furnace.

The cost of power at Trollhattan is stated to be (in 1911) 30s. 3d. per h.p. year, or 41s. 3d. per kw. year. In the opinion of Mr. Harboard the weak part of the process is the large amount of metallic powder produced in proportion to the solid metal, but it is considered that the practice then in use can be greatly improved, leading to decreased consumption of energy and reduction of labour costs.

SNYDER FURNACE.[1]

This process was devised for the treatment of complex lead-zinc ores and has a number of interesting features though, as far as is known, it is not now in operation. The ore is roasted to convert the sulphides into oxides and mixed with fluxes, such as lime and iron (unless already present), in such proportions as when smelted to produce a slag which will form at a temperature between the volatilisation temperatures of zinc and lead (approximately from 1000° to 1100° C.). A slag of this nature is given by a material with 30 per cent lime, 30 per cent iron-oxide, and 40 per cent silica. In the furnace illustrated in Fig. 180 the charge, together with the admixed carbon, is first pre-heated in the chamber *ff* so as to start the process of reduction as far as possible and expel gases and water vapour, without actually vapourising the zinc. The material is then introduced into the reaction chamber *a*. The lower portion of the main chamber *a* of the furnace is divided by a

---

[1] U.S. Pat. 933,133 ; " Electrochem. and Metall. Ind.," 1909, 7, 451.

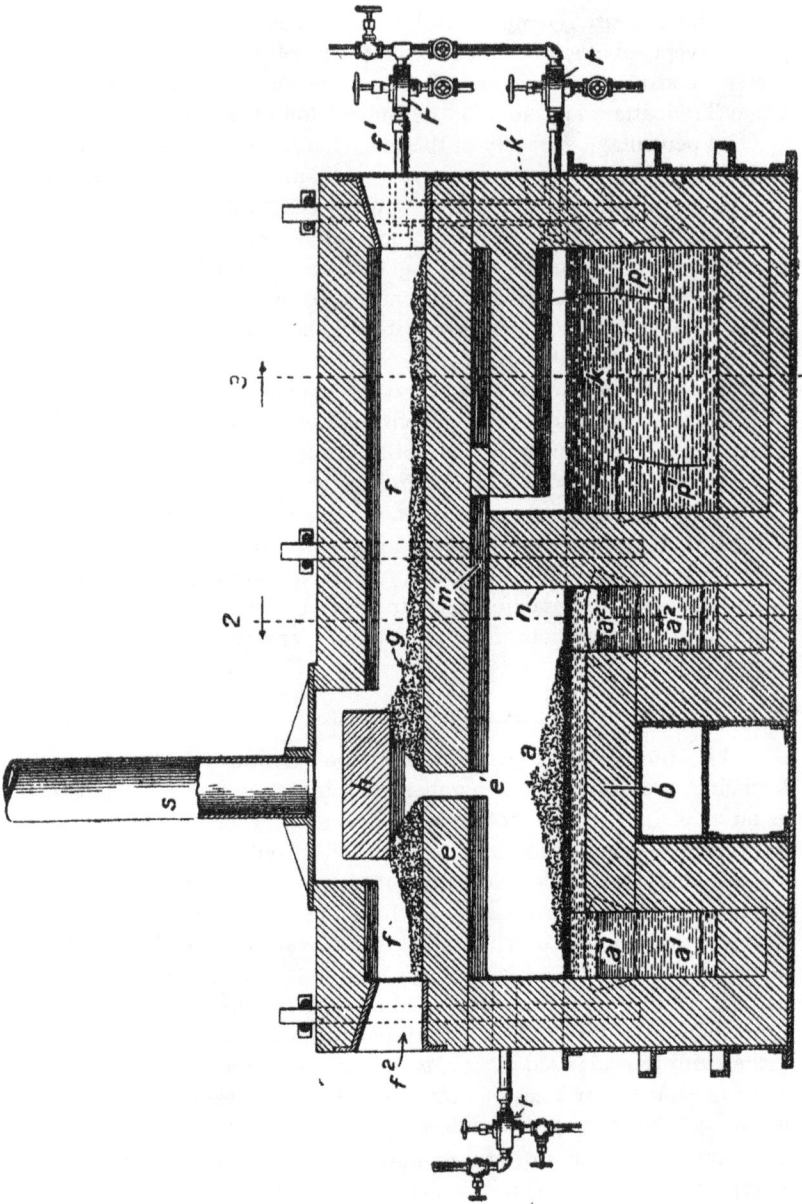

FIG. 180.

bridge b forming two wells a, a^2 which communicate by U-shaped passages with the exterior of the furnace, as shown in Fig. 181, which is a view in cross section on line 2 in Fig. 180. The electric conductors C are dipped into the molten metal in the outer arms of these U-shaped passages, while the molten metal at a and a^2 makes electrical contact with the slag and furnace charge which extends between the two molten electrodes. A layer of carbon is maintained

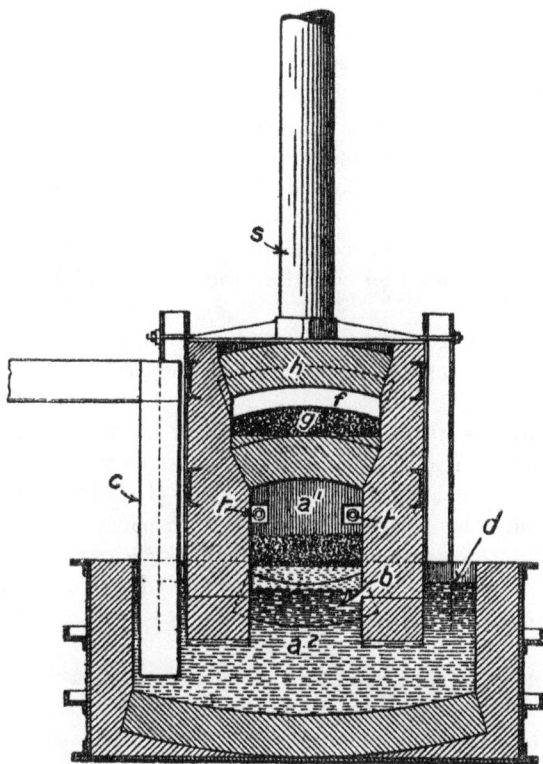

FIG. 181.

on the surface of the molten slag and serves to increase the conductivity. An opening e' is provided in the reverberatory roof of the smelting chamber through which the roasted ore may be fed into the smelting chamber. Zinc which is reduced and volatilised in the chamber a passes through m into the chamber k where it is condensed, while lead accumulates in the electrode wells and can be siphoned off as necessary from d. The current passed through the bath is adjusted so as to raise the temperature sufficiently for the volatilisation of zinc, but not so high as to cause volatilisation of lead. By tapping the slag the process becomes continuous. This process was formerly in operation on an experimental scale at Nelson, British Columbia.

SALGUES FURNACE.[1]

A process for the electric smelting of zinc was devised by M. A. Salgues, in 1903, and installed at Crampagna (Ariège), France. Heating of the charge is, in this furnace, effected by means of two carbon electrodes, one of which is embedded in the base of the furnace, while the second is suspended vertically through an opening in the roof and projects into the material of the bath. The furnace is built of firebrick supported in an iron case, and provided with a tapping hole at the base of the hearth. The top part of the furnace, or roof, can be removed for cleaning or repairing the interior. The furnace is constructed to be completely air-tight, and the roof is provided with openings for admitting the charge and for stirring the bath.

The materials treated are admitted either in the form of oxide or sulphide, and are smelted together with fluxing materials and carbon or iron.

The materials treated include roasted or unroasted blendes, raw or calcined calamine, silicious calamine, and complex ores. In all cases slags are obtained with less than 1 per cent of zinc.

The condensers arranged are of large capacity and maintained at a temperature between 412° and 930° C., while particular regard is paid to avoid inleaks of air which, in contact with zinc vapour, might lead to violent explosions. The methods of smelting adopted are as follow :—

1. Submitting the charge to a preliminary heating, which is sufficient to bring about reduction of the zinc oxide and expulsion of the gases, but not high enough to fuse the product, and afterwards smelting the material in an electric furnace and expelling the zinc as vapour in a concentrated form.

2. Conducting the electric furnace stage at a temperature sufficiently low to cause the zinc to accumulate as liquid under the slag. The slags for this purpose must be arranged to be highly fusible.

3. In a third method which is partly electrolytic, it is proposed to heat the ore without the addition of carbon in presence of carbon monoxide, thus reducing the oxide but leaving the silicate. Heating is

[1] " Bull. Soc. Ing. civ. d. France," 1903 (2), 64.

then applied electrically by direct current through the resistance of the charge, and at the same time electrolysis of the silicate is brought about.

Production.—With a furnace of 100 kw., fed with a cold mineral of 40 to 45 per cent zinc, the output of metal is stated to amount to 5 kg. per 24 kw. hours or 1·8 ton per kw. year.

CÔTE-PIERRON PROCESS.[1]

In the Côte-Pierron process, it is attempted to avoid the difficulty in zinc condensation due to the metal vapour being evolved in presence of a large amount of carbon monoxide by smelting a mixture of unroasted zinc blende and iron, when at a relatively low temperature reaction takes place in accordance with the equation

$$ZnS + Fe = Zn + FeS.$$

The zinc is evolved as vapour, together with a certain amount of gases, moisture, and carbon dioxide. A " zinc mist " is formed in the condenser which can, however, be largely obtained in the form of liquid metal. The furnace used has undergone several modifications, but the principle adopted in the condensation consists in passing the zinc vapour through a column of heated carbon which is interposed between the furnace and the condenser. This process is specially adapted for the treatment of complex zinc-lead ores. The crushed ore, mixed with iron particles and a suitable flux, is introduced into the furnace, the walls of which are heated from the previous run and serve to raise the temperature of the charge sufficiently to cause the lead sulphide to react with liberation of lead which flows to the bottom of the furnace and can be tapped off. Electrical heating is then applied, when a reaction occurs whereby the iron displaces the zinc, which is set free in the form of vapour, and is condensed in the form of vapour, practically free from lead. The iron sulphide collects on the bottom of the furnace in the form of a liquid bath, covered by a fused slag. By increasing the heating, the slag is rendered liquid, and most of the contained zinc is expelled. The fused iron sulphide and slag are finally tapped from the furnace, and a new charge admitted.

A type of furnace which has been adopted is shown in Fig. 182 in vertical and horizontal sections. A is the crucible, the bottom of which contains in its centre a projection *b*. The crucible consists in its lower part and side walls of graphite encased by sheet

[1] " Metall. and Chem. Engineering," 1909, 7, 469.

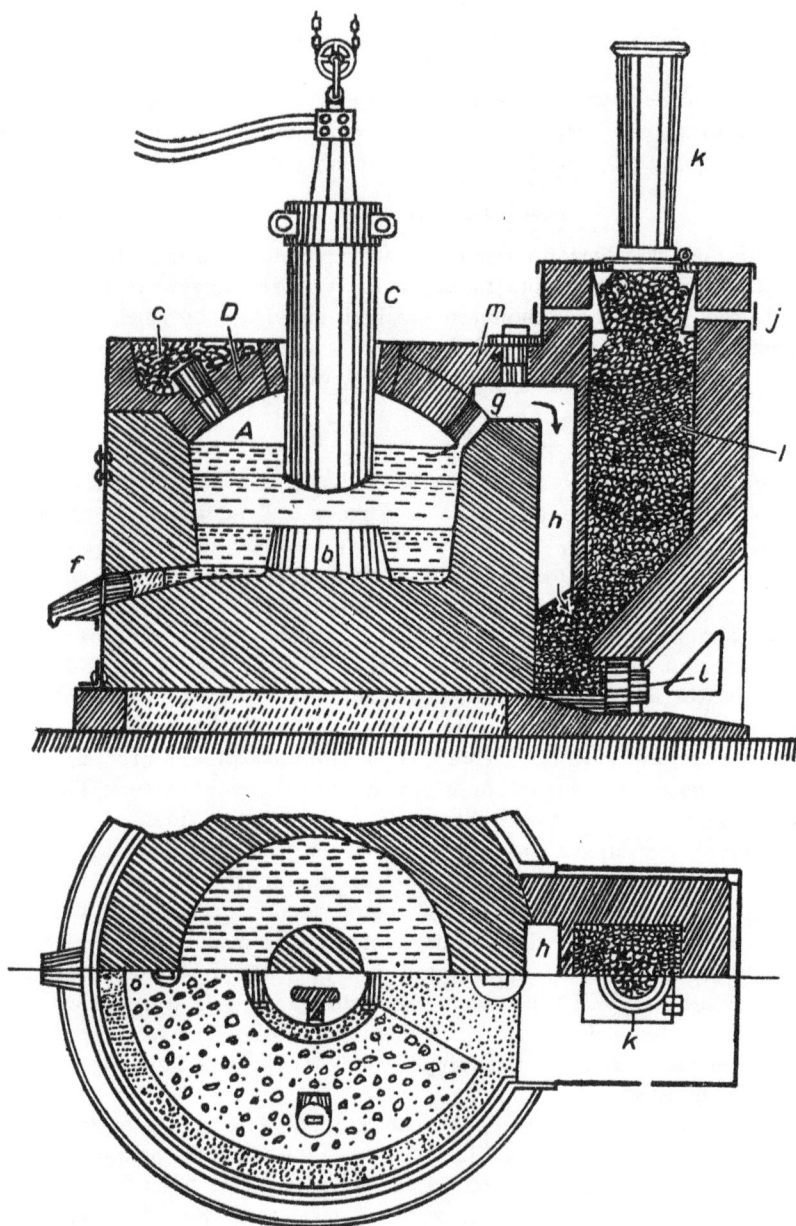

FIG. 182.

metal and is connected to one pole of the electric circuit. The other pole is connected to the vertical electrode C, the lower end of which is above the projection *b*. The roof is formed by the arch D of refractory brick. The charge is introduced through openings *c*, which are closed during operation by means of refractory stoppers ; the tapping hole is shown at *f*. The zinc vapours leave through *g* and pass through a long channel *h* to the condenser I, which is formed of a chamber of vertical refractory walls, filled with granular carbon maintained at about 900° C. The condenser is tapped through the hole *l*. Air can be admitted at the top through *j*, and issuing at the stack *k* makes it possible to bring the carbon in the upper part of the condenser to red heat. The stack can also be heated electrically by shunting off part of the main current and leading through the carbon. The opening *m* is provided for cleaning the channels *g* and *h* in case of clogging.

The charge to be treated is placed on the top of the furnace D, where it is dried before admission to the furnace. The zinc vapour leaves the crucible at a temperature of 1200° or 1300° C., and passing from the channel *h* into the condenser I, serves to heat the carbon granules to the temperature of reduction of zinc oxide which accompanies the zinc vapour. On contact with the carbon surface the condensation of the zinc takes place gradually. The liquid zinc which percolates down the column of carbon forms thereby new condensing surfaces for the zinc vapour. The liquid zinc collects on the bottom behind the door *l*, through which it is withdrawn from time to time.

A further design of furnace by Côte and Pierron is illustrated in Fig. 183.[1] Two electrodes *bb'* are introduced through the roof of the cylindrical enclosure *a* formed of refractory bricks encased on the outside with a covering of sheet-iron.

The charging of the furnace takes place continuously from the hopper B through the opening *i*. The residual slag and matte from the smelting are withdrawn through *t*. A carbon rod *u*, connected to the lever *v*, serves as an auxiliary electrode, and is employed immediately before tapping for the purpose of fusing any congealed material in the neighbourhood of the tapping-hole. A further opening in the walls of the furnace is provided at *w*, normally closed by a plug of refractory earth, to give access to the interior for the purpose of cleaning or repairs. According to the position of the electrodes, the heating of the furnace contents can be brought about either by arc or resistance action.

[1] Brit. Pat. 5100 of 1907.

The zinc vapour evolved passes out through the opening t' to the condenser.

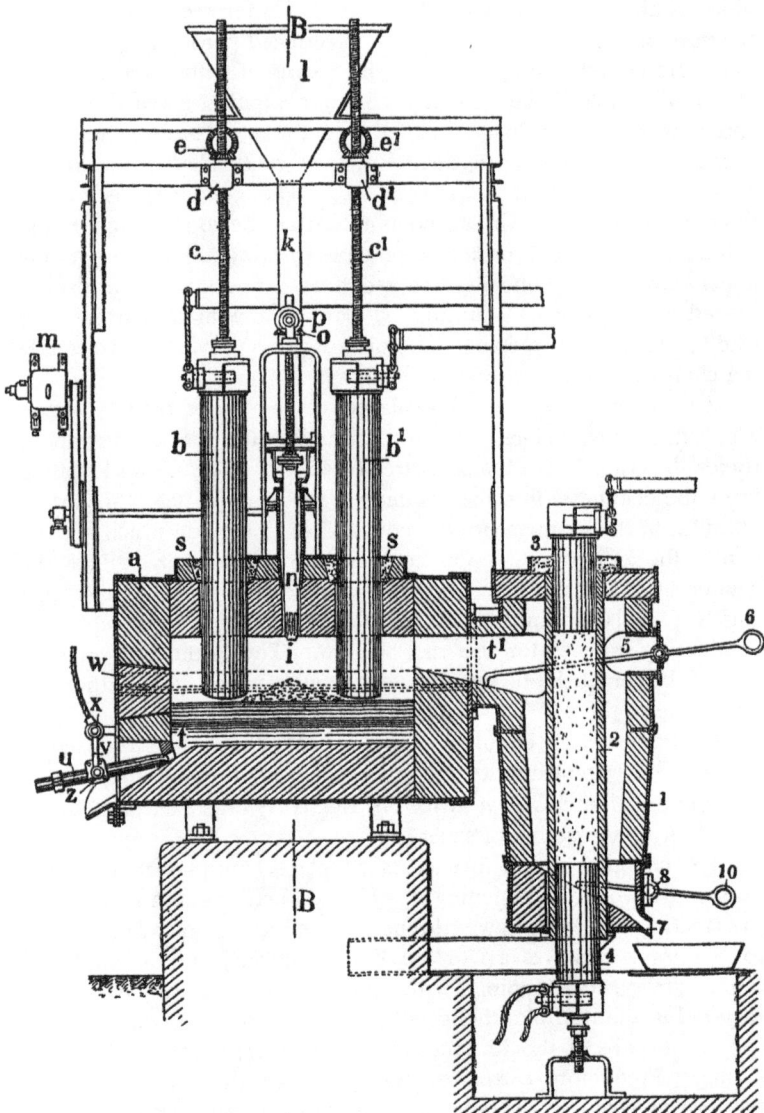

FIG. 183.

The condenser consists of a vertical cast-iron cylinder 1, with an inner lining of refractory earths. A refractory tube 2, filled with fragments of retort carbon, is arranged centrally along the axis of the condenser. The two ends of the tube are closed by the electrodes 3 and 4 which are held in close contact with the column of retort carbon. A current is then passed through the column and the temperature of the condenser thereby maintained at the required degree.

A rabble is provided at 6 which enables the channel t' to be cleared of any accumulated deposit.

Condensed zinc in liquid form is removed through the opening 7.

The first experiments with the Côte-Pierron furnace were conducted at Lyon in 1906 with an unit of 100 h.p. The furnace consisted of a cylindrical crucible of 40 cm. inside diameter and 45 cm. height, built up of magnesia bricks and held together by a metallic casing. In 1907 the *Société des Fonderies Électriques* was formed to exploit the furnace on a larger scale, and a works was erected at Arugy in the Pyrenees, near important zinc-blende mines. At these works the production of zinc oxide has been engaged in. For this product the use of iron in the smelting process has been abandoned in favour of a process in which the blende is smelted together with lime and carbon, when reaction takes place in accordance with the equation

$$ZnS + CaO + C = CO + CaS + Zn.$$

In 1918 this modification of the Côte-Pierron process was adopted in a works erected at Maurienne near Epierre.[1] The zinc is liberated as vapour and condenses as liquid and zinc dust in an impure condition in an adjoining chamber. This chamber constitutes a subsidiary electric furnace, and, before cooling, the crude metal is again volatilised, passes from here as a vapour, and condenses in a third chamber giving a final product of an alleged purity of 99·93 per cent. The ore used has the composition—

Zn, 30 to 40 per cent; Pb, 5 to 15 per cent; Fe (as oxide or sulphide), 8 to 12 per cent; S, 20 to 30 per cent; CaO, SiO_2, or BaO, 10 to 15 per cent; F, 3 to 5 per cent; foreign metals 2 to 3 per cent; As, 2 to 3 per cent.

Only 1·5 per cent Zn remains in the slag, while the total losses of zinc vary from 6 to 9 per cent. The electrode consumption amounts to 12 kg. per ton of ore treated.

The equipment at Maurienne consists of four furnaces each of 500 h.p., and with a capacity for each furnace of 4 metric tons of zinc-ore per twenty-four hours. The distribution of the power is

[1] G. Flusin, " Bull. Tech. de la Suisse Romande," 1917, **43**, 233 ; " Met. and Chem. Engineering," 1918, **18**, 17.

350 to 375 h.p. in the fusion bath and 125 to 150 h.p. in the condenser.

JOHNSON PROCESS.

A continuous zinc-smelting furnace of the smothered-arc type has been developed by W. McA. Johnson[1] which, in a recent design, is arranged as shown in Fig. 184. The main feature of this process is that the iron of the ore is first reduced and then the zinc, the reduced iron forming a resistance medium to the passage of the current through the charge. Ore containing, for example, 40 per cent zinc, 20 per cent iron, 30 per cent sulphur, and small quantities of copper sulphide, lead sulphide and precious metals, is roasted at 850° to 900° C. The sulphur need be reduced to 3 per cent only so that a concentrated sulphur dioxide gas is

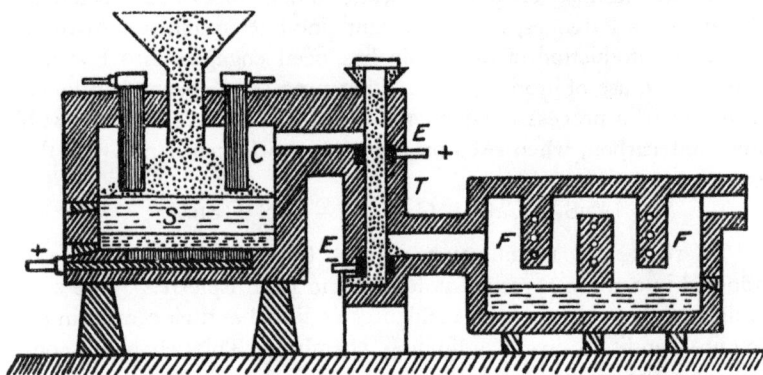

FIG. 184.

obtained. The residue which contains the greater part of the zinc and the iron in the form of oxides is heated with 35 per cent of carbon to 850° to 950° C. Spongy iron is formed while the zinc oxide remains unreduced. The heating is then continued in the electric furnace (Fig. 184) using the iron sponge as resistance, and taking the temperature of the charge to 1000° to 1150° C. The zinc oxide is thereby reduced, the zinc distilling and the iron reacting with any unconverted blende to form a matte which is removed by tapping. To obviate the formation of "blue powder" through the reaction between carbon dioxide which is evolved and the zinc, the furnace gas and vapour are passed through a filter of granulated carbon at T, which is heated by the passage of a current between the electrodes EE. The temperature of this column is maintained sufficiently high to prevent condensation of the zinc and to bring about reduction of

[1] U.S. Pat. 964,268.

carbon dioxide to monoxide. The mixture of zinc vapour and carbon monoxide gas then passes into the condenser F, which is designed so as to bring the zinc vapour into contact with the surface of liquid zinc and to preserve a suitable temperature. A furnace of 25 kw. capacity has been constructed on this principle.[1] The slag aimed at melted at 1100° to 1175° while the working temperature of the slag body was 1250° to 1300°. The gases passed into the condenser at a temperature of about 850°. The rate of smelting per cubic foot of reaction zone is stated to be fifty times as rapid as in the fireclay zinc retort, and five times that of the ordinary lead-smelting furnace. It is claimed that the proportion of zinc vapour which is condensed to liquid metal is more favourable than the value of 70 to 80 per cent obtained in usual retort practice, and the losses in the liquid slag are stated not to exceed 0·5 per cent of the zinc content of the charge. When melting a zinc-lead ore, the lead accumulates as bullion at the base of the furnace, and can be tapped, while copper reacts with any sulphur present to form a matte. By combining the electric furnace with a pre-heater, use is made of the fact that the thermal efficiency of fuel-heating as in the old retort method is high at first, when the charge is readily reducible, while at the end the efficiency is low.

It is estimated that it will be possible to treat an ore containing 35 per cent of zinc with a power consumption of 600 kw. hours per 2000 lb. of ore. On the basis of the above data, this will give 3·4 tons of liquid zinc per kw. year of 8000 hours.

The advantages which it is claimed will be obtained are as follows :—

1. Purer and more uniform product.
2. Higher recovery of zinc, exceeding 95 per cent.
3. Wider range of ores that can be treated.
4. Lower labour charges as units are enlarged.
5. Higher thermal efficiency.
6. Recovery of lead, copper, silver, and gold values.
7. Decreased capital outlay when combined with a pre-heater or where power is rented.

RUTHENBERG PROCESS.

In this process an attempt was made to conduct the smelting of complex zinc ores by heating the sulphide ore mixed with suitable fluxes in an oxidising atmosphere, volatilising the zinc as oxide, which is passed together with the gaseous products of reaction,

[1] Cf. J. W. Richards, "Trans. Amer. Electrochem. Soc.," 1911, **19**, 311.

20 *

including sulphur dioxide, into absorption towers where, in contact with water, a solution of zinc sulphite (partially oxidised to sulphate) is yielded. The heating of the charge is brought about in an electric furnace by radiation from a three-phase arc maintained between carbon electrodes arranged some distance above the charge, as in the Stassano furnace. Lead bullion, and, in the case of copper-containing ores, a copper matte, can be tapped from the furnace, and similarly the fused slag. The earlier experimental trials with this process were carried out at Loughborough. Later the North Eastern Electric Smelting Company was formed to develop the method, and installed the process on a commercial scale at Wallsend-on-Tyne, Newcastle. An electrolytic process for treating the zinc solutions was adopted. Many practical difficulties were encountered in the economic application of heat to the smelting, in the solution of the zinc oxide volatilised, and in the electrolytic treatment of the solutions obtained. The process was not brought to a successful issue and is no longer in operation. The experience gained in this undertaking, however, should serve as a valuable contribution in the development of the electro-chemical treatment of complex zinc ores.

FULTON PROCESS.

In a process devised by C. H. Fulton,[1] oxidised zinc ore is mixed with crushed coke and coal-tar pitch and formed by means of a hydraulic press into briquettes 9·25 ins. diameter by 21 ins. long, in a manner similar to that used in the manufacture of graphite and carbon electrodes. After an initial heating or baking in a fuel-furnace at about 500° C., the briquettes are connected as resisters in an electric furnace whereby the ore is reduced and zinc together with carbon monoxide distilled. In the case of complex zinc-lead ores, the distillation is conducted at a carefully regulated temperature so that a large proportion of lead is retained in the briquette residue which may then be smelted for this metal.

In a plant installed at St. Louis, U.S.A., the furnace holds a charge of thirty-six briquettes, each of ninety pounds weight, and arranged in twelve columns of three each. The columns are set within a circle enclosed in a surrounding retort and operated on a three-phase circuit. Four columns, connected in series, are placed in each phase by the usual star-connexions. The contacts at the ends are made by graphite connecting blocks.

The time required for completion of the distillation is about six hours. The capacity of the furnace is 5100 lb. of zinc concentrates

1 " Metall. and Chem. Engineering," 1920, 22, 73, 130.

per twenty-four hours. When commencing with hot retorts, the voltage varies from 150 at the beginning, with an energy consumption of 120 to 150 kw., to 75 to 90 volts, with an energy consumption of 122 to 184 kw. at the end.

The condenser consists of a rectangular chamber, 42 ins. diameter by 9 ft. high and lined with firebrick. For the initial charge, the condenser is heated by an oil-burner to 940° C. During subsequent operation, the temperature maintains itself at about 700° C. at the inlet and 500° C. at the outlet. The recovery of the metal is as follows :—

Spelter tapped	42·7 per cent.
Scrapings, rechargeable	30·8 ,,
Zinc in brick from condenser	3·9 ,,
Zinc in briquettes residue	9·1 ,,
Unaccounted for	13·5 ,,
	100·0

No "blue powder" is obtained in the product.

The power consumption is given as 2576 kw. hour per ton of ore.

IMBERT ZINC PROCESS.

The Imbert process consists in the smelting of zinc blende together with iron in the Thomson and Fitzgerald type of furnace.[1] The reaction which takes place is in accordance with the equation

$$ZnS + Fe = Zn + FeS$$

but is incomplete unless a large excess of iron is present or unless the temperature of the reaction is very high. Imbert discovered that a mixture of 1 part of ferric oxide and 3 parts of ferrous sulphide forms a very fluid bath at a temperature of 1000° to 1100° C. and that this fused bath will contain 6 parts of blende, whereby, on the addition of iron, the above reaction is able to proceed and the zinc distils. The most satisfactory temperature for the reaction is found to be 1350° to 1400° C.

The Imbert process has been installed at Hohenlohehutte in Upper Silesia.

Recovery of Zinc from Blast-furnace Fumes in Lead Smelting.—A consideration of the situation at the present day with regard to complex zinc-lead ores would appear to show that with the progress recently made in the hydro-metallurgy of zinc, it should be advantageous to apply an electrolytic process to recover zinc from blast-furnace fumes evolved at a high temperature, and under other

[1] Cf. p. 290.

conditions ensuring a good expulsion of the zinc as oxide from the slag. The zinc oxide might be brought into solution by means of spent electrolyte containing free acid. With an electric smelting furnace a higher economy would be expected by applying some method of liberating and condensing the zinc directly in the metallic state, whereas the recovery from zinc oxide would, under usual conditions, be most advantageously conjoined with the blast furnace. Valuable metals carried together with the zinc would also be recovered.

B. Electrolytic Separation from a Fused Electrolyte.

The most notable example of this class is the process devised by Swinburne and Ashcroft with a view to treating the Broken Hill complex zinc-lead ore. In this method the sulphide ore is crushed and fed into a "transformer," consisting of a fireclay tower, resembling a blast furnace. A stream of chlorine is blown in at the bottom through a carbon tube, and bubbling through the fused chloride bath reacts with the metal sulphides to form chlorides with the liberation of sulphur. As heat is evolved by this reaction the "transformer" is self-heating and the temperature controlled by the rate of admission of the chlorine.

The fused mass from this reaction contains chlorides of all the metals present and is run into water, which, when cool enough, is filtered. Lead (and silver) chloride, together with the gangue, remain as residue, and can be readily reduced to their respective metals, while the filtrate contains a little lead and silver in solution, and copper, iron, manganese, and zinc. The lead and silver are displaced by spongy copper, and copper by zinc, leaving iron, manganese, and zinc chlorides. The iron is chlorinated to the ferric state, and precipitated as hydrated ferric oxide by adding zinc oxide. On further chlorination in presence of more zinc oxide, the manganese is precipitated as peroxide. The solution of zinc chloride is then evaporated down carefully and fused. A small amount of oxychloride which is formed can be removed by giving a preliminary electrolysis in a separate vat, after which the fused chlorides are passed into the final electrolytic vats. The anodes are composed of carbon and the cathode of fused zinc.

The temperature is maintained without external heating by the passage of the current. With small vats a current of 3000 amps. at 4 volts per cell is employed, and with a larger vat 10,000 amps. at 3 volts. The vats consist merely of iron cases lined with firebrick. The chloride soaks into the porous brick and solidifies, ' leaving solidified charge itself to serve as a lining.

The Swinburne-Ashcroft process was erected at the Castner Kellner Works at Weston Point, Cheshire, the plant having a capacity of 30 tons of ore per week. The chlorine used was the by-product from the electrolytic alkali plant. The electrolytic treatment of the zinc chloride was later abandoned, the compound instead being sold as such.

C. HYDRO-METALLURGICAL METHODS.

The earliest experiments on the electro-metallurgy of zinc related to the electrolytic extraction of the metal from aqueous solutions. The pioneering work in this field was that of Nahnsen, who, about 1890, devised a process for the electrolytic refining of impure spelter in Upper Silesia; Ashcroft, who installed a large plant at Cockle Creek, New South Wales, which, however, was not continued, for the electrolytic extraction of zinc from leached ores; and Hoepfner who developed a process which was brought into operation at Fuhr-fort on the Rhine and at the works of Brunner, Mond & Co., at Winnington in Cheshire. After many years of successful operation the Hoepfner process is no longer worked in this country.

The separation of zinc by electro-deposition from aqueous solu-tions obtained from the treatment of ore has now found a successful industrial application in several different processes which may be divided into two main classes.

(*a*) Dissolving the zinc in the form of sulphate, by leaching the oxide with a sulphuric acid solution, or by leaching with water ore which has been roasted to sulphate.

(*b*) Dissolving the zinc in the form of chloride by roasting the sulphide ore with sodium chloride and leaching, or transforming into chloride after roasting to oxide.

Recent developments which have now brought processes for the electrolytic recovery of zinc into the first line in the metallurgical field are based on sulphate processes, and it is to these particular cases that the present consideration of the subject will be limited.

The hydro-metallurgy of zinc is concerned with two distinct operations :—

1. The transfer of the zinc from the ore to an aqueous solution.

2. The electrolytic extraction of the zinc from the solution.

1. The solution of the zinc is generally brought about by roasting to oxide and extracting with a solution containing sulphuric acid. The main difficulty which is encountered here in the case of a

ferruginous blende is to roast to oxide without causing the formation of a ferrite of zinc, which cannot readily be dissolved in acid. On this account the solubility of zinc varies largely with different ores. Ores so far applied for electrolytic treatment have consisted of complex zinc-lead sulphides. With zinc concentrates from these ores, containing 45 per cent of zinc, an extraction of as much as 95 to 97 per cent of the zinc is in many cases obtained.

A further difficulty in handling the solutions prepared by leaching the roasted ore is due to gelatinous silica, which passes into the solution and impedes subsequent filtration.

2. In preparing the solutions for electrolysis a difficulty presented is that of removing certain deleterious impurities without elaborate chemical treatment of the solutions.

The main obstacle in the electrolytic separation of zinc from aqueous solutions lies in the fact of the metal being more electro-negative than hydrogen, consequently, in an acid solution which contains many hydrogen ions, these will ordinarily separate in preference to the zinc. Further, largely on account of the hydrogen which is at the same time separated, the deposited zinc has a tendency to assume a spongy condition instead of a coherent form. A further objectionable feature is the tendency, in many cases, of the zinc to be deposited in large crystalline growths which, at an early stage in the electrolysis, become detached or bridge over to the anode.

Over-voltage.—On account of its high electro-negative nature, the possibility of the separation of zinc from aqueous solution with any efficiency is, in the presence of free acid, mainly brought about by the operation of "over-voltage". This phenomenon may be defined as a passive resistance which is offered by a surface of pure zinc to the formation of gaseous hydrogen from the ionic condition. In the presence of impurities, whereby the deposition of other metals together with the zinc occurs, the magnitude of the over-voltage is considerably lessened and evolution of hydrogen facilitated. It is in order to provide a high value for the over-voltage that a thorough purification of the solution is usually necessary with solutions for the deposition of zinc.

A further factor which controls the preferential deposition from an electrolyte containing a mixture of dissolved substances is the velocity of reaction of certain intermediate changes which are incompletely understood, but which attend the change from the ionic condition to that of the free element. This property can be utilised by adjusting the current density, for example a high-current density will give a relatively smaller proportion of an element which has a slow transition velocity than will a low-current density.

In the earlier processes which were developed for the electro-deposition of zinc, such as those of Cowper-Coles, Mylius and Fromm, Langbein and Pfanhauser, Hoepfner, the Bradley-Williams process which was developed at Newcastle-on-Tyne and the Lasczynski process which was installed in Austria and Sweden, it was taken as a maxim that the essential conditions for successful operation included the adoption of small concentration of acid, low-current density, and purity of solutions. As a departure from this rule Siemens and Halske devised a process by which with specially purified solutions, electrolysis could be conducted to yield hard adherent deposits of zinc in presence of a concentration of acid amounting to 10 to 12 per cent and using the usual low-current density of 1 to 3 amps. per square dcm. However, the special purification of the solutions which is necessary limited the commercial application of the process. It has subsequently been found [1] that by simultaneously increasing the acidity and current density, particularly in presence of small quantities of colloidal matter, the efficiency in the deposition of zinc rises and enables the production of smooth coherent deposits of considerable thickness from solutions not of a high degree of purity in presence of a high concentration of free sulphuric acid.

PROCESS OF THE ANACONDA COMPANY. [2]

The largest plant now in use for the electrolytic extraction of zinc is a process which has been developed by Laist and Frick,[3] and brought into operation at Anaconda and at Great Falls, Montana, U.S.A. The production of 5 tons per day was obtained in 1915, and in 1918 the plant was enlarged to a capacity of 150 tons per day. The process consists in concentrating the zinc ore, preferably by flotation, so as to make a concentrate with as little insoluble matter as possible and as rich in zinc as the nature of the ore allows. The concentrate is roasted so as to produce a calcine containing from 2 to 3 per cent of sulphur, most of which is present as sulphate. The temperature is not allowed to exceed 1350° F. (732° C.), under which conditions the formation of zinc ferrite is not pronounced. The

[1] Pring and Tainton, "Trans. Chem. Soc.," 1914, **105**, 710; Brit. Pat. 7235 of 1911.
[2] W. R. Ingalls, "Eng. and Mining Jour.," 1916, **101**, 425; "Trans. Amer. Electrochem. Soc.," 1916, **29**, 347; "Chem. and Metall. Engineering," 1921, **24**, 245.
[3] U.S. Pat. 1,167,700.

calcined product after cooling is treated with a solution consisting partly of spent electrolyte with free sulphuric acid, which dissolves the zinc to the extent of from 85 to 95 per cent together with a little of the iron. Manganese dioxide is added for the purpose of oxidising the iron which is then precipitated by means of a small amount of powdered limestone, which at the same time carries down any arsenic or antimony. The resulting solution containing zinc, cadmium, and copper is separated by filtration from the residue which contains the lead, silver, gold, and part of the copper originally present. The solution is treated with metallic zinc to precipitate the copper and cadmium, and is then pumped through a clarifying filter press into a storage tank where it is mixed with spent electrolyte from the end cells of the series so as to give a content of about $2\frac{1}{2}$ per cent of free sulphuric acid and 5 per cent of zinc, and is then passed to the electrolytic cells. Manganese can be separated at the lead anodes and recovered as dioxide. The zinc is deposited on aluminium cathodes with a current density of 23 amps. per square foot, and the cathodes are removed after forty-eight hours and the zinc peeled off. The residue or gangue remaining from the filtration and containing iron, lead, silver, etc., is then treated in the blast furnace.

Process at Trail, British Columbia.[1]

An electrolytic process for the extraction of zinc was brought into operation in 1916 at the Trail Smelting Works. The ore treated for zinc extraction has the following approximate composition :—

Lead .	14	per cent.
Iron .	23 to 31	,,
Insoluble residue	4	,,
Alumina .	3	,,
Lime .	2	,,
Magnesia .	2	,,
Sulphur .	24 to 29	,,
Zinc .	19 ,, 24	,,
Cadmium .	0·04	,,

The ore is crushed in tube mills to a fine powder, roasted in Wedge roasters at a temperature of 950° to 1100° F., and treated with spent electrolyte until the solution is neutral. After agitating with "atomised" zinc to precipitate copper and cadmium, the crude electrolyte is filtered and delivered to tanks which supply the electrolytic tank room. The electrolytic vats are made of

[1] E. H. Hamilton, "Trans. Amer. Electrochem. Soc.," 1918, **32**, 317.

concrete, lined with asphalt, and arranged in cascades of 8 vats in series. Each vat is 82 inches long, 27 inches wide, and 42 inches deep. Each tank has 17 lead anodes, each of 9 square feet (0·8 square m.) area, and 16 aluminium cathodes, of 10 square feet (0·9 square m.) area. The current density employed is 24 amps. per square foot (265 per square m.), the voltage across the vats is about 3¼, the electrolyte contains about 7·5 per cent zinc, 0·014 per cent iron, 0·003 to 0·008 per cent cadmium, 0·0003 per cent copper, and sometimes manganese, lime, magnesia, and other impurities entering the solution from the ore. The average spent electrolyte from the electrolytic tanks is about 3 to 5 per cent zinc, 0·013 per cent iron, and 5 to 7 per cent free sulphuric acid. The temperature of the solution in the tanks varies from 30° to 45° C.

The cathode zinc is stripped from the aluminium cathodes and is melted in a reverberatory furnace. The production is about 50 tons per day, of a product containing 99·92 per cent zinc.

HIGH-CURRENT DENSITY PROCESS.

The process of J. N. Pring and U. C. Tainton [1] employing a high-current density and high acid concentration is now being applied on a commercial scale, and consists in extracting the roasted ore with spent electrolyte recovered from the end stage of the process and containing 15 to 18 per cent sulphuric acid. A given volume of the solution is treated with the roasted ore until it becomes neutral. By adding lime or finely divided zinc oxide, ferric oxide can be precipitated. The remaining solution is then filtered, and by adding a further volume of spent electrolyte the concentration of free acid is raised to the required degree, viz. 6 to 8 per cent of sulphuric acid or upwards. During the subsequent electrolysis, manganese is separated at the anode as peroxide which adheres to the electrode or falls as a powder to the bottom of the electrolyte. The solution can be submitted to a preliminary electrolysis for a short time under conditions more suited for the separation of manganese peroxide and removing the impure zinc formed on the cathode before applying the electrolyte for the extraction of the zinc.

The wide range of acid concentration permissible in this process enables the extraction of the zinc to proceed with a given volume of electrolyte to a high degree.

In presence of 12 grm. sulphuric acid per 100 c.c. and with a current density of 18 amps. per square dcm. (or 170 amps. per square

[1] Loc. cit.

foot), a current efficiency of 95 per cent is obtained, while in presence of 0·05 per cent gum tragacanth and 14 per cent sulphuric acid, a current efficiency of 97·1 per cent is obtained when using a current density of 60 amps. per square dcm. (560 amps. per square foot).

In commercial operation, with a solution containing as much as 275 grm. sulphuric acid, and as little as 32 grm. zinc per litre, when employing a current density of 100 amp. per square foot, at a voltage of 4·0, with lead anodes, zinc is deposited in a smooth form up to a thickness of ¼ inch with a current efficiency of 89 per cent. This output corresponds to 1·65 kw. hour per lb. of zinc, or 2·4 metric tons of zinc per kw. year of 8700 hours.

Good deposits of zinc can be obtained in presence of small amounts of iron and other impurities. In the case of iron this possibility is largely due to the fact that with a high-current density, on account of a delay in intermediate reactions accompanying the change from the ionic condition to that of free element, the proportion of iron separated together with the zinc is much less than the ratio these elements bear to each other in the electrolyte. The main advantages gained by this process are that a very exhaustive purification of the solution is obviated, a large saving of vat space and time for the deposition of a given amount of zinc is obtained by the abnormally high-current density. On the other hand, the power consumption is no higher for a given output than the slower methods, since, through the high conductivity of the electrolyte, on account of the large amount of free acid, electrode potentials of 4·0 to 4·5 volts are employed. The spent electrolyte, containing up to 28 per cent sulphuric acid, is in a suitable form for the leaching of roasted ore in a continuous process.

CONCENTRATING SMELTING OF ZINC.[1]

A method which should be appropriate for combining with an electrolytic extraction process for zinc is that practised at Colorado, U.S.A. High-grade complex zinc-lead copper ores are smelted directly without preliminary roasting in a blast furnace, when the main part of the precious metal values collect in a copper matte and the zinc and part of the lead is driven off as a fume and collected in a bag-house.

PROSPECTS OF PROCESSES FOR ELECTROLYTIC EXTRACTION OF ZINC.

The conditions essential for the successful operation of the electrolytic extraction of zinc from ores would appear to be the following :—[2]

---

[1] R. G. Hall, " Trans. Amer. Inst. Min. Engineers," 1917, 57, 713.
[2] Compare W. R. Ingalls, loc. cit. ; R. G. Hall, loc. cit.

1. Cheap electrical power such as from water power or large central steam-power stations.

2. For the treatment of complex ores such as the Broken Hill deposits which cannot be efficiently treated by ordinary fuel-heated processes.

3. For the treating of a high-grade ore that will give up a high percentage of zinc by lixiviation with sulphuric acid and containing a high value of silver and possibly lead. In the pyro-metallurgy of zinc, the extraction of silver is in the neighbourhood of 65 per cent, whereas by hydro-metallurgical treatment this should amount to 90 to 95 per cent.

4. A further consideration in the economy of these processes is the high grade of purity attainable with electrolytic zinc which makes it of particular value for purposes such as the manufacture of special brasses. According to the relation of supply and demand, electrolytic zinc commands a higher price than spelter.

Cost of Operation of Electrolytic Zinc Extraction Processes.[1]

The following costs are estimated for the treatment of an ore containing: Zn, 30 per cent, Pb, 15 per cent, and silver, 12 oz. per ton. The power required to deposit a ton of zinc is 3000 kw. hours measured at the vats, or 4000 kw. hours to include generator and other losses and energy for the general operation of the factory.

Total costs per ton (2000 lb.) of zinc on basis of \$5 = £1.

	£.
Power, 4000 kw. hours at 0·125d.	2·5
Labour, at 12s. per diem, and management . . .	2·5
Fuel and supplies	0·5
Roasting, including labour and fuel	1·0
Insurance, maintenance, and depreciation	2·0
Total	8·5
Freight charge at average rate of £3 per ton . .	3·0
Total	11·5

At a price of £30 per ton, this leaves a balance of £18·5 for the cost of ore and profit on zinc per ton. In the case of an ore with the same content of lead and silver, but with 25 per cent zinc, the cost of production would be increased to the extent of 4s. per ton of zinc. Complex ores of this composition are very plentiful, and are of

[1] T. French, " Trans. Amer. Electrochem. Soc.," 1918, **32**, 321.

no value for lead smelting by ordinary processes unless the zinc is first extracted when a product is left, which, after elimination of sulphur and zinc, has about double the percentage content of lead and silver and is in a very suitable form for smelting.

It is estimated that with lead at £20 per ton, and silver at 2s. 6d. per oz., the further profit in the recovery of these metals, allowing for freight and all charges, for treating the residue (about 2 tons) from ore yielding 1 ton of zinc, amounts to £12.

The British Dominions, notably British Columbia, Australia, Burma, and Rhodesia are very rich in zinc ores and complex zinc-lead-silver ores. The deposits are in many cases not far removed from large water powers, and should permit of favourable development by electro-metallurgical methods.

The Electrical Production of Zinc in Tasmania (cf. p. 436).— The electrical extraction of zinc is being developed in Tasmania, and has been attempted along three main lines, viz. :—

(*a*) Partial roasting of the ore followed by leaching, purification, and electrolysis of the solution. The chief difficulty encountered has been the accumulation of impurities in the electrolyte.

(*b*) The electrolysis of fused zinc chloride. The chief difficulty has been in the production on a large scale of a salt free from oxychloride.

(*c*) Smelting with coal in an electric furnace.

The method which it has been finally decided to use is that of the Anaconda Copper Company (cf. p. 313).

In 1919 a works was in operation at Risdon, near Hobart, which was producing zinc electrolytically to the amount of 11 tons daily.

Hydro-metallurgy of Copper.

A very extensive development has in recent years been made in the process of extracting copper from its ores by converting the copper of the ore, such as by roasting, into a form which is soluble in water or acids, leaching to form a sulphate or chloride solution, and separating the copper by electro-deposition. It is estimated that the total amount of ore now treated by this process exceeds 1,000,000 tons per annum. At the Rio Tinto Company in America, ore is submitted to heap-roasting and the sulphate formed is extracted by leaching with water. The solution is then electrolysed, using anodes of lead which become coated with peroxide, and during the electrolysis sulphuric acid accumulates in the electrolyte. In other processes the copper is brought into solution in the form of chloride. In continuous processes of leaching and electrolysis, the accumulation of large quantities of iron in the electrolyte is prevented by agitating the spent electrolyte with an excess of roasted ore when a large amount of iron is precipitated as ferric oxide and impurities such as manganese, arsenic, antimony, and bismuth are largely removed.

In other processes the sulphide ore is roasted to a mixture of sulphate and oxide and leached with a solution of sulphur dioxide when copper is obtained in solution as sulphate and sulphite. After filtration the clear solution is electrolysed with a voltage on each vat of 1·8, and a current destiny of 3 amps. per square foot when copper is deposited on the cathode and sulphuric acid formed at the lead anode. Electrolysis is continued until the copper is reduced to from 0·25 to 0·5 per cent when it is used for further leaching. The output of copper amounts to 6·6 tons per kw. year.

SECTION XV.

ELECTRICAL SMELTING OF COPPER AND TIN ORES.

EXPERIMENTS have at different times been carried out on the smelting of copper ores in furnaces of the type designed for the treatment of iron ores. Tests were originally made by Keller on behalf of the Chilian Government with a low-grade copper-ore.[1] The furnace used was the plural-hearth form designed for iron-ore smelting (cf. p. 168).

The reactions undergone in this process are identical with those obtained with the blast furnace, but the heat is derived from the passage of the current in place of from coke, which in the blast furnace is admixed with the charge. On account of the higher temperature attainable with electrical heating, more latitude is offered in the composition of the slags, which may be more refractory than those permissible in the blast furnace, or, for the same slag, the higher degree of fluidity permits of a more complete separation of slag and matte. A further advantage obtained by electrical smelting is that finely divided materials can be employed to a larger extent than is possible in a blast furnace. The operation of the furnace in Keller's experiments consisted in bringing the ore and charge to a state of fusion, whereby a matte is formed consisting of copper and iron sulphides which sinks to the bottom, and a fused slag which forms above. The slag and matte were then withdrawn through separate tapping holes. The furnace used had a capacity of about 500 kw. and was operated at 4750 amps. at 119 volts. The ore used contained 6 to 7 per cent copper, and 25 tons were melted in twenty-four hours giving directly a high-grade matte containing 50 per cent copper. The slag contained 27·2 per cent silica and 32·5 per cent iron, and was thus suitable for the manufacture of ferro-silicon. With regard to the economy of the process, the production of 1 ton of copper from 16 tons of a 6 to 7 per cent ore by the blast-furnace may be taken to require the consumption of 3·2 tons of coke, while the electrical energy required for the smelting of this

[1] C. Vattier, " Bull. de la Soc. d. ingenieurs Civ. de France," 1903 (2), 19.

amount of ore was found to be 1·25 kw. years.   The consumption of electrodes amounted to 75 kg. per ton of copper produced.   It would follow from these results that the electric furnace could successfully compete with the blast furnace in districts where coke is scarce and water power can be cheaply developed.   From the standpoint of fuel and power the economies of the two methods become equal, when 1·25 kw. year of electrical power can be generated for the same cost as 3·2 tons of coke.

It has been further shown by the results of extended experiments made by Keller that in the case of complex ores containing both copper and zinc, much greater scope is offered by the electric furnace for successful operation in presence of this metal on account of the higher temperature and wider range of slags possible.   The zinc oxide, being expelled in a comparatively concentrated form, is moreover suited for recovery.

Similar experiments to the above have been carried out successfully in the Stassano and Girod furnaces.   In the case of the former [1] a furnace consuming 4000 amps. at 110 volts had a capacity of 1 ton of ore per hour.

SMELTING OF COPPER ORES IN THE GIROD FURNACE. [2]

Experiments have been conducted in a Girod furnace on the smelting of an ore containing from 5·7 to 21 per cent copper oxide, 28 to 78 per cent silica, 4 to 13 per cent alumina, and 4 to 16 per cent ferric oxide.   The procedure adopted consisted in smelting directly to metallic copper with the use of either charcoal, coke, or anthracite as reducing agents.   The furnaces were provided with base electrodes of steel cooled by internal water circulation and vertical carbon electrodes made at the Girod works (cf. p. 223).   A power of 200 kw. was applied at from 30 to 150 volts.   The ore used is difficult to smelt in a blast furnace, but was conveniently treated in these experiments.   The slag produced containing 52 per cent silica was found to be sufficiently fluid at 1400° C. to allow the reduced copper particles to settle and coalesce, though a temperature of 1500° C. is required to enable the slag to flow freely.   A temperature of 1920° C. was found necessary to render the slag liquid when the highly-acid ore was melted without fluxes.

The pig-copper produced contained from 65 to 95 per cent copper and 1 to 21 per cent iron, the purity of the product being higher the lower the melting-point of the slag.

---

[1] Cf. Vanoy, " Elektrochem. Technik," 1904, 103.

[2] " Met. and Chem. Engineering," 1913, 11, 22.

In continuous operation, producing a slag with 52 per cent silica and 0·46 per cent copper oxide, a power consumption of 1000 to 1200 kw. hours per metric ton of ore was required. With an easily fluxible ore, the power consumption amounted to about 500 kw. hours The electrode consumption averaged 17·6 lb. per ton of ore and was operated with a current density of 25·8 amp. per square inch of electrode section. The furnace lining found most suitable consisted of fireclay with 80 per cent silica and 15 per cent alumina.

Experiments of Lyon and Keeney.[1]

It is found from experiments conducted by D. A. Lyon and R. M. Keeney that fine native copper concentrates of between 25 and 40 per cent copper (as oxide) can be smelted in an electric furnace of large capacity with a power expenditure of 750 kw. hours per ton of ore. A high grade of black copper is produced and the percentage of copper in the slag does not, with a suitable slag, exceed 0·50 per cent of the total copper charged. Other losses should not exceed 1·0 per cent, giving a total loss of 1·5 per cent of all the copper charged. The loss of copper by volatilisation is found to be high if the slag is much more acid than a monosilicate and if it contains much alumina. When operating at a low temperature on a monosilicate slag, the reduction of iron by the carbon electrode is not excessive or sufficient to result in a product containing less than 95 per cent metallic copper. With care, a product containing 98 per cent copper can be produced. A continuous type of furnace is recommended in which the mixture of ore and flux is charged at intervals into the top of a short stack above the smelting crucible. The charge is thus pre-heated before it reaches the smelting zone.

Ores and concentrates which cannot be successfully treated in a blast furnace are normally smelted in a reverberatory furnace, and on account of the low thermal efficiency of these furnaces should, as a rule, with economy be replaced by the electric furnace. An important advantage gained with the electric furnace over the blast furnace in the treatment of finely divided ore is that, in the former, the material can be smelted without first agglomerating or briquetting.

The Electric Blast Furnace or Combination of Electric and Pyrite Smelting.

In pyrite and semi-pyrite smelting, which is carried out in a blast furnace, the reaction temperature is produced by the heat of oxida-

[1] "Trans. Am. Inst. Min. Eng.," 1913, **47**, 233; "Met. and Chem. Engineering," 1913, **11**, 522.

tion of the copper sulphide, or partly by this means and partly by the addition of coke. It is proposed by Lyon and Keeney to employ a type of furnace in which electric heat is applied to replace the use of coke in pyritic smelting. The suggested construction of the furnace is that of a modern copper blast furnace with the usual tuyères, while below these, spaced equidistantly around the circumference, electrodes extend down into the crucible. By this system, in which the use of coke is obviated, the difficulties ordinarily encountered in pyrite furnaces through the presence of either a deficiency or excess of coke, are avoided. Heat is, moreover, introduced into the furnace without depriving the combustion zone of oxygen. A similar method has been proposed by S. B. Ladd.[1]

ELECTRIC SMELTING OF TIN-ORE.[2]

The electric smelting of tin-ore has been carried out as a commercial experiment in Cornwall with a plant designed by Härden and the Gröndall Kjellin Company.

The reduction of stannous oxide by carbon requires a comparatively high temperature, when considerable volatilisation of the metal takes place. With the reverberatory furnace a large loss of metal thus occurs through vapour carried away by the furnace gases. These losses may be largely reduced by the use of the electric furnace.

The tin-ore consists of stannic oxide, SnO_2, and is reduced by carbon in accordance with the equation

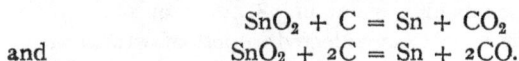

$$SnO_2 + C = Sn + CO_2$$
and
$$SnO_2 + 2C = Sn + 2CO.$$

A furnace of 40 kw. capacity was first applied. This was made of shaft form with a bottom chamber of magnesite bricks rammed with calcined magnesite and tar. A charging shaft rested on a vault or arch. The inner diameter of the hearth was about $5\frac{1}{2}$ inches, and the total height about 14 inches. Two lateral and inclined carbon rods about $2\frac{1}{2}$ inches diameter were used as electrodes and reached into the centre of the hearth about $2\frac{1}{2}$ inches from the bottom. The charge consisted of

Tin-ore.	20 lb.
Coke powder	5 ,,
Flux	5 ,,

The power expenditure in one series of experiments corresponded to about 4900 kw. hours per ton of metal.

[1] "Met. and Chem. Engineering," 1910, 8, 7.
[2] J. Härden, "Met. and Chem. Engineering," 1911, 9, 453.

A metal was obtained with over 90 per cent tin and contained very little iron and no arsenic. The recovery amounted to 90 per cent.

These experiments were continued on a larger scale with a unit of 100 kw. Three-phase current was applied, and a furnace of the shaft type was used with three electrodes 8 inches × 8 inches square and 4 feet 6 inches long, cooled by water jackets. The charge was arranged to form a cone-shaped heap on the hearth covering the reacting zone. Heating was brought about by resistance, the formation of an arc being avoided. Arrangements were made to prevent the escape of volatilised metal, and only 0·5 per cent or less of the tin was lost through volatilisation. Tap-holes on different levels were provided for drawing off the metal and slag separately. When frozen at the tap-hole, this was cleared by means of an auxiliary electrode or an oxy-hydrogen flame.

At the beginning of a smelting process the empty furnace was pre-heated by a fuel fire. A charge was then admitted of a mixture of 100 kg. of ore with 14 kg. of anthracite, while a current of about 1000 amperes per phase at 60 volts was applied. As the reaction commenced the volts were gradually lowered to the working condition of 40 volts and 2500 amps. which was steadily maintained during the whole run.

By the addition at intervals of certain reagents, the sulphides and arsenides of iron were caused to enter the slag instead of alloying with the tin, and enabled a metal of 98 per cent purity to be produced. The liquid metal was collected in pots of wrought-iron, and a stream of compressed air passed through when metal of 99·7 purity resulted.

It was found that a slag containing as low as 0·25 per cent of tin could be produced. The output of this furnace amounted only to about 6 or 8 cwt. per day. The power consumed was 3000 kw. hours per ton of metal.

At the other extreme the operation could be so conducted, by altering the composition of the charge, that a slag containing 17 to 19 per cent of tin was produced. In this case the output increased to 1 ton 11 cwt. per day with a power consumption of 1300 kw. hours per ton of metal.

The method of operation found to give the greatest economy with an ore containing 57 per cent of tin was to obtain a yield of 96 to 98 per cent with a power consumption of 2220 kw. hours for the reduction and an electrode consumption of 28 lb. per ton of metal, while the amount of carbon necessary for the reduction amounted to 14 per cent.

A number of advantages are offered by the electric furnace method, and in cases where power can be obtained at a reasonable rate it appears quite feasible to smelt electrically. Charcoal may also be used as the reducing agent.

Electrical Reduction of Tin Dross.—An electric furnace has been applied for the recovery of tin from tin dross.[1]

In this case the furnace enclosure contains a lower stationary carbon electrode and a movable upper electrode. Slag is first admitted to the furnace and heated internally by the current passed between the carbon electrodes. The dross is placed on top of the slag and thus comes into contact with the latter only at the point of reduction. The liberated gases filter through the dross while any tin oxide which is volatilised is condensed in the colder portion of the dross. The globules of tin produced pass downwards through the slag and lose most of the impurities so that very little refining of the product is necessary. The tin is tapped from time to time. The furnaces used are about 44 kw. capacity and are operated at the beginning at a voltage of about 80, which towards the end decreases to 45 to 50 while the current is maintained constant.

[1] R. S. Wile, " Trans. Amer. Electrochem. Soc.," 1910, 18, 205.

SECTION XVI.

CARBORUNDUM, ALUNDUM, AND ARTIFICIAL GRAPHITE.

CARBORUNDUM is the name which has been generally adopted for the electric-furnace product, consisting of carbide of silicon in a crystalline condition, which has now acquired very large industrial importance.

Silicon carbide was first prepared by Despretz in 1849, in experiments in which carbon electrodes were mixed with various oxides to ascertain if a lowering in the melting-point of carbon could be obtained. Sand, when used as an admixture, was found to volatilise and lead to the formation of hard crystals which would scratch ruby or chrome-steel.

Moissan, in 1893, described several ways of preparing this carbide :—

1. By heating silica and carbon in an electric furnace.

2. Direct union of silicon and carbon by solution of the latter in the former at the temperature of the blast furnace.

3. Direct union of silicon and carbon in presence of molten iron.

4. Union of the vapours of silicon and carbon.

Acheson, in 1891, conducted experimental work with a view to preparing crystallised carbon. It was thought that alumina or aluminium silicate might form a suitable solvent for carbon at a high temperature. A blue crystalline substance of great hardness obtained in this way was considered to be a compound of carbon and alumina, and accordingly given the name carbo-corundum or carborundum. Acheson immediately engaged in the manufacture of this product industrially, and on ascertaining its true composition, replaced corundum by sand for its preparation. In the earliest furnaces the heating was brought about by means of an arc formed between carbon electrodes in the middle of the charge, and then gradually withdrawing the electrodes so that the current passed through the charge. In order to obtain a better regulation of the heating, the method was changed by placing a rod of carbon in

326

between the electrodes and afterwards a core of coke, to conduct the current through the furnace. However, on account of an earlier patent held by the Cowles brothers on the use of granulated coke as a heating medium in electric furnaces, it was for some time found necessary by the Acheson Company to use rods of carbon in place of coke as the heating medium.

The Acheson Company began operations at Monongahela, Pa., U.S.A., using first a furnace of 6 kw. capacity, which was increased to 55 kw. then 100 kw. With this last a production of 136 kg. of carborundum was obtained per twenty-four hours, representing a power expenditure of 17·6 kw. hours per kg. This enterprise was followed by the formation of the Carborundum Company at Niagara Falls, in 1895, and a factory was also erected at Barmen, in Switzerland, and at Dresden.

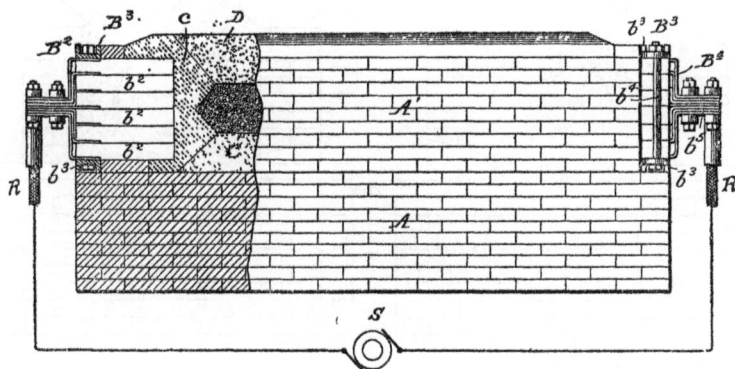

FIG. 185.

The economy of the process was found to be raised by the adoption of larger units. Thus, with the original size units at Niagara Falls of 746 kw. or 1000 h.p., a production was obtained of 3150 kg. of carborundum, in the crystalline form (excluding the amorphous material), in thirty-six hours, which corresponds to a power expenditure of 8·5 kw. hours per kg. carborundum.

The furnaces of 1000 h.p. originally in use at Niagara Falls are built up as shown in Figs. 185 and 186, showing vertical sections taken lengthwise and crosswise respectively.

The case of the furnace is made of fireproof stone and brick. The bed is of stone (granite), as careful construction is necessary, on account of the possibility of the production of silicon locally, which runs to the base in a molten condition, and is highly reactive.

Contact with the stone bed is avoided by always providing a sufficiently thick layer of unaltered material underneath, and below this a layer of quartz *i*, while below the quartz is a layer of sawdust *h*, which ensures electrical insulation and porosity so that gases will not accumulate underneath. The larger the furnaces the greater the care that must be taken to allow gas generated to escape freely.

The core C has a diameter of 40 to 54 cm. and is composed of pieces of coke of about 2 cm. diameter. The current is led in by twenty-five carbon electrodes, placed in five horizontal rows *b*. Copper plates are placed in between each row of carbons, to ensure contact with the current terminals. The junction of the electrode plates with the refractory walls of the furnace is made air-tight by

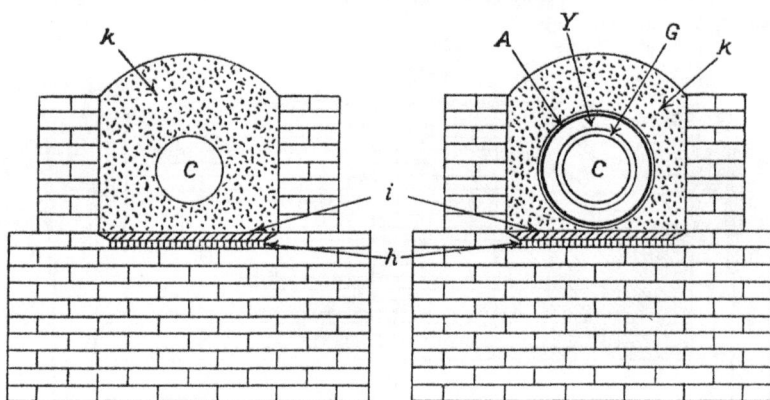

FIG. 186.   FIG. 187.

applying a mixture of cement, graphite, and coke. The carbon electrodes which project into the air are protected by coating with a layer of amorphous silicon carbide mixed with water glass. The masonry around is composed of material containing a large amount of silicon carbide. The total length of the furnace (1000 h.p. size) is about 23 feet, and the internal dimensions: length, 16½ feet, breadth, 6 feet, height, 5 feet 9 inches. The brick walls of the furnace are loosely laid, without mortar, and for the most part rebuilt after each run. The carbon electrodes are each 86 cm. long and 10 × 10 cm. cross section.

The charge is made up of the following percentage composition :—

FIG. 188.

(From Borchers' "Electric Furnaces".)

Sand										52·2
Coke										35·4
Sawdust										10·6
Salt										1·8
										100·0

Sawdust is added to confer porosity and enable the escape of gases, whilst the salt facilitates the removal of iron and aluminium as volatile chlorides.

The raw materials used are in as pure condition as practicable. The carbon used is in the form of anthracite, petroleum coke (retort carbon), or the best furnace coke, which contains 10 per cent ash. If too high in sulphur, noxious gases are evolved during the operation of the furnace. The carbon is crushed and sieved so as to be between certain limits of size. The powdered material is not used in the charge, but utilised for making a good connexion between the electrodes and core.

Sand is used in the form of quartz or white sand, containing 99·5 per cent SiO_2, which is pulverised so as to give lumps 2 to 4 mm. diameter. It is essential that the ingredients of the charge should be very well mixed. During the filling of the furnace, iron plates are inserted so as to leave a clear space around the electrodes, and similarly around the central channel to be occupied by the heating core. The space around the electrodes is afterwards filled in with finely divided coke. Coke which has been used in a previous heating conducts very much better a second time, on account of graphitisation and through having lost impurities and occluded gases. A mixture of old and new coke is usually employed, and a layer of paper generally used to separate the core from the surrounding charge.

During the progress of the heating the resistance of the furnace falls considerably, and a regulation of the voltage is accordingly necessary. Current is taken from the Power Company at 2200 volts, and is transformed from 75 to 210 volts according to the resistance of the furnace.

On beginning to heat, gases from the decomposing sawdust are first evolved. This is soon accompanied by the generation of carbon monoxide, which burns with a blue flame in the crevices of the furnace walls as seen in Fig. 188, giving a view of a furnace in operation. After about two hours from the start the furnace is completely enveloped with carbon monoxide flames, which last throughout the whole process (thirty-six hours). A sinking in of the top takes place to some extent due to shrinkage, and occasionally a sudden eruption of vapour, known as "blowing," takes place through the sudden formation of silicon, which is partly volatilised, and on account of insufficient porosity raises a portion of the charge. The formation of silicon takes place either through the core being too hot and decomposing the carbide, or where excess of silica is present. If the blowing becomes too violent the current is interrupted for a time and fresh charge pressed in. After several hours the furnace reaches its maximum conductivity. From this stage onwards the current only increases slightly due to the formation of the conducting cylinder of carborundum. The 1000 h.p. furnaces at this stage usually take 10,000 amps. at 75 volts. The roof of the furnace reaches a dull red heat. The end of the reaction is generally indicated by a gradual falling off of the gases evolved, and by a yellow coloration of the flame due to silicon. At the conclusion of the heating, means are taken to cause as rapid cooling as possible. The furnace is dismantled by taking down the side walls to about half way; loose unaltered mixture is removed, and the furnace then allowed to cool for several hours. A cross section showing the furnace products after heating is illustrated in Fig. 187. The core C is graphitised, and surrounded by an arrow zone of graphite G resulting from decomposed carborundum and retaining its crystalline form. Outside this is a zone of crystallised carborundum Y 35 to 40 cm. thick. Beyond this is a layer of amorphous carborundum at A, and outside this the unaltered charge at K. The core is removed, and after sieving off the fine powder, and mixing with some new material, is used for another run.

A type of furnace subsequently introduced has double the above capacity, or 2000 h.p., and is 30 feet in length and 12 feet in width. The end walls containing the electrode terminals are built of concrete and the side walls are composed of several sectional units, each section being built of firebrick set in iron frames. This facilitates the discharging and reloading of the furnaces and saves considerable labour. The furnaces are loaded by means of an overhead conveyer. Each furnace consumes 1600 kw. with a maximum current of about 20,000 amperes, and produces about 15,000 lb. of crystalline carborundum in each run. The transformers receive the primary current at 2200 volts and 25 cycles and are provided with regulators which, as before, vary the secondary voltage from 200 to 75 volts.

Analysis of Product.—The product obtained has the following composition :—

	Crude Crystals.	Washed Crystals.	Amorphous Carbide (Purified).
Si	63·5	69·1	65·4
C	34·0	30·2	27·9
FeO . . .	1·5 ⎫	0·5	5·1
Al$_2$O$_3$	0·9 ⎭		
CaO	0·1	0·1	0·3
MgO	—	—	0·2

The crystals of carborundum, if allowed to cool away from air, are green or dark grey. If cooled in air, however, they have a beautiful prismatic surface. The crystalline masses are crushed in a machine and washed for some time in dilute sulphuric acid to remove admixed impurities, then with water, dried, and finally graded by sieves.

Efficiency of Carborundum Furnace.—A calculation of the thermal efficiency of the carborundum furnace has been made by Prof. J. W. Richards, by taking into account the specific heat of the materials and the heat absorbed in the chemical reaction. The results showed that 42 per cent of the applied electrical energy is consumed in raising the materials to the temperature of reaction, and 34·5 per cent in the endothermic chemical reaction, thus giving a thermal efficiency of 76·5 per cent. The balance, 23·5 per cent, represents the loss caused by radiation, and heat necessary to expel water vapour and decompose the sawdust, etc. A certain amount of heat, not accounted for here, however, is derived from the combustion of the carbon monoxide on the outside of the furnace.

The 1000 h.p. furnaces producing about 3150 kg. silicon carbide liberate some 3400 kg. (or about 3200 cubic metres) carbon monoxide during each run of thirty-six hours.

Temperature of Formation and of Decomposition of Carborundum. —An estimate of the temperatures which are obtained in the different zones in the carborundum furnace was made in a laboratory furnace with the use of an optical pyrometer.[1] The reduction of silica by carbon was found to begin above about 1615° C., and the transition from amorphous into crystalline silicon carbide between 1920° and 1980° C., whereas decomposition of the carbide into silicon and carbon took place between 2200° and 2240° C.

Measurements made at low gaseous pressures [2] have shown that at a pressure of about 10 mm. the reduction of silica by carbon

---

[1] Tucker and Lampen, "Journ. Amer. Chem. Soc.," 1906, **28**, 853.
[2] Greenwood, "Trans. Chem. Soc.," 1908, **93**, 1483 and 1496.

begins at 1460° C., whereas in presence of iron the reduction point is, by catalytic action, lowered to 1200°, and with manganese to 1100°. The presence of metals similarly accelerates the decomposition of silicon carbide.

The union of silicon and carbon in a vacuum has been found[1] to commence at temperatures between 1250° and 1300° C., and to proceed rapidly above 1400° C.

Uses of Carborundum.—By far the main application of carborundum is as an abrasive, and for this purpose it is made up chiefly in the form of grinding wheels of various sizes and forms. In this process the substance is carefully graded, mixed with kaolin and felspar, and then baked in an oven similar to that used in porcelain manufacture. The fine crystals are used for the preparation of carborundum cloth and paper. Carborundum is of great hardness, will scratch ruby or corundum, and readily abrade chrome-steel. It is, however, somewhat brittle, and will not withstand the application of great pressure. The substance is chiefly suited for polishing cast-iron and steel, sharpening saw-teeth, and for working copper, brass, and other metals, also for marble and even granite. However, for grinding materials of high tensile strength, such as steel and malleable iron, a tougher abrasive is preferable.

Silicon carbide, either in the crystalline or amorphous form, has also a large application as a refractory, and is used for the lining of various types of furnaces where high temperatures are developed. In addition to its heat-resisting qualities, carborundum has a high thermal conductivity which makes it of value in the construction of furnace muffles which are required to transmit heat.

Refractory bricks are supplied under the name of "Carbofrax" which are composed of carborundum crystals bonded together with a small percentage of refractory clay. These can be used at very high temperatures, and through possessing a low coefficient of expansion, can be submitted to sudden changes of temperature without cracking. A further material which has valuable heat-resisting properties, known as "Refrax," which is made up into bricks, is composed of carborundum crystals which are fused together in an electric furnace at a temperature of 2100° C., the use of any added bonding material being avoided. This material is also used for the preparation of tubes to serve as the outer protection of thermocouples. These tubes are generally built into the furnace, such as those for steel and glass and annealing and tempering furnaces,

---

[1] Pring, "Trans. Chem. Soc.," 1908, 93, 2101.

the thermo-couple itself being further protected by an iron, silica, or porcelain casing. A further application of carborundum is for a material known as " Silfrax ". This is used in the form of tubes which consist of graphite, the surface of which has been converted into carborundum for a thickness of $\frac{1}{32}$ inch to $\frac{1}{16}$ inch. This coating gives the tube very high resistance against oxidation without impairing any of the other properties of graphite. The tubes are very resistant to rapid changes of temperature, and suited for sudden immersion in molten metal in making temperature estimations. A material similar to the above was formerly prepared in Germany, under the name of "Silundum" by exposing the carbon or graphite to the action of silicon vapour at a high temperature.

Carborundum is used to some extent in place of ferro-silicon in steel manufacture for deoxidising the metal, and is valuable on account of the complete absence of sulphur and phosphorus. The carbide decomposes in presence of iron at a lower temperature than in air. It has also been applied for the reduction of metallic oxides in accordance with the reactions

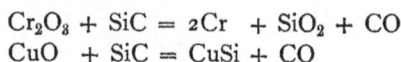

$$Cr_2O_3 + SiC = 2Cr + SiO_2 + CO$$
$$CuO + SiC = CuSi + CO$$

the latter reaction taking place at a temperature of 800° C.

Secondary Reactions in the Carborundum Furnace.—The formation of carborundum takes place in accordance with the equation

$$SiO_2 + 3C = SiC + 2CO.$$

The element silicon is obtained by lowering the carbon content of the mixture, also by reaction between the carbide and silica, as in the equation

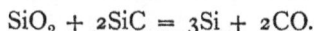

$$SiO_2 + 2SiC = 3Si + 2CO.$$

Compounds of carbon, oxygen, and silicon have been known for some time, thus carbon dioxide, when passed over heated silicon, reacts as follows :—

$$3Si + 2CO_2 = SiO_2 + 2CSiO.$$

Compounds have also been obtained which have been given the formulæ $SiCO_3$, Si_2C_2O, and $Si_2C_3O_7$. In the carborundum furnaces a substance known as "Siloxicon" is formed at a temperature several hundred degrees below that needed for carborundum, and consequently is yielded in a zone outside the layer of amorphous carborundum. The material has been given the formula Si_2C_2O, though it is not clear that this is not composed of a mixture of SiO and

SiC, and at a temperature of 1900° to 2000° C. decomposes into Si, CO, and SiC.

Silicon Monoxide, SiO, which may also be formed during the manufacture of carborundum, is now prepared commercially in special electric furnaces, and occurs as a red powder, which is used as a paint pigment, for polishing, and as a reducing agent.

Silicon is prepared in special furnaces by the Carborundum Company, from a charge of silica and carbon in suitable proportions, and by arranging for the product to flow from the heating core.

The reduction of silica to give silicon was first effected by resistance furnaces, but at present suitably designed arc furnaces are found preferable. According to A. Stansfield,[1] the output of silicon at the Carborundum Company at Niagara Falls amounts to 1500 tons per annum. The material is used in steel making and in the chemical industry owing to its power of resistance to acids. Silicon has been employed instead of aluminium for the reduction of metals by the Goldschmidt process and has also been applied for the production of hydrogen.

ALUNDUM.

Alundum, or artificial corundum, has an extensive use as an abrasive, to replace natural corundum. The preparation of this material was first undertaken by the Norton Emery Company of Worcester, Mass., which erected a works for its manufacture on a large scale at Niagara Falls. The process employed consisted in fusing bauxite (alumina) in an electric furnace and allowing to cool slowly. The material is ground, and moulded to form grinding wheels, to which in some cases, natural corundum is added. The furnace product is less brittle than natural corundum, through its less crystalline structure, and is especially adapted for the grinding of hardened steels and malleable iron.

A type of furnace which was designed by A. C. Higgins[2] in 1904, for the preparation of alundum is shown in Fig. 189. The two electrodes 10, 11 are held by means of the clamps 12, 13 to the conductors 14, 15. The furnace crucible consists of an iron shell 16, fitting into the base ring 17 which is mounted on the hearth or base 18. While the electrodes are immovable the base of the furnace can be raised or lowered according to the position of the molten

[1] "Cassier's Magazine," 1916, **49**, 386.

[2] U.S. Pat. 775,654; "Mineral Industry," 1911, **20**, 31; "Electrochem. and Metall. Ind.," 1909, **7**, 223.

charge. A hood 20 with suitable openings for feeding the furnace and for observation is provided to confine the gases and to direct their escape up the chimney 21. During the heating of the furnace, cooling of the case is provided by allowing a stream of water to flow over its outside surface. The operation of the furnace consists in first raising the base plate until contact is made with the electrodes and an arc formed. Alumina is then fed into the furnace and the base gradually lowered as the fusion proceeds until the cylindrical case is filled when the furnace contents are removed away from the

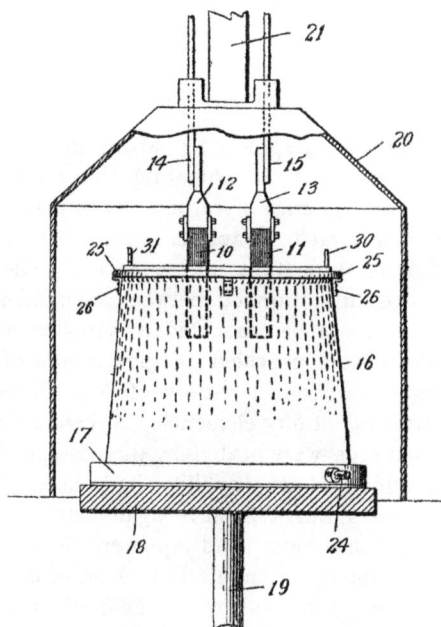

Fig. 189.

electrodes and the iron case withdrawn from the charge, which is facilitated by its conical shape. After removal of the fused product, the case is replaced and the furnace assembled for a further run.

The above type of furnace has been installed by the Norton Company, the crucible being 4 feet in diameter and 5 feet high and giving an ingot of alundum weighing 2·5 tons. A furnace of similar design employing four electrodes has also been adopted.

Aloxite.—The manufacture of fused alumina was commenced by the Carborundum Company at Niagara Falls in 1910, and is placed on the market under the name of " Aloxite ". The material is claimed to possess an irregular crystalline fracture which causes it to break into fragments, which at all times offer a sharp cutting edge in contact with the work. The hardness is stated to be greater than that of ruby combined with a high degree of toughness.

Fused alumina is receiving an increasing application as a refractory on account of its high melting-point (about 2050° C.) and for this purpose is moulded from the finely powdered substance in the form of crucibles, tubes, etc., and baked in an electric furnace. This refractory withstands sudden temperature changes without cracking.

GRAPHITE.

The production of graphite by artificial means was first obtained by Despretz, when investigating the effect of high temperatures on different varieties of carbon. The materials examined were retort carbon, anthracite, graphite, sugar charcoal, carbon from the decomposition of turpentine, and diamond. The heating was effected by means of an arc formed between electrodes enclosed in an iron case, which could be evacuated or supplied with gas under pressure. Windows were provided to give a view of the interior. The current was taken from a battery of 600 Bunsen elements, arranged in twelve rows of fifty elements. A bending of the carbon was observed when rods were heated by the passage of the current and it was thought that its fusibility had been demonstrated. Powdered carbon was thought to have agglomerated into globules. With anthracite, intumescence and apparent local graphitisation occurred. Hydrocarbons, when used as a surrounding gas, were found to decompose. Diamond was converted into graphite. It was observed in general that each variety of carbon loses its hardness in proportion to the length of time for which it is exposed to the high temperature and is finally converted into graphite.

Berthelot conducted the most extensive series of experiments on the different varieties of carbon, and found a basis of distinguishing and separating the different allotropic forms. The method consists in repeatedly heating the finely divided material for several hours with a mixture of nitric acid and potassium chlorate. Amorphous carbon is transformed into a brown substance which is soluble in water. Graphite is finally converted into a yellow substance which Berthelot termed graphitic oxide. If this is heated to about 250° C., it decomposes with intumescence and is transformed into a deep black, flocculent substance which Berthelot named pyrographitic oxide. This, on further treatment with the reagent, behaves like amorphous carbon. Diamond undergoes no change by this treatment. Berthelot proposed to give an exact definition for graphite by applying the term to those varieties which yield graphitic oxide. Coke and carbon from the decomposition of hydrocarbons, retort carbon, and anthracite were found to contain no graphite, while considerable amounts were obtained by the decomposition of carbon tetra-chloride, and by the action of sodium on sodium carbonate and chlorine on boron carbide.

Moissan investigated different varieties of carbon and found great differences in their densities, oxidation temperatures, and reaction temperatures with chromic acid and other oxidising agents. It was not found possible to obtain pure carbon. Moissan described three methods for preparing graphite :—

1. Heating carbon in an electric furnace.

2. By separating from solution in metals by cooling.

3. By dissolving carbon in iron, and then throwing out as graphite by the addition of silicon.

By cooling rapidly a solution of carbon in iron, which is saturated at a high temperature, and then plunged into water or molten lead, solidification of the outside is obtained and a high pressure subsequently generated in the interior. In this way the production of microscopic diamonds resulted.

Girard and Street[1] devised and patented a furnace for the preparation of graphite which was mainly applied for the conversion of rods of amorphous carbon with a view of applying them for use as electrodes. In this apparatus (Fig. 190) the carbon rod arranged vertically was slowly passed through holes in one or two carbon slabs arranged horizontally which were connected with the source of current. An arc was formed between these poles and the surface of the carbon rod undergoing treatment. The arc was caused to rotate by means of a magnetic field placed outside the furnace so as to heat uniformly all portions of the surface. The atmosphere of gas could be selected as desired. In hydrogen the conversion to graphite seemed to take place rather more readily than in air. An admixture of about 2 per cent of the oxides of silicon, boron, or iron, also facilitated the change. The conductivity of electrodes treated in this way was found to increase in the ratio of 1 : 4, and the density from about 1·98 to 2·6, while a content of graphite amounting to 85 per cent was obtained.

---

[1] " Manuel d'Electrochimie," 1898, 447.

The resistance of these electrodes against decomposition when employed in the electrolysis of fused alkali was very largely increased. In a different type of furnace, in which the electrode was replaced by a heated carbon tube, graphite was prepared from granulated carbon.

The process of Girard and Street was worked in France by the Le Carbone Company at Levallois-Perret, near Paris.

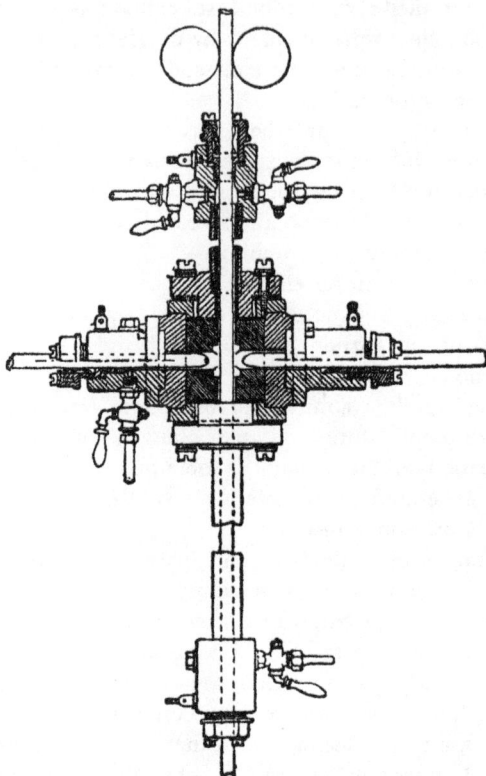

Fig. 190.

THE ACHESON PROCESS.

In the early experiments on the preparation of silicon carbide in the electric furnace, it was noticed by Acheson that graphite was obtained in the part of the furnace which had been taken to a very high temperature. This graphite possessed a crystalline form similar to that of carborundum indicating that its origin lay in the decomposition of the carbide in accordance with the equation

$$SiC = Si + C.$$

In normal operation of the carborundum furnace the layer of graphite occurs between the central heating core and the zone of carborundum crystals (cf. Fig. 187). This graphite is found to possess a high degree of purity, the impurities being more or less thoroughly removed by volatilisation according to the temperature employed and the duration of the heating.

In an exhaustive investigation made by Acheson on the production of graphite, it was found that the presence of admixed oxides, such as those of silicon, aluminium, magnesium, calcium, or iron, played an important part in the process, and indicated that the reaction took place through the intermediate formation of carbides.

Furnace Construction.—In the process which was installed at Niagara Falls the furnaces employed are of similar design to those used in the manufacture of carborundum. The two different types are—

1. For the graphitisation of masses of carbon, usually in the form of anthracite coal ; and

2. The graphitisation of moulded carbons, such as electrodes.

1. For the preparation of bulk graphite the furnace, of 800 kw. capacity, is 9 metres long, and 50 × 35 cm. cross section. A core of granulated coke or carbon plates extends along the furnace, makes contact with the end electrode plates, and is surrounded by the charge. The walls of the furnace are constructed of stone lined with carborundum to protect from fusion. The charge generally consists of anthracite coal, which, with a content of about 5 per cent ash (SiO_2, Al_2O_3, and Fe_2O_3), is specially suitable for the production of graphite. The charge is covered with a mixture of sand and coke to protect from oxidation. It is necessary that the temperature should be taken above the volatilising point of the elements other than carbon present. The resistance of the furnace is at first high, and as the heating and graphitisation proceed gradually falls. A quantity of carbon monoxide is generated which at first burns with a flame-coloured yellow by the volatile impurities, and afterwards with a blue flame. At a later stage the flames diminish and again become coloured yellow, indicating the decomposition of carbides and volatilisation of silicon. The duration of the heating depends on the grade of graphite required.

It is important for many purposes that the graphite should be quite free from sulphur. To safeguard this, it is necessary to ensure that this element is absent from the carbon taken. A higher grade of graphite is prepared by taking retort carbon, petroleum coke,

22 *

or charcoal, and after mixing with the necessary amount of ash giving the charge a more prolonged heating.

2. *Graphitisation of "formed" Articles.*—Moulded articles, such as electrodes, crucibles, dynamo brushes, etc., are prepared of artificial graphite by taking finely divided carbon such as retort carbon, mixing with suitable oxides, such as silica and iron oxide, and a binding material such as tar or sugar molasses. After thoroughly incorporating to form a thick paste, the mixture is compressed into a mould of the desired form and then forced or

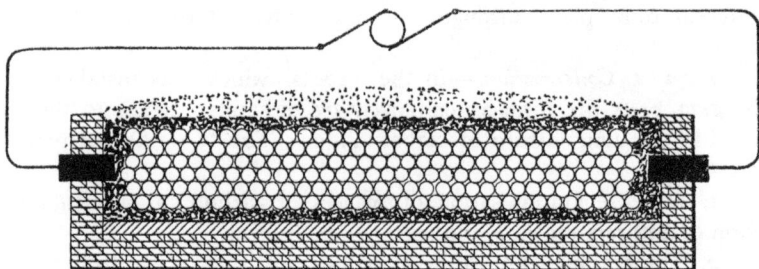

FIG. 191.

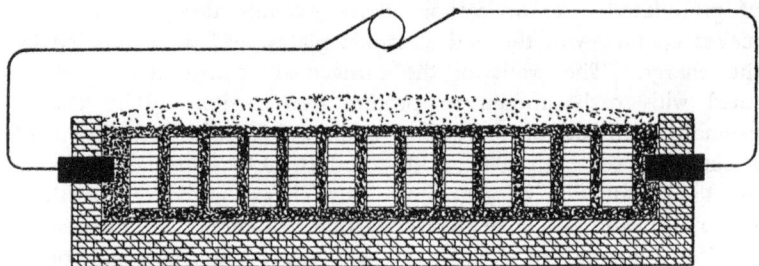

FIG. 192.

squirted from the open end by hydraulic pressure. The articles are then stoved at a moderate temperature, which renders them compact, and arranged in an electric furnace as shown in Figs. 191 and 192, which illustrates the arrangement of a furnace for the treatment of electrodes of either rectangular or circular cross section. The base of the furnace is composed of stone, covered with a layer of suitable refractory material. The end walls are composed of stone and hold the electrodes which consist of carbon rods and are in connexion with the source of current. On the refractory bed of the furnace a layer of granulated coke about 5 cm. thick is placed, and above this the electrodes are arranged in heaps transversely to the longitudinal axis of the furnace. The spaces in between the layers of electrodes are about $\frac{1}{8}$ of the breadth of each heap. These spaces and a layer above are filled with granulated coke consisting of pieces about 2 mm. thick. The space between this layer of granulated coke and the walls of the furnace is filled with a refractory material, consisting of a mixture of coke and sand. During the filling the coke core and the refractory mixture are kept separate by the interposition of plates of sheet iron. A layer of the refractory mixture is finally placed on top.

The overall length of the furnace is about 30 feet. The current terminals each consist of nine carbon rods covered on the outside of the furnace with brass caps, and not provided with water-cooling.

At the beginning of the heating the resistance of the furnace is high. With a 1000 h.p. unit a voltage of 200 is applied, and a current of about 1400 amps. passed. With the progress of heating the resistance falls at first rapidly and later more slowly, until it finally becomes constant, denoting completion of the change into graphite. At this stage the current passed amounts to about 9000 amps. at 80 volts. The duration of each run is from twelve to twenty-four hours according to the degree of graphitisation required. Small articles to be graphitised, such as thin carbon rods, are placed inside larger tubes for protection.

For most purposes it is found desirable to leave a proportion of the carbon unconverted, the product thus obtained being harder than pure graphite.

Yield of Furnaces.—Units of 1000 h.p. in a run of twenty to twenty-four hours give 6 tons of bulk graphite from anthracite, the product containing less than 2 per cent of soluble carbon. With formed articles the yield is 3 to $3\frac{1}{2}$ tons.

It has been estimated by J. W. Richards that in these furnaces about 82·5 per cent of the energy employed is consumed in raising the charge to the graphitising temperature and 17·5 per cent radiated during the run. A heat evolution of 8 kilogram-calories per kilo, which accompanies the change of amorphous carbon into graphite, amounts to about 10 per cent of the heat supplied by the current. This item, however, is approximately balanced by the heat absorbed in volatilising the 5 to 10 per cent impurities (ash) in the charge.

Application of Acheson Graphite.—The value of electric furnace graphite lies in its high resistance to oxidation, not alone at high temperatures, but also at low temperatures, e.g. to oxidising agents such as nitric acid. Graphite can be used as electrodes in the electrolysis of fused alkali and aqueous solutions. Its electrical conductivity is some four times as great as that of the densest form of retort carbon. Articles made from natural graphite which it is necessary to powder and agglomerate with some bonding agent,

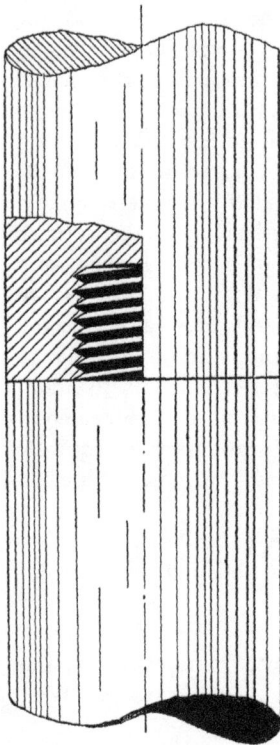

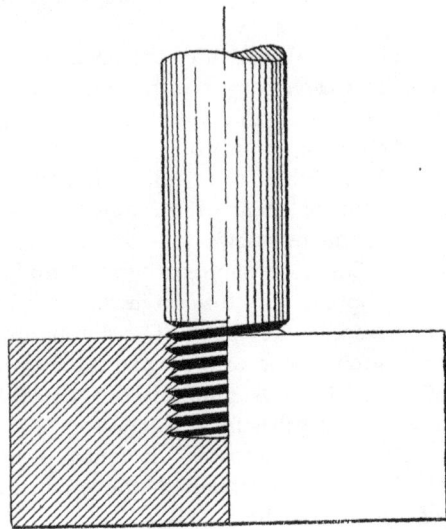

FIG. 194.

FIG. 193.

through not being homogeneous, do not possess these favourable properties. A further valuable property of furnace graphite is the readiness with which it can be machined, giving accurate threads by means of which it can be connected up for structural purposes. Examples of objects of graphite which have been connected in this manner are seen in Figs. 193 and 194.

Unctuous Graphite.—A soft or unctuous form of graphite is prepared in special furnaces arranged with a starting core of graphite rods. The charge consists of carbon together with ash or silica in a larger amount than that employed in the manufacture of ordinary graphite. The heating is more prolonged and the temperature also taken higher. The variety of graphite thus produced has valuable properties as a lubricant as it does not coalesce under pressure.

Colloidal Graphite.—A deflocculated or colloidal form of graphite is prepared by treating the above unctuous variety with a solution of gallo-tannic acid. A colloidal form of graphite is thus produced which remains permanently suspended. In a medium of either water or oil the material is applied as a lubricant under the trade names of "aqua-dag" "or oil-dag".[1]

PHOSPHORUS.[2]

The preparation of phosphorus is now conducted almost exclusively by an electric furnace process. The necessity of protecting the vapour of this element from air during distillation from the reacting materials makes a method in which the heating is applied internally by electrical means particularly suited for this operation.

Readman Process.—This method was introduced in 1889 and was brought into commercial use in most countries. Bone-ash or crude phosphoric acid is mixed with powdered coal or charcoal, or mineral calcium phosphate is roasted, crushed, and mixed with charcoal and silica or some basic salt. The mixture is reduced in a continuously operated electric furnace in a reducing atmosphere by passing the current from carbon electrodes through the mass which is heated to incandescence. The silica combines with the calcium to form a calcium silicate slag, and phosphorus distils and is evolved together with carbon monoxide. The distillation begins at 1150° C. and requires a temperature of 1400° to 1500° C. to complete the process. The chemical reaction is given by the equation

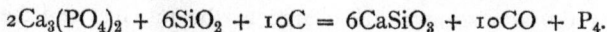

$$2Ca_3(PO_4)_2 + 6SiO_2 + 10C = 6CaSiO_3 + 10CO + P_4.$$

Irvine Process.—The Readman process was modified by the Irvine furnace in 1901. The charge used is the same as in the earlier method, although either aluminium or calcium phosphate can be used with the silica or basic salt flux. The two carbon electrodes are suspended vertically from above, and to commence the heating coal is used to connect the path of current.

After the charge melts, the slag forms on top and serves as a

[1] "Electrochem. and Metall. Ind.," 1907, **5**, 452.
[2] G. W. Stose, "Electrochem. and Metall. Ind.," 1907, **5**, 407.

conductor for the current between the electrodes. The fusion is continuous, and the excess of slag is tapped off gradually so as not to expose the ends of the electrodes.

The Readman process originated at the Albright and Wilson factory at Oldbury, England. In 1897 this company built a factory at Niagara Falls under the name of the Oldbury Electro-chemical Company, and the Readman process was installed with an initial power consumption of 300 h.p. The Irvine process was later adopted by which 80 to 90 per cent of the phosphorus is reported to be extracted from the raw material consisting of a high-grade phosphate rock. The furnaces were made in units of 50 h.p., each with a capacity of 170 lb. of phosphorus daily or 12,000 kw. hours per ton. The Oldbury Electro-chemical Company now employs a total of 8600 h.p. for the manufacture of phosphorus at Niagara Falls.

Landis Furnace.—A furnace which has been devised by G. C. Landis[1] is described as shown in Fig. 195 as consisting of a rectangular enclosure contained in a metallic casing B, and provided with a lining A, formed of non-absorbent or vitrified bricks cemented together by a mixture of silicate of soda and powdered asbestos. This is further lined by blocks of carbon placed in two layers so that the inner layer can be renewed as it becomes worn out without disturbing the outer blocks. The carbon lining E is in contact with a base-plate C of conducting material which has a rod c projecting through the walls of the furnace and leading to a terminal block c' to which one of the electrical conductors is attached.

The roof of the furnace is formed by a cap-plate G, an upward flange to form a receptacle for the circulation of water for the purpose of cooling, and a downward flange which fits in an annular chamber containing water to serve as an air-tight seal. The furnace cover is provided with a central vertical shaft J for the admission of the electrode F, and two inclined channels H and I, the former serving for the admission of the charge to the furnace, and the latter as outlet for phosphorus and other vapours resulting from the reaction.

The charge is admitted to the furnace through the hopper H^2 closed by a valve. The shaft J has mounted upon it a sealing vessel K, which is insulated from the furnace body by the interposition of a mass of non-conducting material. The annular vessel contained between the walls of K is filled with water, and receives the inverted cup M, which is secured by means of transverse screws to

[1] U.S. Pat. 842,099 (1907).

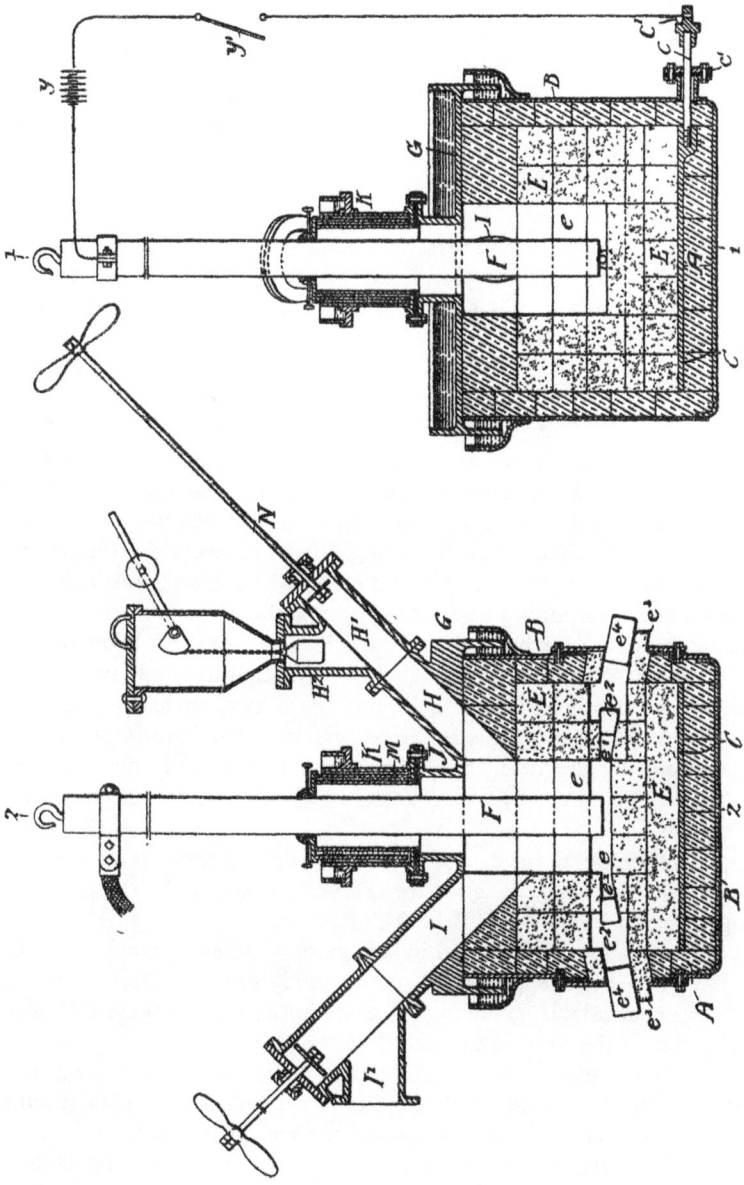

FIG. 195.

the carbon electrode F. The joint between the electrode and the cup M is closed by means of a luting of fireclay.

The furnace is continuous in operation, the molten slag being at intervals tapped through the holes e_1 which are closed by wooden plugs at e_2 and carbon plugs at e_4. The tapping holes are provided in duplicate in case it becomes necessary to plug one permanently through the action of wear.

The Landis process is said to be in operation at the works of the American Phosphorus Company at Yorkhaven, Pa., where water power is employed. The materials utilised are wavellite (aluminium phosphate) and calcium phosphate, which are roasted, mixed with silica and charcoal, and smelted in the electric furnace. The slag is drawn off every three or four hours and the phosphorus condensed under water.

CARBON BISULPHIDE, MANUFACTURE OF—IN ELECTRIC FURNACE.

The manufacture of carbon bisulphide is an instance of the successful application of the electric furnace to a process which was formerly carried out by fuel-heating, the temperature of the reaction being well within the limits of such heating. This compound is formed by the action of sulphur on charcoal at a red heat, and was formerly conducted by heating these materials together in small retorts and condensing the carbon bisulphide which is evolved. The manufacture on a large scale involved the multiplication of retorts, as these necessarily had to remain limited in size, on account of the difficulty of causing the heat to penetrate. The retorts are easily destroyed by the heat, and their replacement and the constant attention required by their large number involved considerable expense.

An electric furnace process for the manufacture of carbon bisulphide has been designed by E. R. Taylor, and brought into operation at Penn Yann, N.Y., U.S.A.[1]

The advantages obtained in this process are as follows :—

1. The heat is generated in the body of the material undergoing chemical change, being led there without any appreciable loss of power in the form of the electric current.

2. The temperature can be regulated as desired, and the radiated heat is almost completely utilised in raising the temperature of the raw materials which surround the reaction zone.

3. The use of retorts or any parts requiring renewing is dispensed with.

[1] " Trans. Amer. Electrochem. Soc.," 1902, I, 115 ; 2, 185.

The furnaces designed by Taylor are shown in Figs. 196 and 197. Fig. 196 shows a section in elevation through the electrodes, and Fig. 197 a cross section in elevation at right angles. The furnace is 40 feet in height, with a diameter decreasing slightly towards the bottom. Charcoal is fed into the centre chamber *y*, and sulphur in the surrounding annular chamber *z*. The four electrodes, consisting of short carbon bars, are arranged as at C and D, and at right angles to these a second pair. The space between the electrodes is filled with granular coke or pieces of carbon rods, and serves to

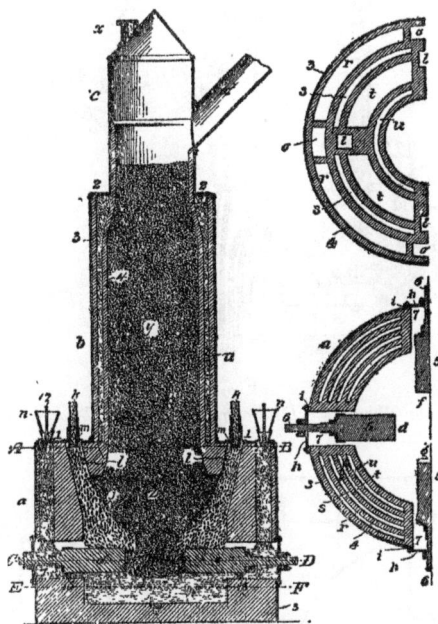

FIG. 196.
(*From Thorpe's "Dictionary of Applied Chemistry".*)

conduct the current, and by the resistance offered provides the heated zone. This granular carbon is replenished as becomes necessary by a supply admitted through *kk*. Sulphur is supplied to the annular chamber through *z*, Fig. 196 (or 17, Fig. 197), melted by the radiant heat, and flows along channels to the reservoirs AB, and from here to the well at the base of the furnace, until the liquid sulphur comes into contact with the heated carbon. The temperature in this zone is too high to permit of reaction, the sulphur volatilises, and the vapour passes upwards in the column of charcoal until it encounters a zone of temperature suitable for reaction, when carbon bisulphide is formed and passes away as a gas.   The con-

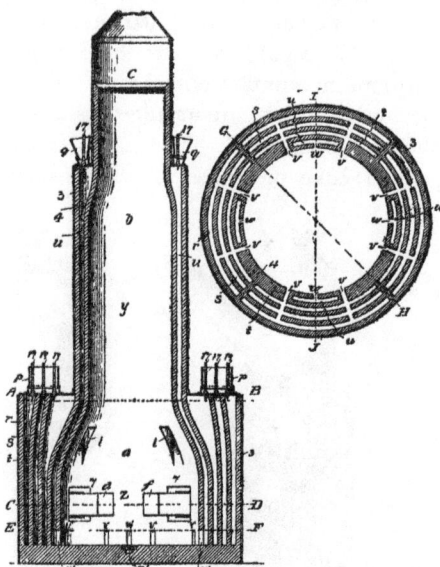

FIG. 197.
(*From Thorpe's " Dictionary of Applied Chemistry ".*)

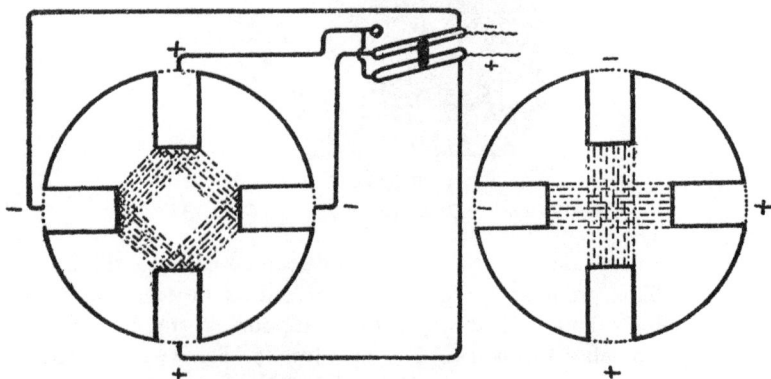

FIG. 198.

trol of this furnace is largely automatic, as if the heat becomes too intense more sulphur melts, and rising up around the electrode surface cuts down the current on account of its insulating properties.

Consequently the only labour involved in operating this furnace is that of feeding the shaft and annular chamber periodically with sulphur and charcoal, and a run of fourteen to seventeen months duration without any other attention is usually possible.

The furnaces are supplied with four-phase alternating current, with 4000 amps. at from 50 to 70 volts, amounting to a power of about 250 kw.

By changing the position of a switch, the current can be passed between the electrodes crossways or diagonally, as shown in Fig. 198, and in this way the path of heated carbon is renewed. To condense the carbon disulphide the furnace gases are passed through a row of tubes, arranged vertically in a cylinder, through which water circulates, and any gas still uncondensed is passed over charcoal and thus absorbed. For this purpose use is made of the charcoal which is later to be employed in the furnace, and the gas is passed over it while it is undergoing a drying treatment by heating. Hydrogen sulphide, which is formed to some extent during this operation, is absorbed with iron oxide.

The carbon bisulphide which is condensed is redistilled, and dissolved sulphur remains as a residue. The output of one of these furnaces when operated with 250 kw. is 5 tons per day, or 5·6 tons per h.p. year (7·3 tons per kw. year). Each furnace would permit of operation at double the above capacity by applying more power.

This process may be regarded as providing a considerable advance in the manufacture of carbon bisulphide, as with it no escaping fumes, which are highly noxious and dangerously inflammable, are apparent, and the amount of labour is reduced to a minimum. When operating with one furnace the whole factory at Penn Yann only requires two or three attendants during the day, and one or two during the night.

The main uses for carbon bisulphide are as a solvent of rubber and sulphur, in agriculture for destroying bacterial and other noxious forms of life, and as a solvent in the manufacture of a form of artificial silk (Viscose).

The plant at Penn Yann utilises a water power, where a fall of 32 feet in the river is available, and the factory supplies the whole demand of the United States for carbon bisulphide, the older processes having been thereby replaced.

SECTION XVII.

MISCELLANEOUS ELECTRIC FURNACES.

High-Frequency Induction Furnace. The "Pinch" Effect Furnace.

Inductive Heating with High-Frequency Currents.[1]—In a type of furnace developed by E. F. Northrup for heating by induction currents, the difference in principle from that of the ordinary induction furnace is that in the latter the metal under treatment is contained in an annular crucible and forms the single secondary turn or loop of a transformer, while interlinking of the magnetic circuit with that of the primary is arranged by an iron core or yoke. In the Northrup furnace the material to be heated is placed in a cylindrical enclosure in the centre of a solenoid or coil through which current of a very high frequency is passed. Currents which are generally termed eddy currents are induced in this internal conducting mass or tube, and lead to a heating of the mass while the coil itself remains cold. By surrounding the material by an efficient heat insulator the temperature attainable with a suitable current supply is only limited by the fusing-point of the refractory. To cause a rapid generation of heat in the conducting mass, it is necessary to employ a large number of "ampère-turns" in the solenoid and to use current of a frequency exceeding 10,000 cycles per second, while frequencies as high as 100,000 may be used and voltages of 6000 to 8000.

In the usual type of induction furnace the self-induction in the primary winding increases as the closeness of coupling between the primary coil and the single-turn closed secondary is diminished. Consequently, with increasing size of furnace the frequency of the e.m.f. impressed must be reduced to avoid an excessive lag of the primary current behind the impressed e.m.f. In some cases it has thus been necessary to employ a periodicity as low as 5. With low-frequency currents it is impracticable to utilise condensers in order to improve the power factor by introducing negative

[1] E. F. Northrup, "Trans. Amer. Electrochem. Soc.," 1919, **35**, 69; "The Iron Age," 1919, 1294; "Trans. Far. Soc.," 1917, **13**, 213.

reactance on account of the large capacity which would be required.

In the method applied by E. F. Northrup the other extreme
is made use of by employing very high frequencies and dispensing
with all magnetic circuits.

A suitable arrangement for this purpose is illustrated in Fig. 199.
The inductor coil which surrounds the chamber containing the
crucible to be heated is, in a typical small-size unit, composed

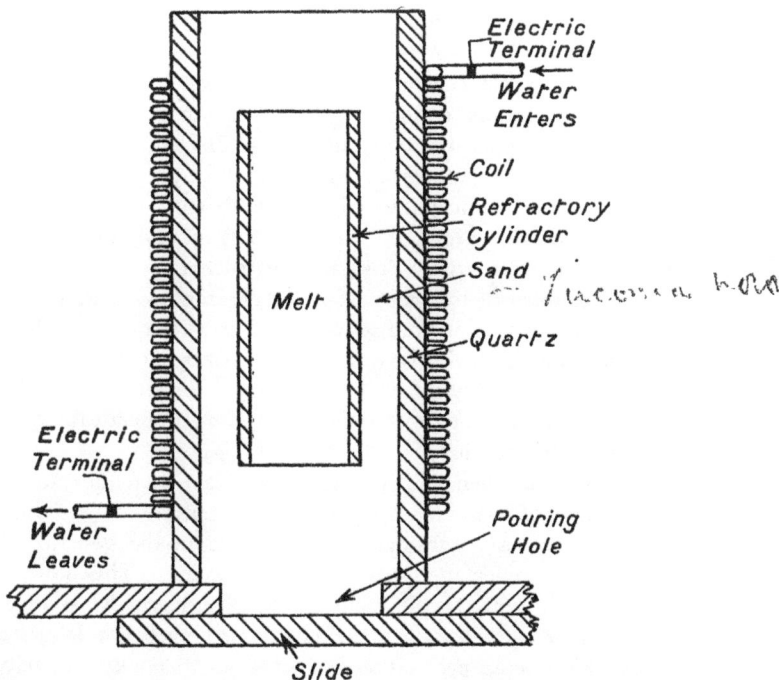

FIG. 199.

of fifty-two turns of flat copper strip.   The copper is wound
edgewise in order to have as many windings as possible in a
given length, and to give the conductor sufficient cross section
to carry the current.   As the peak voltage applied to the coil
may be as high as 7000 to 8000 volts a quartz cylinder is used
to insulate electrically the inductor coil from the crucible which
is on the inside of the quartz cylinder.   The crucible is 14 cm.
diameter in outside, and the space between the crucible and the
quartz cylinder is filled with suitable heat-insulating material.   The

refractory enclosure can be earthed, and the inductor coil surrounded by a metal cage to avoid danger from the high potential.

It is not essential that the metal used in the melting chamber be in the form of a cylinder. Small broken masses may be placed in a moulded cylinder of suitable material, inserted in the inductor, and surrounded by powdered refractory. The temperature of the mass is raised rapidly and, in the case of, iron, melting occurs in a few minutes.

An efficient circulation of the molten material is brought about by the action of electro-magnetic forces. Currents for the operation of this furnace have so far been obtained either from

(a) The Tesla oscillatory current circuit. In this method the circuit of a supply of 8000 volts, which may be either direct or alternating current, is connected in parallel with large condensers with a discharge gap consisting of graphite poles suspended over mercury in an atmosphere of alcohol vapour. The circuit leads from the discharge gap to the terminals of the solenoid. If the energy is not too rapidly dissipated an oscillatory discharge is given at each rupture of the gap. Under favourable conditions a power factor can be obtained up to 60 or even 75 per cent. Oscillatory currents produced in this way have been applied for the operation of laboratory furnaces of 60 kw. capacity. On account of the large condenser capacity required the method is impracticable for units of much larger capacity.

(b) High-frequency alternators such as used in radio work. Small units of these have been used for current of 25,000 cycles, and in 1919 tests were made at the works of the general Electric Company, Schenectady, with a unit of 200 kw. The power factor is adjusted to unity by the inclusion of a condenser (cf. p. 60).

Furnaces of this type have been introduced for laboratory or small-scale operations in units of from 20 to 60 kw. capacity, and have found a useful application in the melting of metals of the platinum groups so as to avoid contamination; in preparing carbon-free alloys such as those of electrolytic iron with tungsten, molybdenum, copper, and in the heat treatment of steel articles. The furnace has also been used for the melting of glass and refractory oxides and heat treatment of porcelain, by surrounding the containing crucible in a cylinder of nickel or graphite.

In place of the coil of copper strip a hollow copper tube can be used and cooled by air or water circulation. In the latter case the outflowing stream of water is arranged to issue from the outlet in the form of a spray, so as to avoid an electrical short-circuit between the ingoing and outflowing water.

Power Consumption and Efficiency.—In the preparation of carbon-free ferro-alloys in a small furnace, melting and thorough mixing of the product has been obtained of 5 lb. of metal with 18 kw. in less than one half hour.

With a 200 kw. unit, 8·4 lb. of copper have been melted per kw. hour as measured at the terminals of the supply line, representing an efficiency of about 80 per cent. The application of this type of furnace on a large scale would appear to depend only on the development of less expensive and more efficient means of obtaining in large amounts currents of 10,000 cycles or more.

THE "PINCH" PHENOMENON. THE HERING FURNACE.[1]

When a current is passed through a liquid-conducting channel the accompanying electro-magnetic field causes the surface layers to be attracted towards the centre, so that a force is exerted acting from the circumference to the centre of the conductor in a direction perpendicular to the axis. This force tends to cause a contraction of the cross section, which is liable to result in a separation of the column at some part of the channel where the section is narrower than the remainder. On account of this so-called "pinch" phenomenon, the current which can be passed and the temperature which can be reached in a channel of liquid

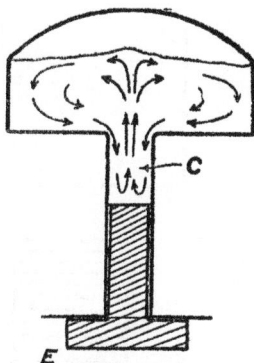

FIG. 200.

metal is limited. If, as in Fig. 200, a cylindrical channel C is arranged vertically under a wider bath and current is passed from the solid electrode E which closes the base of the channel, then, on passing a current, the force which acts radially from the circumference to the centre leads to a difference of pressure between the periphery and the axis which causes the central portion of the channel to be propelled forward in the direction shown by the arrows, while liquid from the outside layers enters to replace that leaving the centre. This propelling force, which increases as the square of the current and diminishes as the cross section increases, causes a circulation of the liquid in the direction shown by the arrows.

[1] C. Hering, "Trans. Amer. Electrochem. Soc.," 1911, **19**, 255; "J. Faraday Soc.," 1911, **7**, 202.

23

By arranging this column as the resister for the generation of heat in an electric furnace an active circulation of the contents of the bath can be brought about, the heat uniformly distributed, and intimate contact of the metal with the covering slag ensured.

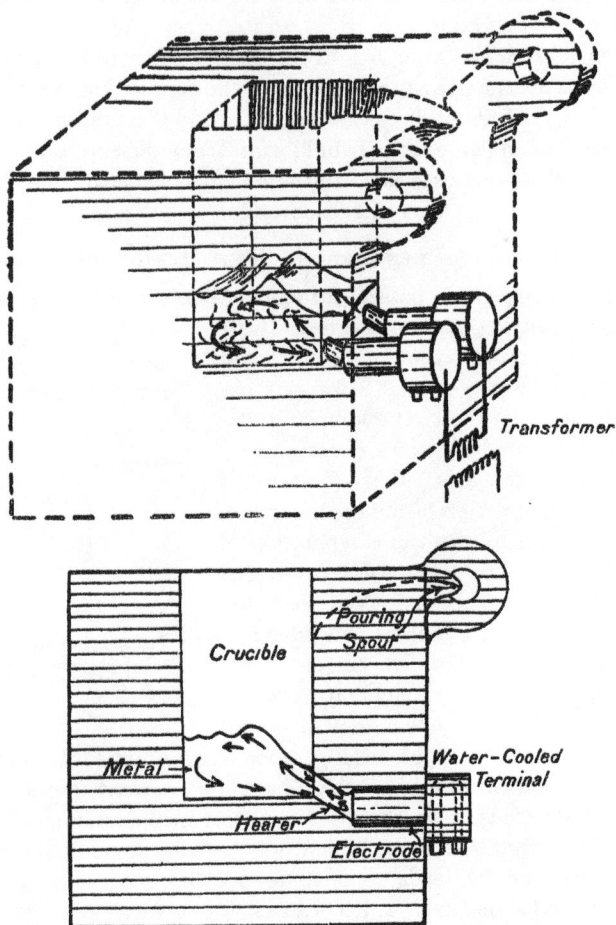

FIG. 201.

In furnaces constructed on this principle two liquid resister channels are used with single-phase current as illustrated in Fig. 201 and three channels for three-phase current, each channel being connected with an electrode.

The electrodes are made of metal, preferably of the same metal as that being melted in the furnace. With a resister ¾ inch diameter and 4 inches long, a current of 3000 to 5000 amps. is used at 5 volts.

A tilting arrangement and pouring spout can be arranged for emptying the furnace as shown in Fig. 201.

When in operation the top surface of the metal is visibly agitated with the production of wave motion as illustrated in the diagrams.

In addition for use in the treatment of metals, it is proposed to apply this furnace for the melting of non-conductors such as glass or for treating granular ores. In these cases a suitable metal would be used as heating medium on the bed of the furnace.

SECTION XVIII.

ELECTROLYTIC PROCESSES WITH FUSED ELECTROLYTES.

THE production of the alkali and alkaline earth metals and aluminium, on account of their powerful chemical affinities, can only with great difficulty be brought about by purely chemical agencies. In all cases, from the time of the first isolation of the alkali metals by electrolytic processes, the technical preparation of these metals has remained in the domain of electro-chemical methods. The procedure in all instances with this class of materials consists in utilising the property possessed by compounds of these elements, when in the fused condition at high temperatures or when contained in fused solvents of other compounds, of becoming conductors of the second class. Under these circumstances, electrolytic dissociation is brought about similarly to the behaviour at ordinary temperatures of aqueous solutions and, by the application of a suitable potential, the cation is separated in the metallic condition on the surface of the cathode. During the electrolysis, the necessary high temperature is maintained through heat generated by the passage of the current.

ALUMINIUM.

The industrial preparation of aluminium was first brought to a practical success in methods devised during the years 1886 to 1890 by Héroult, Kiliani, Hall, and Minet. The Héroult process was brought into operation at Froges (Isère) and later at Champ, la Praz and Gardanne, by the Société Electrométallurgique Française, and in Scotland by the British Aluminium Company. In America the Hall process was installed by the Pittsburg Reduction Company in 1888. The Minet process was operated in Paris in 1887 and a process of Kiliani, working in conjunction with Héroult, at Neuhausen, Switzerland, in 1890.

In the case of all methods which have so far received practical application, the preparation of aluminium is brought about by dissolving alumina in the halogen compounds of the alkalis, alkaline earths or of aluminium, and separating the aluminium by electrolysis.

356

The raw materials employed consist almost exclusively of bauxite, which is a hydrate of alumina of the formula $Al_2O_3 . 3H_2O$ and cryolite, which has the composition $3NaF . AlF_3$.   An addition is generally made to the bath of an excess of aluminium fluoride, calcium fluoride, sodium chloride or other compounds in order to reduce the melting-point of the electrolyte and diminish its density so as to facilitate the sinking of the aluminium to the bottom of the bath.

It is necessary to submit the alumina employed to a careful purification and to prepare it in a physical condition which enables solution to take place rapidly in the molten bath.

During the electrolysis, aluminium is liberated at the cathode

FIG. 202.

and oxygen at the anode which reacts with the carbon to form carbon monoxide.   At the same time small amounts of fluorine are sometimes liberated, the decomposition voltage of the fluoride being only slightly above that of alumina.   It is necessary to adjust any loss of fluorine from the bath by the addition of aluminium fluoride. Alumina is replaced as the electrolysis proceeds and the composition of the bath remains constant.

Héroult Process.—The type of furnace at first employed by Héroult at Froges, in 1888, was constructed as shown in Fig. 202.   At first electrolysis was attempted by the use of alumina alone as the charge. In this case the high temperature necessary for the reaction resulted, when using a carbon cathode, in the formation aluminium carbide. Cathodes of copper and other metals were accordingly applied, and the preparation of aluminium alloys brought about. Later by the addition of cryolite as a flux or solvent, the separation of aluminium metal was found possible when using a cathode either of iron or carbon. The anode, in this apparatus, consisted of a bundle of carbon poles fastened together by copper bands so as to form a block 1 metre long and 0·25 square metre section. The type of furnace adopted later and installed at Neuhausen is shown in Fig.

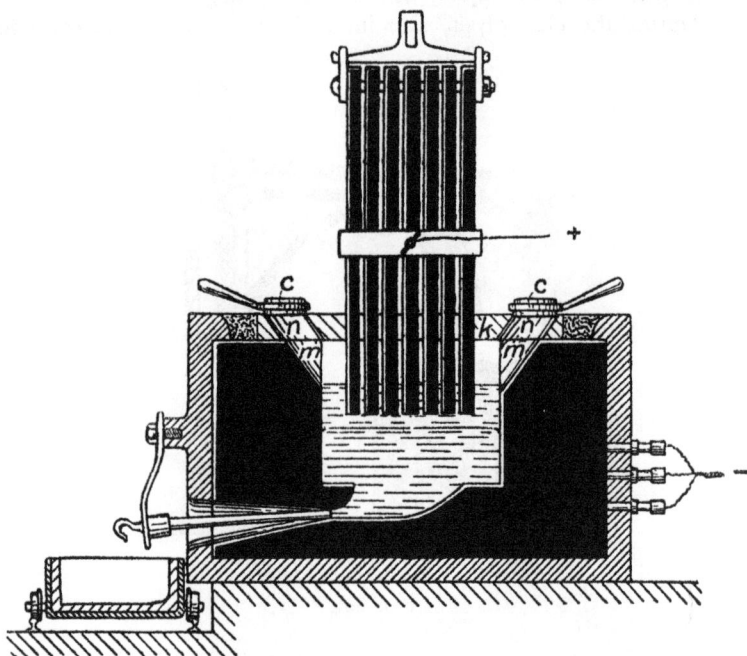

FIG. 203.

203. The anode consists of a bundle of rectangular plates of carbon which are cemented together by means of a mixture of carbon with tar and molasses or glucose solution. The cathode consists of a carbon block contained in an enclosing metal case. The roof of the furnace is covered with graphite plates k, furnished with openings for the admission of the anode in the centre and the charge at C.

On account of the high melting-point of the charge, the furnace was originally used for the production of aluminium alloys, the pro-

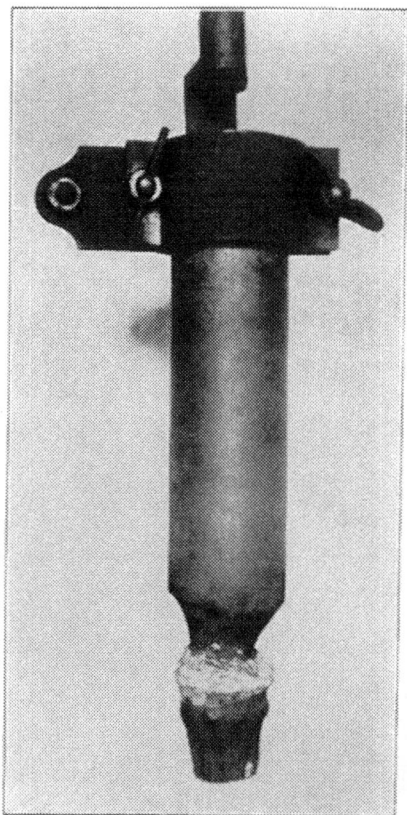

FIG. 204.—LEIDENFROST PHENOMENON.

(From Borchers' "Electric Furnaces".)

cedure consisting in first adding granulated copper which was fused by lowering the anode so as to make electrical contact and then adding alumina. A current of 13,000 amps. for about 5 square feet section was employed at 12 to 15 volts.

By adding a large excess of cryolite together with the alumina, it was later found possible to conduct the electrolysis at a much lower temperature and obtain metallic aluminium.

One of the main difficulties at first experienced with the Héroult furnace was due to overheating at the surface of the anode through the high resistance caused by the formation of a film of gas. One of the effects of this high temperature is to cause an erosion of the carbon immediately above the surface of the bath which, as illustrated by Borchers, is shown in Fig. 204, and is known as the Leidenfrost phenomenon.

Attempts to overcome this defect were for a time directed by Héroult and Kiliani to the rotation of the anode, and subsequently by employing an electrolyte of lower fusing-point and increasing the anode area to reduce the current density.

According to Winteler,[1] in the type of aluminium furnaces subsequently adopted, the current density amounts to about 7000 amps. per square metre of bath section or 650 amps. per square foot section and the potential about 8 volts. The power expended for 1 ton of aluminium is given as 1500 to 1700 e.h.p. for twenty-four hours or 3·35 to 3·8 kw. years of 8000 hours. The present (1920) yield usually obtained amounts to 1 ton per 1000 e.h.p. for twenty-four hours. The temperature of the bath is from 800° to 1000° C., and the most favourable composition, a mixture of cryolite with 10 per cent alumina. The aluminium collects on the bed of the cell from whence it can be removed by tapping. The specific gravity of the metal is 2·7, while that of the fused bath is 3·0.

A suitable form of furnace for large-scale operation, as described by Winteler, is illustrated in Figs. 205 to 207.

The dimensions of the bath are 3 feet 6 inches long, 1 foot 10 inches wide, and 1 foot high. The current employed is 3200 amps. The case is made of wrought-iron and protected against the corrosive action of the bath by means of a solidified layer of the charge itself. The floor of the bath is lined with carbon plates which serve as cathode, the layer of metal which collects here itself eventually forming the cathode. The anodes consist of carbon electrodes with copper connecting-bars which are attached by screw terminals to copper rods. The most favourable distance between anode and

[1] " The Aluminium Industry," Brunswick, 1903.

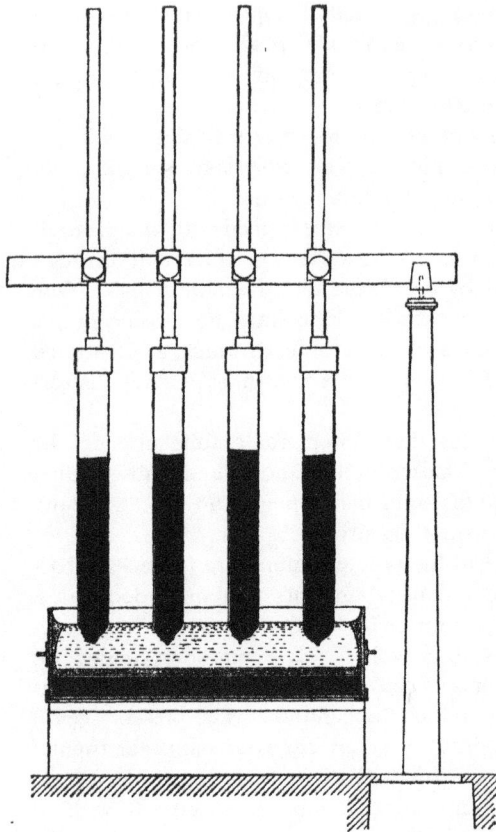
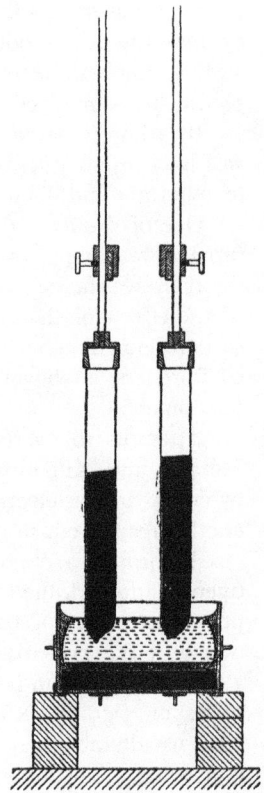

FIG. 205.　　　　　　　　　FIG. 206.

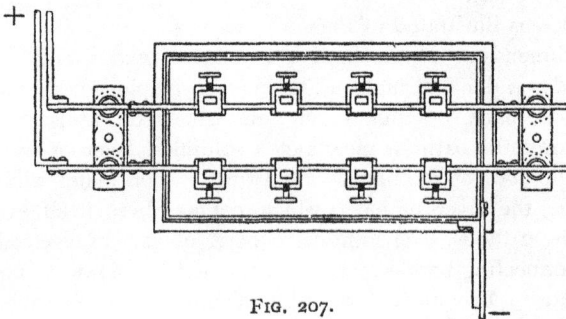

FIG. 207.

cathode is about 6 cm., and the electrodes are adjusted to this distance as consumed.

Héroult Installations in France.—The works erected at Froges (Isère) formed the first electro-metallurgical factory in France. A few hundred horse-power was generated here from water power. A second works was built at La Praz, near Modane (Savoy), where 13,000 e.h.p. is available during most of the year. In 1894 a works for the preparation of alumina was taken over at Gardanne (B. du Rhône), and in 1903 a further works was built at Saint-Michel de Maurienne, where 17,000 e.h.p. is developed.

A works at Argentière, in the Durance valley, near Briançon, has also been constructed and takes 35,000 to 45,000 e.h.p. The company thus utilises during the greater part of the year a total power of 65,000 to 70,000 e.h.p.

In addition to aluminium and its alloys in the form of ingots, bars, cables, tubes, etc., the Société Electro-métallurgique Française engages in the production of steel, ferro-alloys, and carbon electrodes.

Furnaces of British Aluminium Company.—The Héroult furnaces employed by the British Aluminium Company, consisted, in the original installation at Foyers,[1] of boxes lined with carbon which formed the cathode, while the anode consisted of a bundle of carbon rods suspended within the case and reaching nearly to the bottom. With natural cryolite as used in this country, furnaces cannot be worked below 950° C., and the voltage, which was originally 6½-7½, cannot be reduced below 5¼. A current density of 600 amps. per square foot of cathode and 1200 amps. per square foot of anode or 8000 amps. per cell is used. The yield of metal amounts to 1 lb. per 12 e.h.p. hours, or 0·5 metric ton per kw. year of 8000 hours.

The Héroult process is now operated by the British Aluminium Company at their works at Kinlochleven and Foyers, in Scotland and at Stangfjord near Bergen and Vigeland, in Norway, and by the *Aluminium Corporation* at Dolgarrog, North Wales.

The Minet Aluminium Furnace.—The type of furnace adopted by Minet is illustrated in Fig. 208 and contains two electrodes suspended vertically in the bath. In some types the walls of the crucible are also connected through a resistance to the negative terminal, so as to cause the separation of a regulated small quantity of aluminium on the walls of the vessel and thus protect them against corrosion from the bath. The advantage claimed to be obtained in the Minet furnace is that less contamination is imparted to the aluminium, which forms on the surface of the

---

[1] Cf. pp. 11, 413; " J. Soc. Chem. Ind.," 1898, **17,** 310.

cathode and flows to a basin hollowed in the base of the furnace from whence it is tapped.

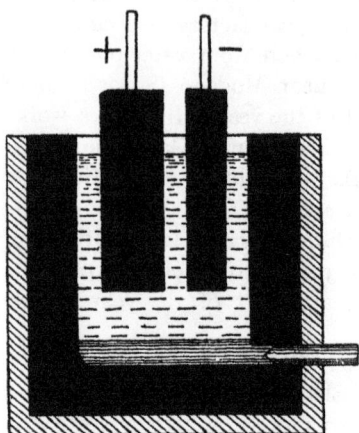

A disadvantage with the system, however, is that the metal globules are said to diffuse through the electrolyte to some extent instead of coalescing, while, if the electrodes are placed too low in the bath, a danger arises of the aluminium collected at the base of acting as an intermediate electrode.

The electrolyte employed in the Minet process consists of seventy parts of sodium chloride, and thirty parts of sodium aluminium fluoride to which alumina is added. The bath is electrolysed by a current of 4000 amps. at 7·5 volts. The temperature employed is 800° C., when the specific gravity of the bath amounts to 1·76. At 1055° C. volatilisation commences. A current efficiency of 70 per cent is stated to be obtained.

FIG. 208.

HALL PROCESS.[1]

The original type of the Hall process consisted, as shown in Fig. 209, of a rectangular iron case 5 feet long and 3 feet wide, lined, in the first place, with a layer 4 inches thick of wood charcoal to serve as heat insulation. Above this was placed a second lining of iron plates with a rim 4 inches high which was connected with the outside walls and served for the introduction of the current to the bath. Above this and around the walls a carbon lining was tamped in to an average thickness of 16 inches. The bed of the furnace was shaped to allow the aluminium produced to flow into a channel from which it could be tapped through an opening in the base. The carbon lining was composed partly of old electrode ends, in order to increase the conductivity, and the whole bed was baked at a high temperature before use. As the result of much investigation on the production of a compound of low melting-point, the flux finally adopted to serve as solvent of the alumina has the following composition :—

$$Al_2 F_6 . 6 Na F + Al_2 F_6 + Ca F.$$

[1] Eng. Pat. 5669 of 1889; "Electrochemical Industry," 1903, 160.

Alumina is added to a maximum proportion of 15 to 20 per cent when the solution is saturated. The anodes consist of four rows of ten to eleven carbon plates of circular section, which with a length of 18 inches have a surface of 7 square inches.

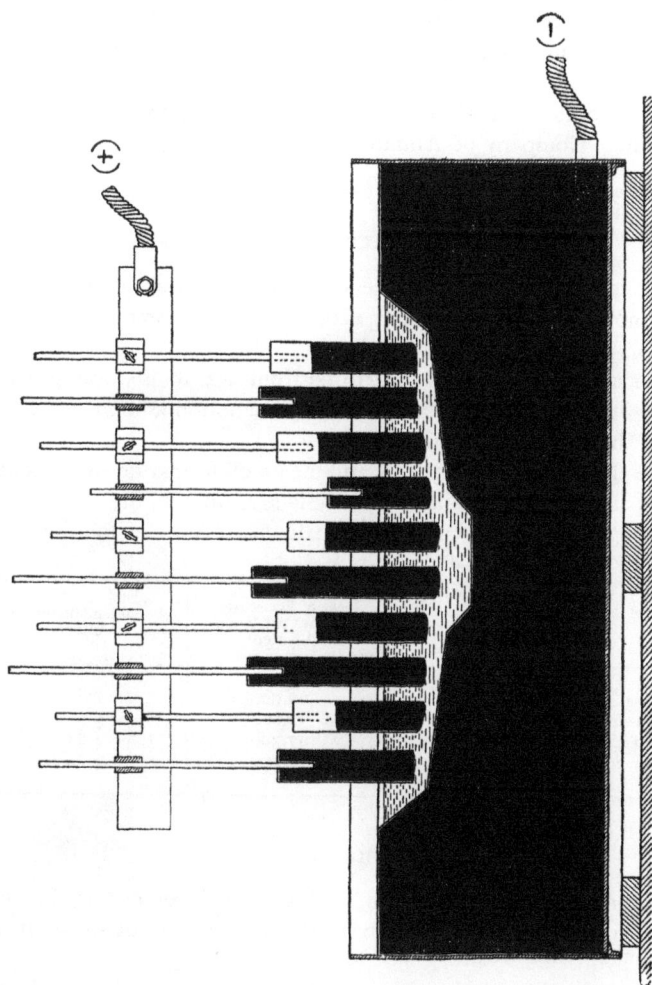

FIG. 209.

A potential difference of 7 to 8 volts was formerly employed between anode and cathode but, by employing larger baths, this has been reduced to 5 to 6 volts. The temperature employed was originally about 900° C., but it has been stated that, in modern practice, the process is worked at a temperature of from 750° to 800° C. The contents of the bath solidify at the edges, and the top. A layer of powdered alumina is placed on the crusted surface of the electrolyte and thus undergoes a final drying and pre-heating before admitting to the bath by pressing under the surface.

A rise in voltage denoting exhaustion of the alumina from the electrolyte is indicated by a lamp placed in the circuit over each bath.

The process of Hall is now in operation at the works of the Aluminium Company of America at Niagara Falls and Shawinigan Falls in Canada.

EFFICIENCY OF ALUMINIUM PROCESSES.

The recognised normal current efficiency in aluminium electrolysis amounts to 75 to 85 per cent, though in some cases, falling as low as 60 to 65 per cent.

The average output is said to be from 0·4 to 0·5 ton per kw. year of 8760 hours. The net anode consumption amounts to about 0·75 of the weight of aluminium.

The decomposition voltage and heats of formation of the compounds used in aluminium electrolysis are, from indirect data and neglecting temperature coefficients, calculated to be as follows :—

Compound.	Heat of Formation. Calories per Gram Equivalent.	Decomposition Voltage.
Alumina	65,000	2·82
Aluminium fluoride . . .	70,000	3·03
Sodium chloride	97,300	4·21
Sodium fluoride	110,800	4·80

SODIUM.

The industrial preparation of sodium was developed by Castner through investigations begun in 1888 on the electrolysis of fused sodium hydroxide.

The main difficulties were found to consist in the solubility, which increases with the temperature, of the sodium and oxygen in the fused hydroxide. At temperatures not far removed from the melting-point of the hydroxide, the solution and diffusion followed by recombination of the separated products may proceed as quickly as the materials are liberated so that no yield is obtained. The successful operation of the process has, on this account, been found to consist in conducting the electrolysis at as low a temperature as possible and quickly removing the separated sodium from contact with the electrolyte.

The highest temperature found satisfactory is 330° C., or 20° above the melting-point of the hydroxide (310°). The apparatus designed by Castner consists, as shown in Fig. 210, of a cast-iron case A of about 14 inches diameter, the upper part being 24 inches high and provided at the base with an extension tube B of 3¼ inches diameter and 32 inches long. The cathode H consists of a copper bar secured at the base of the surrounding case by a wooden plug, in the space above which fused sodium hydrate is filled and allowed

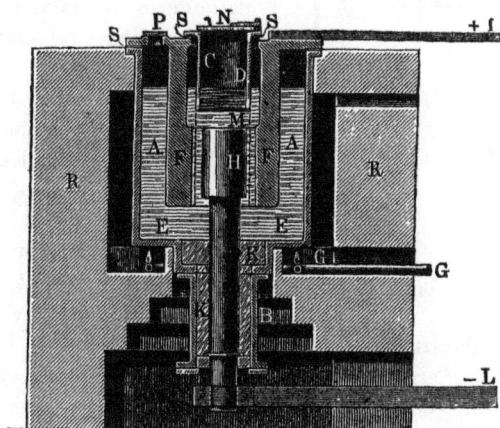

FIG. 210.

to solidify. The electrolyte in the space above can be heated by means of the gas-ring burner G.

Directly above the cathode is suspended an iron cylinder, closed by the lid N which is provided with an opening for the escape of hydrogen, while from the lower part, a wire gauze cylinder extends so as to form a diaphragm between the electrodes and prevent the sodium after separation from diffusing to the anode. The anode E was originally made of an alloy of nickel and silver which was found resistant against corrosion.

The sodium liberated at the cathode H together with the hydrogen rises into the cylinder N, and is periodically removed by means of a perforated ladle which allows its separation from the fused electrolyte to take place.

In practice, twelve to twenty of these cells are arranged in series and supplied with a current of 500 amps. at 110 volts, the current efficiency amounting to about 70 per cent.

The Castner process is mainly operated by the Castner-Kellner Alkali Company at Weston Point, St. Helens, and Newcastle-on-Tyne in England, and the Niagara Electro-chemical Company in America which now employs a total of 13,000 e.h.p.

The Castner-Becker Process. —A modification of the Castner cell introduced by Becker[1] is illustrated in Fig. 211. The cathode B is made of metal or retort carbon and connected to a rod b which passes through an extension tube a in the base of the furnace, the remaining space in the tube being sealed by a ring a' of porcelain or fire-proof stone. The extension tube a is, for cooling purposes, surrounded by a double-walled shell h, for air or water circulation, whereby the electrolyte at the base is kept solid.

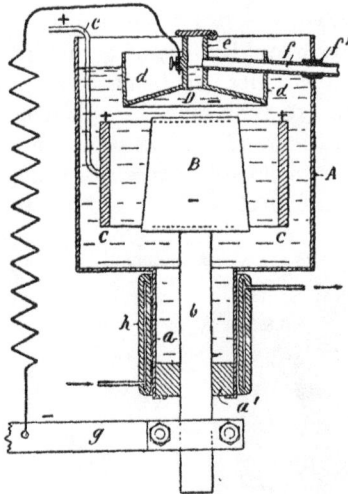

FIG. 211.

The annular anode C is made of metal or retort carbon and suspended from rods c so as to surround completely the cathode. A metal cone D is suspended above the cathode and serves to collect the liberated metal which rises as globules to the surface of the electrolyte. A rim d is provided at the periphery so as to furnish a cooling surface for the roof of d, which may be still more efficiently cooled by attaching an annular cover as shown in Fig. 212, through which air or water is circulated.

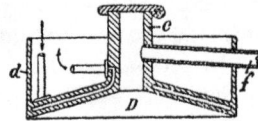

FIG. 212.

The liberated metal rises to the level of the discharge pipe f, through which it can be tapped in liquid form into moulds.

The collecting head D is connected electrically through a resistance to the negative pole, so that a small proportion of the current passes through this vessel and the sodium collected here is thus protected from dissolving in the electrolyte.

[1] Eng. Pat. 11,678 of 1899.

The temperature of the electrolyte is maintained by the passage of the current whereby only the central portion need be kept liquid. The electrolyte used differs from that in the Castner cell in consisting of a mixture of sodium hydroxide and carbonate. The advantages thus obtained are greater economy, and in avoiding explosions with hydrogen and oxygen which occur in the original Castner cell.

The yield of sodium by this process is said to amount to 0·44 metric ton per kw. year of 8000 hours.

CALCIUM.

In a process in operation at the Bitterfeld electro-chemical works in Germany, calcium is prepared from the fused chloride by a process in which, as shown in Fig. 213, the cathode *a* which is kept cold, is, at its base *d*, arranged to make contact with the surface of an electrolyte of fused calcium chloride. On passing the current, calcium is separated in globular form, adheres to the cathode and solidifies. The electrode and adhering calcium are slowly raised from the bath so that a cylindrical rod of calcium *c* is gradually built up, and by means of the thin coating of chloride which adheres, is protected against oxidation.

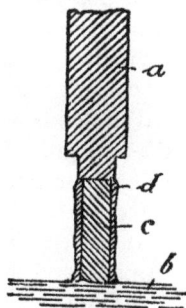

FIG. 213.

The conditions for conducting the electrolytic preparation of calcium on a laboratory scale are pointed out by P. Wöhler.[1] The electrolyte to be employed consists of 100 parts of calcium chloride and twelve parts of calcium fluoride, a mixture which has a melting-point of 660° C. Calcium chloride alone melts at 780° and metallic calcium at 800° C., thus allowing only a very small margin of temperature for the solidification of the metal.

An iron crucible is used to contain the bath. The cathode consists of an iron wire of 8 mm. diameter, and a current is applied of 40 amps. at 38 volts, giving a current density of 100 amps. per square cm. of cathode surface.

[1] "Zeit. f. Elektrochem.," 1905, **36**, 612.

SECTION XIX.

REFRACTORIES.[1]

THE part played by refractories in electric furnace practice is in most cases as a heat-insulating material, and the physical properties of these substances is of first importance in determining the efficiency and economy of furnace processes. The limiting temperature attainable in a furnace enclosure is determined by the melting-point of the surrounding refractory and by the loss of heat through conduction. The consumption of power or thermal efficiency is again proportional to the heat losses through the conductivity of the electrode- and enclosure. Further, a frequent source of loss of power which occurs with electric furnaces is due to the electrical conductivity of the furnace roof or walls which at some points establish contact with the electrodes or with the charge in the neighbourhood of the electrodes, and thus cause stray currents.

The main characteristic of electrothermal methods, and the chief basis of their economy, is that heat is applied locally in a comparatively small zone, just where required, and the efficiency of the operation will frequently be determined by the extent to which thermal losses to the outside air can be minimised.

While for most purposes the highest possible insulating properties are essential in refractories, there are some instances, which are more common in fuel-heated furnaces, where the converse applies, and the highest thermal conductivity is desired. These include cases where the heat is required to penetrate the walls of muffles or crucibles. Though a large amount of work still remains to be conducted on this subject, it has been well established that the properties of refractories such as alumina, magnesia, and magnetic oxide of iron vary greatly on exposure to high temperatures. Apart from the thermal and electrical conductivity, an important feature in refractories is the ability to withstand sudden changes of temperature.

---

[1] R. S. Hutton, " Trans. English Ceramic Society," 1905-6, **5**, part 2, 110; F. A. J. FitzGerald, " Met. and Chem. Engineering," 1912, **10**, 129; R. Hadfield, " Trans. Faraday Soc.," 1917, **12**, 86; Bywater, *ibid.*, p. 116.

MEASUREMENTS ON THERMAL CONDUCTIVITY.

A series of experiments carried out by R. S. Hutton and J. R. Beard[1] with a number of refractory materials in a finely divided condition, at temperatures up to 1000° C. showed the following order of conductivity, beginning with the best insulator :—

Kieselguhr (infusorial earth).
Magnesia (Pattinson's light calcined).
Firebrick.
Lime.
Magnesia (calcined "Veitsch").
Quartz ("Enamel").
Quartz (fused).
Retort Carbon.
Magnesia (calcined Greek).
Magnesia (fused).
Magnesia ("Mabor" brick).
Carborundum (fine).
Carborundum (coarse).
Sand (white Calais).

CHARACTERISTICS OF COMMON REFRACTORIES.

In applying materials for furnace insulation at high temperatures it is necessary to pay regard to the influence of the conditions of use on their properties such as stability, resistance against oxidation, melting-point, shrinkage, disintegration, and flaking.

Wood Charcoal, Acetylene Black, and Carbonised Cotton have very high insulating properties and, in absence of air or an oxidising atmosphere, can be used at the highest producible temperatures.

Fused Magnesia and Alumina.—In cases where it is necessary to avoid the presence of carbon, these substances are pre-eminently suited for application at very high temperatures in the absence of contact with carbon or other bodies with which they react chemically. When in the fused or "electrically shrunk" form, these oxides are capable of withstanding very sudden changes of temperature. Alumina can be used at temperatures up to 2100° C. and magnesia up to about 2400° C., when marked volatilisation takes place prior to melting. The unique properties of this form of magnesia was first pointed out by E. K. Scott[2] and R. S. Hutton,[3] and in view

---

[1] "Trans. Faraday Soc.," 1905, **1**, 264. Cf. Appendix I.
[2] *Ibid.*, 289.   [3] Loc. cit.

of the possibility of cheaply producing electrically shrunk magnesia on a large scale, the extension of its use as a special refractory may be expected. Formed articles, such as crucibles, bricks, and muffles, can be readily manufactured from the fused oxide by mixing the finely powdered material with a small quantity of magnesium chloride solution, pressing in a mould, and baking at a temperature of about 1800° C. when the material becomes agglomerated to a solid form. The magnesium chloride which first forms oxychloride is decomposed to oxide, and thus leaves no foreign substance to contaminate the magnesia. In the use of refractories in electric furnaces, apart from the influence of actual solid materials with which they may come into contact, regard must be paid to the influence of volatilised substances. For instance, in electric steel furnaces, metallic oxides including lime and magnesia are volatilised, and coming into contact with the heated roof of the furnace, are liable to impair considerably its durability.

Alundum.—Though so far the production of fused magnesia has not been undertaken as a commercial process, that of alumina, which was originally prepared as an abrasive, by the Norton Company at Niagara Falls, has been applied by this firm for the manufacture of refractory ware.[1] The fused alumina is known as alundum, the trade name applied to the abrasive.[2]

The main application which has so far been made of this refractory is to the construction of laboratory apparatus, such as muffles, crucibles, combustion boats, cores for laboratory electric furnaces, combustion tubes, etc.

The heat conductivity of the material is much higher than that of firebrick. Alundum bricks have been made with a view to their employment for the roofs of electric furnaces. The durability has been found to be very much greater than that of silica bricks, but in the case of arc furnaces for steel, the vapours of lime and other oxides were found to react with the alumina causing the material to break off in layers and impair its durability. For this application, though the roof of alundum bricks would still have a considerably longer life than that of silica bricks, the difference is not considered sufficiently great to warrant the extra cost of the manufacture.

Fused Silica ("*Artificial Quartz*" or "*Silica Glass*").—The preparation of fused silica in the form of "quartz fibres," was first accomplished by Boys, and in the form of small vessels by Shenstone. In both cases the oxy-hydrogen blowpipe served as the source of

[1] "Trans. Amer. Electrochem. Soc.," 1911, 19, 333.   [2] Cf. p. 334.

heat. The electric furnace was later applied for the preparation of tubes and other articles of silica by R. S. Hutton,[1] who moulded relatively long tubes from quartz crystals or Calais sand, both by indirect heating with the electric arc and also by passing the electric current through a carbon core surrounded by sand. A method was then developed by Hutton, Bottomley, and Paget[2] for blowing and shaping vessels from the semi-fluid material produced around an electrically heated carbon core.

The apparatus used consists, as shown in Fig. 214, of an iron case, through the ends of which the graphite electrodes EE are introduced. A thinner carbon rod is fitted in the ends of the electrodes and is raised to a high temperature by the passage of a current. The space in the box around the electrodes is filled with silica which is raised to its melting-point (about 1750° C.) in the neighbourhood of the carbon rod, and by means of the carbon monoxide evolved is dis-

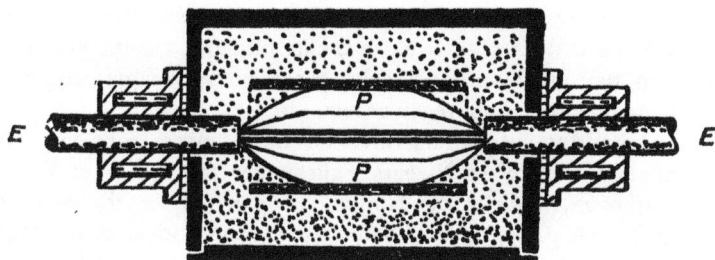

FIG. 214.

tended as shown in the figure. The carbon plates shown at PP serve as a mould, and after a certain interval the case is opened and the pasty mass rapidly removed and blown by means of compressed air into the desired form. In order to keep the tube of uniform cross section during the heating, the furnace case is rotated.

With other improvements the Thermal Syndicate of Wallsend-on-Tyne engaged in the production of large pipes, bricks, dishes (for use in processes such as sulphuric acid concentration) insulators, pyrometer tubes, and other articles. At first, on account of the inclusion of small air bubbles in the plastic mass, a translucent material only was obtained, but developments of the process now enable the production of a transparent product.

[1] "Mem. Manch. Lit. and Phil. Soc.," 1901, 46 (vi), 1; "Trans. Amer. Electrochem. Soc.," 1902, 2, 105.
[2] Eng. Pat. 10,670 of 1904; "Zeit. fur angewandte Chemie," 1912, 25, 1845

24 *

The most valuable properties of this fused silica are its high melting-point, hardness, and its ability to withstand extreme and sudden changes of temperature.

Devitrification of Silica.—When silica is maintained for long periods at temperatures between 1150° C. and 1400° C., devitrification occurs, consisting probably in the formation of tridymite crystals. This devitrified material is brittle, has a thermal coefficient of expansion from fifteen to thirty times as much as vitreous silica, and cracks if suddenly cooled after heating. This phenomenon of devitrification can usually be counteracted by raising the temperature of the silica above 1400°, and cooling rapidly to below 1150° C.

Coefficient of Expansion.—The coefficient of expansion of crystalline silica is 780×10^{-8} on the main axis, and 1420×10^{-8} in the cross section. For fused silica the coefficient of expansion at 200° C. has been found to be 518×10^{-9}, and at 1100° C., 583×10^{-9}.[1]

Carborundum is applied as a refractory in cases where a high conductivity is desired such as in muffle furnaces, retorts, crucibles, partition walls in by-product coke ovens, pottery saggers and bats, kiln floors, recuperators, regenerators, etc.

Carborundum is manufactured for these purposes by the Carborundum Company of Niagara Falls under the names of Refrax and Carbofrax, and by the Norton Company under the name of Crystolon (*vide* p. 373). Carbofrax in the form of bricks or muffles is made from pure carborundum crystals, bonded together with a small percentage of refractory clay. Refrax articles consist of a dense mass of carborundum crystals which are held together by recrystallisation. No binding agent is used, the recrystallisation being carried out in an electric furnace. The main characteristics of these materials are the high thermal conductivity, low coefficient of expansion, resistance to chemical action, and ability to withstand temperatures up to about 2200° C. without change.

The linear coefficient of expansion of carborundum from 200° to 900° C. is stated to be $4\cdot7 \times 10^{-6}$, as compared with $5\cdot6 \times 10^{-6}$ to $1\cdot6 \times 10^{-5}$ for firebrick, and $8\cdot6 \times 10^{-6}$ for fused alumina. The thermal conductivities of the above forms of carborundum, and other materials for comparison are given in the table below :—[2]

---

[1] "Met. and Chem. Engineering," 1920, **22**, 593.

[2] "Carborundum Refractories," The Carborundum Co., Ltd., U.S.A. Cf. Appendix I.

Material.	Conductivity (Calories per Deg. C. per Centimetre Cube per Second).
Firebrick	0·0034
Magnesite . . .	0·0071
Chrome	0·0067
Silica	0·0020
Refrax	0·0275
Carbofrax . . .	0·0243

Crystolon, as manufactured by the Norton Company, consists of carborundum either in crystalline or amorphous form. The amorphous material is a greenish-coloured product which is formed at a temperature below that required for the crystalline form (*vide* p. 330). The material has been applied to a considerable extent in furnace construction in the form of bricks, cement, or for lining, and has been marketed under the name of "carborundum fire-sand". Articles of crystallised carborundum are made by a process of FitzGerald in which the carborundum grains or powders are mixed with a temporary binder, such as a solution of glue or dextrine; the mixture is then moulded in the desired form and heated in an electric furnace to the temperature at which silicon carbide is formed. This causes a crystallisation or recrystallisation of the silicon carbide and a strong very refractory article is obtained.

Crystolon in the form of bricks has been applied with satisfactory results in the construction of the roofs of electric steel furnaces. On account of the high thermal conductivity of the material, however, it is necessary in this case to apply an outside layer of fireclay bricks, and through the electrical conductivity at high temperatures, to avoid contact of this refractory with the electrodes.

Siloxicon.—In addition to amorphous carborundum, a product of importance as a refractory is obtained during the manufacture of carborundum, consisting of a compound of carbon, silicon, and oxygen, or an oxycarbide of silicon. It was first found by Schutzenberger in 1881,[1] that crystallised silicon when raised to a white heat in an atmosphere of carbon dioxide absorbed that gas and was converted into a greenish product which, after purification, left a compound corresponding to the formula $SiCO$.

Acheson,[2] in 1903, produced compounds of a variable composition, such as Si_2C_2O and Si_7C_7O by heating carbon and silica in appropriate proportions in a furnace of the carborundum type.

[1] " Comptes rendus," **92,** 1508.   [2] U.S. Pat. No. 722,793.

The name siloxicon was given to these materials, and they are described as amorphous powders, very refractory, indifferent to all acids except hydrofluoric, and capable of being moulded and ignited to yield a coherent mass.

Fibrox.—A variety of silicon oxycarbide in the form of very finely divided fibres, of unique conducting properties, has been prepared by E. Weintraub,[1] and introduced commercially under the name of "Fibrox". The material is prepared by heating silicon in an enclosed crucible to a temperature of 1400° to 1500° C. in a gas furnace. The vapour of silicon, in presence of an added catalyst such as calcium fluoride, reacts slowly with the carbon monoxide and dioxide which permeate the walls of the crucible. A compound is obtained in the form of a mass of exceedingly fine fibres which eventually fill the crucible. The material is removed and successive yields obtained. The diameter of the fibres is said to average about 0·6 micron, or about the wave length of yellow light, and the substance has an apparent density of 0·0025 to 0·0030, or weighs only about 2½ to 3 grm. per litre. The real density of the material, when of the composition SiCO, is found to be 1·81, so that the actual volume of the solid is calculated to amount to only 0·1 to 0·5 per cent, the remainder consisting of enclosed air. The heat resistivity of the material increases with the density of packing, but falls considerably with increasing temperature. However, even at higher temperatures, the material when packed to the same degree of compression has a resistivity which is considerably higher than that of any other substance with which comparative measurements were made. At a temperature of 1400° C. "fibrox" is slowly oxidised by air to silica. In distinction from its high thermal insulating properties the material is a relatively good electrical conductor.

Commercial Furnace Refractories.—The application of the above substances, which may be regarded more as ideal refractories, has up to the present been limited to laboratory furnaces, except in some instances for special purposes. For regular furnace operation no special refractory has yet been applied on any considerable scale for resisting higher temperatures other than materials which have been developed in connexion with fuel-heated furnaces. The refractories which are in general use for the linings of both electric and fuel furnaces may be classified under the following three headings :—

---

[1] " Trans. Amer. Electrochem. Soc.," 1915, **27**, 267.

(*a*) *Acid.*—Ganister, Dinas rock, and fireclays including "Schamotte" or "Chamotte" firebricks.

(*b*) *Basic.*—Dolomite, magnesite.

(*c*) *Neutral.*—Bauxite, zirconia, chrome-iron-ore, graphite or carbon, and a few fireclays.

These materials are used either in an agglomerated condition, or after moulding in the form of bricks.

Unconverted Charge as Refractory.—The main feature of most types of electric furnaces is that through the application of heat internally in the body of the raw materials the unconverted charge itself, which is undergoing reaction, forms its own insulating material and thus protects from any great heat the outside walls of the furnace, which may be composed of ordinary firebricks. In the case of steel furnaces, however, it is necessary to apply a refractory which will withstand the high temperature of the metal with which it comes into contact, and will not react prejudicially with this or the covering slag. For this and similar purposes the materials which are most commonly used as refractories are the following :—

(*a*) *Acid Materials.*—*Ganister.*—Ganister occurs in nature and probably results from the disintegration of sandstone. The material has been found of particular value for the lining of Bessemer converters, where in addition to a high temperature the corrosive action of slags has to be resisted. The material has been used from an early date in crucible steel manufacture, a particular variety of Sheffield ganister which occurs in strata immediately below the coal seams in the lower coal measure being of special value in this process. Typical analyses of these different materials are shown in the table below :—

	South Yorkshire Ganister.	Black Ganister (Sheffield).	Dinas Quartz.
Silica	88·7	98·5	97·6
Alumina	7·5	0·3	0·5
Oxide of iron	1·5	1·3	1·5
Lime	0·8	0·2	0·2
Magnesia	1·2	—	—
Potash	—	trace	0·1
Soda	—	—	0·03

Dinas Silica Bricks are made of Dinas quartz, which has the composition given above. In the selection of refractory materials it has for a long time been recognised that the higher the proportion of silica the more refractory the substance, but with increase in silica the cohesion is lessened and cracks develop more readily during heating. Alumina and magnesia render the material cohesive and plastic, but an excess of the former lowers the refractory quality and makes it subject to vitrification at high temperatures. Highly silicious bricks usually expand by the influence of heat, and provision of the necessary spacing should be made for this in furnace construction.

Silica or Dinas bricks are largely used for the construction of the roofs of electric steel furnaces, but the material is unsuitable for use in contact with basic slags, and contact with a basic lining in the lower part of the furnace is necessarily to be avoided.

"*Schamotte*" *or* "*Chamotte*" *Firebricks* are made from a burnt fireclay, and joined by a cement of the same material. These have good heat-insulating properties, are resistant against changes of temperature, and suitable for use in the construction of induction furnaces.

(*b*) *Basic Materials.—Dolomite* consists of calcium carbonate, containing a large proportion of magnesium carbonate, and is largely used as the refractory for the hearths of electric steel furnaces, having formerly been generally adopted as a basic lining for the Bessemer converter. For use in these connexions the dolomite is first crushed to the form of lumps of 3 to 6 inches diameter and thoroughly calcined.

Before any appreciable absorption of water vapour or carbon dioxide from the atmosphere has been allowed to occur, the material is thoroughly incorporated with a small percentage of dry tar or pitch, and the mixture applied to build up the furnace lining by tamping or compressing in the desired position. By a preliminary gradual heating of the furnace to a high temperature the binding material is decomposed and partly volatilised, while the lining remains in an agglomerated condition. Dolomite is also moulded in the form of bricks by a similar process.

Magnesite, or magnesium carbonate, is also used for furnace linings and applied with a binding material in a manner similar to that with dolomite, or moulded in the form of bricks. The main difference in its properties from dolomite is that magnesite is less basic or more neutral, and does not react with water vapour or carbon dioxide by atmospheric exposure to the same degree as dolomite. Except when prepared from " electrically shrunk " material, magnesite is apt to crack through changes of temperature. Magnesite has found a satisfactory application for the construction of furnace roofs, and has a low shrinkage on heating.

(c) *Neutral Refractories.*—*Bauxite.*—This compound is still more refractory than silica, but when ordinarily made up as bricks is excessively porous, and subject to contraction on heating. However, these defects can be mitigated by well calcining the material and particularly by fusing electrically, though this method has not yet been accepted as economical in the preparation of a furnace refractory for general industrial use (*vide* p. 334).

Zirconia[1] is of value in the construction of articles such as crucibles for special purposes. The pure oxide has a melting-point of about 2700° C. and is highly resistant against chemical action, being unattacked for instance by molten silica or fluxes and slags. Zirconia, however, is somewhat easily reduced by carbon at high temperatures giving carbide. Zirconia is obtained as a naturally occurring oxide, baddeleyite, which contains about 96·5 per cent ZrO_2. Another form of natural zirconia contains about 84 per cent ZrO_2, together with 7·7 per cent silica, and 3·1 per cent ferric oxide.

The melting-point of zirconia is largely influenced by the presence of silica and other oxides and by materials which are added to serve as binder in the construction of formed articles. Zirconia has a very low thermal conductivity, and in the form of moulded articles, a high tensile strength, so that crucibles and other formed objects can be made of thinner walls than in the case of other refractories.

The binding material for use in making formed articles, which has been found most satisfactory both in regard to the melting-point and low value for heat shrinkage of the product, is the finely divided oxide made into a thin paste with water.

Chromite or Chrome-iron Ore is largely used for the lining of ferro-alloy furnaces, but is not suitable for steel furnaces where it comes into direct contact with the metal on account of chemical reaction.

Carbon linings consisting of powdered retort carbon (graphite) or coke agglomerated with tar and tamped into position are used with furnaces for the smelting of iron-ores and other purposes, but are not applicable to steel furnaces on account of reaction with the metal. In many furnaces this material serves both for the introduction of the current and as a refractory.

Fireclays.[2]—Fireclays are an impure form of kaolin or China clay, and derived by the decomposition by weathering of felspar or minerals related to it. A typical felspar such as orthoclase is a

[1] H. C. Meyer, " Met. and Chem. Engineering," 1915, **13**, 263.
[2] F. J. Bywater, " Trans. Faraday Soc.," 1917, **12**, 116.

may be partly or entirely replaced by other elements such as sodium
or calcium leaving a material represented by the general formula
$RO, Al_2O_3, 6SiO_2$. Through the influence of weathering, the alkalis,
lime, and silicic acid are carried away, leaving kaolins of the general
formula $Al_2O_3, 2SiO_2 + 2H_2O$. This compound may be regarded as
a true clay substance and is the most refractory combination of silica
and alumina. This product is rarely attained, however, intermediate
fireclays of the general formula $Al_2O_3, 6SiO_2$, together with varying
amounts of water and extraneous materials in the form of lime,
potash, soda, and iron, being more usual. If these oxides are
present to any considerable extent, they exert a deleterious effect on
the refractoriness of the clay, on account of the silicates of these
bases being more fusible than that of alumina, and in some instances
through the formation of multiple silicates which are still more fusible.
The chemical composition of a clay is not in itself a guide to its
suitability for firebricks as this depends on the condition in which
the silica is present in the clay. The refractoriness of the material
is largely affected by the silica being present in the form of quartz,
or in combination with alumina. Alumina has a higher melting-
point than silica, and in mixtures of the two the compound $Al_2O_3,
2SiO$, i.e. kaolin, is the most infusible, the melting-point being (ac-
cording to Le Chatelier) the same as that of silica, 1800° C. Apart
from its composition the relative size of its constituent parts, and the
mechanical condition of the clay as it leaves the grinding mill, have
considerable influence on the refractory character of the manufactured brick. Thus, the melting-point is higher in the case of a
material with a large grain.

Mortar or Cement.—When building a furnace lining or roof from
bricks, it is essential to use some form of mortar as cementing
material. For this purpose, use may be made of tar or tar mixed
with powdered dolomite or magnesite when the bricks are of this
material. With neutral or acid bricks, a small amount of clay mixed
with an excess of silica sand may be used. Magnesia forms a good
cement for silicious materials by a fluxing action, and is highly refractory. However, if added to any considerable extent the melting-
point of the silica is largely lowered.

Linings used with Different Furnaces.—When the process involves
the use of a basic slag the hearth has necessarily to be of a basic
material, and acid with an acid slag. With the direct-arc furnaces
such as those of Héroult and Girod, the roof is generally formed
of silica bricks. With the Stassano furnace, on account of the higher

temperature to which the roof is exposed, both roof and hearth are generally of basic material, the latter being made of magnesite bricks.

Electrical Resistance of Firebricks.[1]—Measurements made on the electrical resistance of firebricks at different temperatures showed that with chrome brick the conductivity becomes appreciable, and begins to increase rapidly with the temperature at 850° C. Magnesia and silica bricks retain their insulating properties up to 1300° C. The conductivity of Caledonia brick first becomes appreciable about 1300° C., magnesia brick about 1350°, and silica brick about 1470°. At the highest temperatures observed, about 1550° C., the resistivities of all the bricks tested are about the same, being about 25 ohms for 1 cm. cube.

[1] A. Stansfield, D. L. McLeod and J. W. McMahon, " Trans. Amer. Electrochem. Soc.," 1912, **22**, 89.

SECTION XX.

HEAT LOSSES THROUGH FURNACE WALLS.[1]

THE loss of heat through the walls of a furnace is, in accordance with the laws of thermal conductivity, determined by the following relations. The flow of heat through a rectangular section of furnace wall is given by the expression

$$H = k(T_1 - T_0)\frac{S}{l}$$

where H represents the flow of heat, T_0 the temperature of the outside surface, and T_1 that of the inside surface of the furnace wall, S the area, and l the thickness of the wall, and k the mean thermal conductivity between the temperatures T_1 and T_0. Taking the furnace as a whole, the area of the outside surface is greater than that of the inner surface in proportion to the thickness of the walls. In accordance with this, the total flow of heat is given by the expression

$$H = k(T_1 - T_0)\frac{\sqrt{sS}}{l}$$

where s is the inner area of the furnace walls, and S the outer area, or the value $\sqrt{sS}$ is the geometric mean of the two areas. The above formula also represents the flow of heat through a spherical enclosure.

Influence of Shape of Furnace on Heat Loss.—The relative loss of heat from a furnace enclosure is determined by the ratio of the area of the radiating surface to the volume of the heated charge or furnace contents. With a rectangular furnace, it follows from geometrical considerations that this ratio will have its minimum value when the furnace is square, while the most favourable shape of enclosure for heat conservation, as having the maximum volume for the minimum surface, is the sphere.

---

[1] C. Hering, " Trans. Amer. Electrochem. Soc.," 1908, **14**, 215; 1912, **21**, 511; " Electrochem. and Metall. Ind.," 1909, **7**, 11, 72; " Metall. and Chem. Engineering," 1910, **8**, 676; 1911, **9**, 13, 652; 1912, **10**, 40, 97, 159; F. T. Snyder, " Trans. Amer. Electrochem. Soc.," 1910, **18**, 235; J. W. Richards, " Metallurgical Calculations," vol. 1.

Influence of Size of Furnace on Relative Heat Loss.—The thermal loss from a heated enclosure is proportional to the area of the walls or to the square of the linear dimensions, whereas the capacity is proportional to the cube of the linear dimensions. It follows from this that for a given temperature and thickness of furnace walls, the loss of heat per unit volume of charge increases in inverse proportion to the linear dimensions or, as would usually be the case, if the thickness of the walls of the furnace is also in proportion to its linear dimensions, then the thermal loss per unit volume of charge is inversely proportional to the square of the linear dimensions of the furnace. By increasing, for example, the linear dimensions of all parts of the furnace ten times, the heat loss per unit volume is diminished in the ratio of $\frac{1}{100}$.

Units.—Thermal conductivities are usually expressed in terms of gram-calories per second per centimetre cube for a difference in temperature of one degree, but in electrical applications the unit known as the *thermal mho* is generally employed to express in terms of watts the heat radiated under the above conditions, the watt being equivalent to 0·239 cal. per second. The reciprocal of this value is the *thermal ohm*, and represents the difference of temperature divided by the flow of heat in watts per centimetre cube.

Radiation and Convection.—While the loss of heat from a furnace enclosure is mainly determined by the thermal conductance through the walls, there is a certain temperature gradient between the outside walls of the furnace and the adjacent layer of air, also between the hot gases in the furnace and the furnace walls. The temperature of the inside walls of the furnace will consequently be below that of the furnace gases, and that of the outside walls above the temperature of the surrounding atmosphere. While the loss of heat through the solid is determined by conduction, that between the outside surface and the surrounding air is controlled by the laws of radiation and convection.

The radiation varies according to the colour and nature of the radiating surface, being low for a polished metal and at a maximum for rough and black surfaces, when the radiation is proportional to the fourth power of the difference in temperature (cf. p. 87).

It is estimated that for a difference in temperature of 100° C., the loss by radiation, in distinction from convection, amounts to 0·015 grm. cal. per second for each square centimetre of radiating surface.

Contact Resistivity.—The loss of heat by convection is measured in terms of *contact conductivity* or its reciprocal, *contact resistance*, which is denoted by R, and represents the difference of temperature in degrees centigrade between the hot body and the surrounding medium, divided by the number of watts or gram-calories per second which flows from each square centimetre or square inch of surface. The value of R has been found to be given by the expression

$$R = \frac{36,000}{2 + \sqrt{v}}$$

where v denotes the velocity of air (which results from the temperature difference) in centimetres per second.

Thus, if the velocity of air is 10 cm. per second, the value of R will be

$$\frac{36,000}{2 + \sqrt{10}} = 6970$$

and the loss of heat per square centimetre surface per second for a difference of temperature of 100°, as in the above case, will amount to

$$\frac{100}{6970} = 0.014 \text{ cal. per second.}$$

For this difference of temperature the loss of heat by convection is seen to be approximately equal to that by radiation, while for greater differences of temperature the radiation loss, on account of the operation of the fourth-power law, becomes proportionately much greater. The term *contact resistivity* is in some cases, as in the table below, used to denote the total transfer of heat from the furnace walls to the surrounding air, both by convection and radiation. The loss of heat from furnaces for varying differences in temperature between the walls and the surrounding air has been estimated to have the following values :—

Material of Walls.	Difference of Temperature. Degrees Centigrade.	Loss of Heat.		Contact Resistivity.
		Kw. per Square Foot.	Watts per Square Inch.	Thermal Ohms per Square Inch.
Brick	70	0.09	0.63	112
,,	130	0.2	1.4	94
,,	200	0.39	2.7	74
Iron	70	0.08	0.56	126
,,	130	0.18	1.25	104
,,	200	0.34	2.4	85

It is further estimated that at a dull red heat the loss is about 7 to 11 kw. per square foot, and at a bright red heat 12 to 15 kw. per square foot.

Calculation of Heat Loss.—The loss of heat from a furnace in watts can be calculated by dividing the difference in temperature between the interior of the furnace and the surrounding air by the sum of the thermal resistance $\left(\dfrac{\text{or thermal resistivity}}{\text{area}}\right)$ of the furnace walls and the contact resistance $\left(\dfrac{\text{or contact resistivity}}{\text{area}}\right)$ of the surface in thermal ohms. The contact resistance in the interior of the furnace, which will generally be of minor importance, is neglected.

SECTION XX1.

ELECTRODE DIMENSIONS AND HEAT LOSSES.

A LARGE source of heat loss in electric furnaces occurs through the electrodes. The two factors which contribute to this loss are firstly the heat which is developed in the electrode through the passage of the current in accordance with Ohm's law and corresponds to C^2r, and secondly the heat transference which takes place between the furnace and the cold end of the electrode, in accordance with the difference of temperature and the thermal conductivity.

The relations obtaining in the flow of heat along furnace electrodes have been analysed in detail by C. Hering,[1] who has derived a number of deductions relating to electrode dimensions which are of considerable interest. It is assumed in the first place that no interchange of heat takes place between the electrode and the furnace walls or space surrounding the electrode, a condition which will be realised when the temperature gradient of the furnace walls is the same as that of the electrode. With reference to the C^2r loss, if the two extremities of the electrode are maintained at a constant low temperature and no heat is given to the surrounding lining, it will follow that on passing a current, the heat conducted to each cold end of the electrode is equal to one-half of the total heat generated in the electrode through the C^2r loss. It follows from this that in furnace operation the two components of heat loss at the cold end of the electrode are given by the expression

$$Q = kT\frac{S}{L} + \frac{C^2r}{2}\frac{L}{S} \quad . \quad . \quad . \quad . \quad (1)$$

where Q is the total energy expressed in watts which flows out at the cold end of the conductor; C the current in amperes; T the difference in temperature in centigrade degrees between the hot and the cold ends; S the cross section in square inches; L the length in inches; k the average heat conductivity in watts per second per degree centigrade for 1 inch length and 1 square inch section for

---

[1] " Trans. Amer. Electrochem. Soc.," 1909, **16**, 265 et seq.

1 degree difference in temperature in the given range ; r the average electrical resistivity or specific resistance in ohms for 1 inch length and 1 square inch section.

It is seen from the above equation that the first term represents the heat in watts that would flow by heat conduction alone, that is, when no current is flowing, and the second term represents one-half of the C^2r loss. According to a mathematical deduction, which assumes that the change of thermal and electrical conductivities with temperature follows a linear relation, Q is found to have a minimum value when the two terms in equation (1) are equal to each other. According to this derivation *the combined loss will be least when the loss by heat conduction alone is made equal to half the C^2r loss ; the total loss will then be equal to the C^2r loss, and no heat will be conducted from the interior of the furnace.*

For a given length of electrode the first term or the heat conduction increases proportionately to the cross section, and the second term varies in inverse proportion to the cross section, while the sum of the two, or Q, has a minimum value when the two terms are equal to each other. It is therefore necessary to adjust the cross section of the electrodes, so that with a given C^2r loss the value of T makes the two terms equal. If the section is made greater than this, the total loss will be increased by the amount of the flow of heat from the furnace, while if the section is made smaller, heat will pass from the end of the electrode into the furnace, and the electrodes will have a higher temperature at some point within the furnace walls.

If the minimum total loss is represented by $Q_{min.}$ and the two terms in equation (1) are made equal, it follows from the equation, by substitution, that

$$Q_{min.} = 1\cdot41\ C\ \sqrt{kr}\ T \qquad . \qquad . \qquad . \quad (2)$$

According to this equation *the condition of minimum heat loss is dependent only on the material, current, and temperature, but not on the absolute dimensions ; it merely determines the relation of the cross section to the length.*

A further practical consideration is that the temperature of the electrode increases with diminishing cross section, and through heating the electrode to a point at which serious loss by oxidation would result, imposes an upper limit to the current density which may be used.

It is seen from equation (2) that the value of $Q_{min.}$ increases in direct proportion with the current and not with its square, is independent of the amount of energy, and becomes smaller if the

25

386   THE ELECTRIC FURNACE

current is decreased and the voltage increased. The formula also shows that the minimum heat loss increases only as the square root of the temperature, so that the temperature of the furnace may be increased considerably without increasing the loss greatly, similarly with the heat conductivity k, and electrical resistivity r, either of which must be decreased considerably in order to diminish the loss appreciably.

Choice of Electrode Material.

For the most favourable economy the two points to be considered are the loss of power and the expenditure of electrode material. In most furnace processes, only carbon and graphite can come into consideration on account of their refractory nature. In considering the relative advantages of these materials, as seen below, to attain the minimum loss with carbon, the electrode, for a given length, requires to be of much larger section than graphite, thus increasing the difficulty with contacts. Other considerations which favour the use of graphite are the greater facility of machining and making joints and terminal connections. Further, in high temperature furnaces, carbon may graphitise partly, so that if correctly proportioned at first, it will evidently not be so at a later stage and the loss will then increase, whereas, with graphite, the values for the thermal and electrical resistance do not permanently change to any considerable extent.

Economy of Power.—In equation (2) the resistivity r may be replaced by its reciprocal, the electrical conductivity K. If the current and temperature are further taken as unity, the result will represent the loss in watts per ampere for one degree of temperature and will therefore be characteristic of the material itself. Making these substitutions, we have

$$Q_{min} = 1\text{'}41\sqrt{\frac{k}{K}} \qquad . \qquad . \qquad . \qquad . \qquad (3)$$

where K is given in reciprocal ohms or *mhos* per 1-inch cube. The minimum loss is thus seen to depend on the ratio of the two conductivities.

The numerical values of this quotient $\frac{k}{K} = kr$ in the case of some typical furnace materials from measurements made by C. Hering [1] are given in the table below :—

[1] "Trans. Amer. Electrochem. Soc.," 1909, **16**, 317 ; "Proc. Amer. Instit. Elec. Eng.," 1910, **29**, 285.

Material.	Thermal Conductivities (*k*). Watts per Degree for 1-inch Cube.	Electrical Resistivity (*r*). Ohms per 1-inch Cube.	kr or $\frac{k}{\mathrm{K}}$.
Copper	9·0	$3·0 \times 10^{-6}$	$2·7 \times 10^{-5}$
Iron	2·0	$3·5 \times 10^{-5}$	$7·0 \times 10^{-5}$
Graphite	3·0	$3·2 \times 10^{-4}$	$9·6 \times 10^{-4}$
Carbon	1·4	$1·4 \times 10^{-3}$	$6·1 \times 10^{-3}$

These data apply to an average temperature of 900° C. between the hot and cold ends of the electrodes. In the case of carbon and graphite the electrical conductivity increases with temperature, the increase with temperature being more rapid with carbon than with graphite, and with the metals diminishes with the temperature. According to these data the thermal conductivity of carbon increases with the temperature and that for graphite and metals diminishes with temperature. It is pointed out by Hansen, the electrical resistivity of carbon electrodes varies very largely with the size which appears to be due to the different degrees of compression to which the granular composition from which they are formed is subjected during their manufacture. Variations were found of from 0·0015 ohm for 1 inch cube for the smaller size to as much as 0·005 ohm for electrodes of 18 inches by 18 inches section.

Cored Electrodes.—In order to diminish the value of $\frac{k}{\mathrm{K}}$ when using carbon electrodes, systems are in some cases being adopted in which the electrode is provided with a central core of metal such as copper or iron. By this means a reduction of both the minimum heat loss and section of the electrodes is effected.

Economy of Electrode Material.—The conditions of minimum heat loss are given when the two terms of equation (1) are equal, i.e.

$$k\mathrm{T}\frac{\mathrm{S}}{\mathrm{L}} = \frac{\mathrm{C}^2 r}{2}\frac{\mathrm{L}}{\mathrm{S}}$$

so that

$$\frac{\mathrm{S}}{\mathrm{L}} = 0·71\mathrm{C}\sqrt{\frac{r}{k\mathrm{T}}} \qquad . \qquad . \qquad . \qquad . \quad (4)$$

According to this equation the conditions for obtaining minimum heat loss are not determined by the absolute dimensions of the electrodes but by the relation of the section to the length.

The rule concerning the economy of electrode material is therefore to make the electrodes as short as practical considerations will permit, and as the cross section also diminishes with the length, the volume or cost diminishes as the square of the length.

25 *

For a given length of electrode, equation (4) may be written in the form

$$S = 0.71 CL \sqrt{\frac{r}{kT}} \quad \cdot \quad \cdot \quad \cdot \quad \cdot \quad (5)$$

so that for a given length, temperature, and material, the section increases in direct proportion with the current and inversely as the square root of the temperature, so that for higher furnace temperaures the section becomes smaller. In order to reduce the current it is therefore desirable to produce the required heat in the furnace by as high a voltage as possible, since the diminution of the current not only enables a smaller section of electrode but, as seen in equation (2), it also diminishes the lowest possible loss of power.

If the current density in amps. per square inch is represented by Cd, from equation (4) we have

$$Cd = \frac{1}{0.71 L} \sqrt{\frac{kT}{r}} \quad \cdot \quad \cdot \quad \cdot \quad \cdot \quad (6)$$

The current density necessary to obtain the minimum heat loss thus varies inversely with the length of the electrode. By water-cooling the electrode at one end, a higher current density may be employed without raising the temperature of the electrode to a point where overheating and oxidation occurs.

From equation (4), making C and T equal to unity, we have

$$\frac{S}{L} = 0.71 \sqrt{\frac{r}{k}} = 0.71 \sqrt{\frac{1}{kK}} \quad \cdot \quad \cdot \quad \cdot \quad (7)$$

where K is the electrical conductivity.

This equation shows that as far as economy of material is concerned, the most favourable conductor is that in which the product of the thermal and electrical conductivities is the greatest.   In this respect, as seen in the table on page 387, metals are superior to graphite and graphite to carbon. Comparing equation (7) with the corresponding one (3) for minimum heat loss, it is seen that the requirements for economy of material and that for power are different and in some respects antagonistic, hence the choice of material is determined to some degree on the relative importance of these two factors. It is seen from (3) and (7) that both economies are improved by an increase of the electrical conductivity alone, while an increase of the heat conductivity alone improves the economy of material but diminishes the economy of power.

Electrode Voltage.—From equation (2), since $Q_{min.} = eC$, the voltage between the hot and cold end of the electrode under conditions of minimum heat loss is given by the formula

$$e = 1\text{·}41 \sqrt{krT} \qquad . \qquad . \qquad . \qquad . \qquad (8)$$

which is independent of the dimensions of the electrodes. According to this, for any given material, the electrode voltage varies only as the square root of the temperature drop. In the case of graphite, for one degree difference in temperature the value of e is therefore 0·0436, and for carbon, 0·1102. Accordingly, in cases of graphite or carbon electrodes of all sizes, if of the theoretically best proportions, the drop of voltage in the furnace will be given by these respective values multiplied by the square root of the temperature drop.

RULES FOR DESIGNING ELECTRODES.

The procedure to be followed in the design of electrodes may be summarised as follows :—

For a given capacity of kilowatts, the current is to be made as small and the voltage as high as practicable.

From equation (4) the proportion of the section to the length is to be calculated. The length is made as short as practicable and generally mainly determined by the width of the furnace walls. The current density is not a factor in calculating the proportions of the electrodes except that it should not lead to a heating of the electrodes to the oxidation temperature. The heat loss from each electrode in watts will be determined by equation (2).

To take a numerical example of these deductions. If the capacity of a furnace is 500 kw. and operated at 50 volts and 10,000 amps., with a temperature of 1700° C. inside and 100° C. at the outside ends of the electrodes. From equation (2), by applying the values given in the table on page 387, the minimum loss in each electrode, in the case of graphite, will be

$$1\text{·}41 \times 10{,}000 \sqrt{3\text{·}0 \times 9\text{·}6 \times 10^{-4} \times 1600} \text{ watts} = 31 \text{ kw.}$$

while, for carbon, the value will be

$$1\text{·}41 \times 10{,}000 \sqrt{1\text{·}4 \times 1\text{·}4 \times 10^{-3} \times 1600} \text{ watts} = 24 \text{ kw.}$$

From equation (5), the most favourable cross section for graphite for unit length is given by the formula

$$S = 0\text{·}71 C \sqrt{\frac{r}{kT}}$$

$$= 0\text{·}71 \times 10{,}000 \times \sqrt{\frac{3\text{·}2 \times 10^{-4}}{3\text{·}0 \times 1600}} = 1\text{·}82$$

while, for carbon, the cross section will be

$$0\text{·}71 \times 10{,}000 \times \sqrt{\frac{6\text{·}1 \times 10^{-3}}{1\text{·}4 \times 1600}} = 11\text{·}63.$$

If, for instance, the length of the electrode is 50 inches and round in section, then the diameter, in the case of graphite, will be 10·8 inches, and in that of carbon 37·2 inches, or in the case of carbon of square section, 24·1 inches wide.

It should be noted that these figures, while serving as a valuable guide in determining the dimensions of electrodes, are subject to correction when more exact data are available on the determining constants.

It must be further remembered that the premises on which many of the deductions are made in the above formulæ do not rigidly hold in practice. The variation of both thermal and electrical resistance with temperature causes deviations which might, however, be adjusted to give results of approximate accuracy by taking a suitable mean value for the conductivity constants which is different from the arithmetic averages of the values at the two temperatures. A further departure is caused by the fact that the heat generated by the C^2r loss in the electrode is not produced uniformly throughout the length but varies with the changing resistance at different temperatures. With carbon, for instance, and to a minor degree, graphite, more heat is generated in the colder parts near the terminals, which leads to a relative disadvantage in the use of carbon. In many cases the absence of interchange of heat between the electrode and the furnace walls or surrounding medium cannot be assumed.

By taking into account these additional factors of the nature of the change of resistance with temperature, and radiation to surroundings, more comprehensive equations have been deduced by E. F. Roeber.[1]

In the case of direct-arc furnaces, such as those used in the preparation of steel, ferro-alloys, and calcium carbide, a source of uncertainty in these deductions would appear to be the indefiniteness of the temperature of the heated end of the electrode, which through the formation of the arc is exposed to a much higher temperature than the main part of the charge and furnace interior.

CURRENT DENSITIES WITH ELECTRODES.

The design of electrodes has, as a rule, been based mainly on the current density which is to be employed.

According to C. A. Hansen,[2] in general work a current density of 50 amps. per square inch may be allowed for carbon, except for furnaces of the Héroult type, where 30 amps. or less is necessary, and 150 amps. per square inch for graphite. In Héroult steel

---

[1] "Trans. Amer. Electrochem. Soc.," 1909, **16**, 363.   [2] *Ibid.*, 344.

furnaces of 20 tons capacity, using currents up to 25,000 amps. on each electrode of amorphous carbon, a current density of 26 to 46 amps. per square inch has been adopted (cf. p. 402).

It is stated by A. Helfenstein that the electrodes used in large calcium carbide furnaces, attain a red heat with a current density of 60 to 65 amps. per square inch.

The following data, published by W. Borchers,[1] give the current density that can be employed in carbon rods of different cross sections :—

Diameter of Carbon. Inches.	Cross Section. Square Inches.	Current Density. Amps. per Sq. In.
2	3	64
4	12·1	54
8	50	32·5
12	111·2	21·5
16	200	10·7

The decrease of permissible current density with increasing diameter of electrode is mainly due to the fact that larger electrodes cannot be made of the same compactness as those of smaller section. On this account, it is usually found preferable to build up the larger electrodes from a bundle of smaller rods.

[1] " Electric Furnaces," p. 201.

SECTION XXII.

THE DESIGN OF ELECTRODE TERMINALS.

FOR the upper or side electrodes of furnaces, carbon, either in moulded form from retort carbon or graphite, is almost universally used. The connexion between carbon electrodes and the current leads is a subject to which a very large amount of work has been

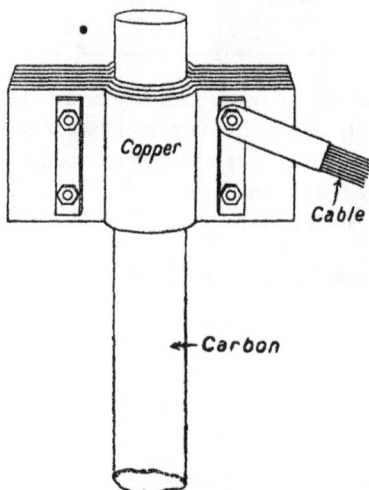 

FIG. 215.   FIG. 216.

devoted as is shown in the variety of designs which have been adopted and the still larger number described in patent literature.

Apart from the heat conducted along the electrode, a small resistance at the junction of the carbon with the metal terminal causes the generation of a large amount of heat. In addition to the loss of energy thus entailed, this resistance leads to oxidation of the metal connecter followed by further deterioration of the contact.

According to C. A. Hansen, the contact losses between brass and graphite with a current of 15,000 amps. amounts to 19·3 kw., and with 20,000 amps. to 24 kw. These data apply to the case of a contact made by clamping a carbon electrode of rectangular section with flat metal holders by means of bolts. With attachments now in use, however, this loss has been very much reduced.

Soldered Connexions.—In laboratory and small types of furnaces a satisfactory connexion of carbon rods and tubes can be made, both in the case of amorphous carbon and graphite, by electro-coppering the ends and soldering into surrounding copper tubes which are themselves enclosed by a water jacket as shown in the instance in Fig. 24, or the carbon can be soldered in a recessed plug enclosing the end of a metal tube through which water is circulated

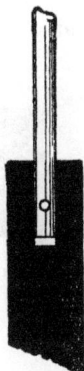 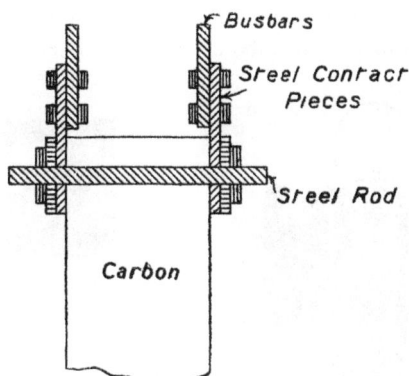

Busbars

Steel Contact Pieces

Steel Rod

Carbon

FIG. 217.   FIG. 218.

as in Fig. 30. With small rods or tubes of carbon, a good contact can also be made by means of a friction contact in holes bored in larger graphite blocks.

Clamp Connexions for Small Electrodes.—For electrodes of circular section up to about 10 cm. diameter, the simplest connexion consists of a number of superposed pieces of pliable copper sheet of rectangular shape with a central curvature. The inner piece which is to be placed in contact with the carbon electrode has a radius of curvature slightly less than that of the carbon. Each succeeding layer of copper is shaped so as to make good contact with the piece below. The metal sheets are firmly clamped around the electrode as shown in Fig. 215. One of the bolts securing the clamps is passed through the copper lug in which the cables are soldered. The air-cooling provided by the copper plates is generally sufficient to prevent a harmful rise of temperature.

FIG. 219.

TYPES OF CONNEXIONS WITH ELECTRODES USED IN INDUSTRIAL FURNACES.[1]

Electrodes in use in industrial furnaces are of two types, those with terminal connexions and those with side holders. The terminal connexions are almost exclusively used in the manufacture of calcium carbide, ferro-silicon, and aluminium, while in electric steel furnaces, side holders are usually employed.

It is to be remembered that metals expand with heat more than

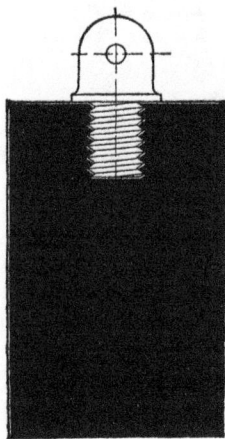

FIG. 221.

FIG. 222.

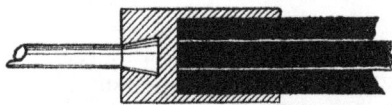

FIG. 220.

FIG. 223.

carbon, so that metal heads clamped on the outside must have provision to prevent them working loose during the heating. In the types illustrated in Figs. 216 and 217, loosening is prevented by providing the carbon with a tapered head.

In the type shown in Fig. 216 conical metal plates inserted between the main clamp and the electrode are firmly pressed against

[1] "Stahl u. Eisen," 1913, **33** (1), 472; " Metall. and Chem. Engineering," 1913, **11**, 321.

the carbon surface by the weight of the electrode. In the type shown in Fig. 217 a cavity with tapering walls is cut in the head of the carbon and a metal rod inserted which widens at the base. When in its lowest position in the bore, the rod is surrounded by metal plates which wedge it against the walls of the cavity.

A transverse hole is in some cases bored in the head of the carbon and a metal rod passed through and secured at the side by clamps as seen in Fig. 218.

On account of the difficulty of machining special shapes of electrodes, the material of the electrode is generally formed before baking. Recesses are in many cases provided in the carbon for the purpose either of making a good contact with the cable, or for join-

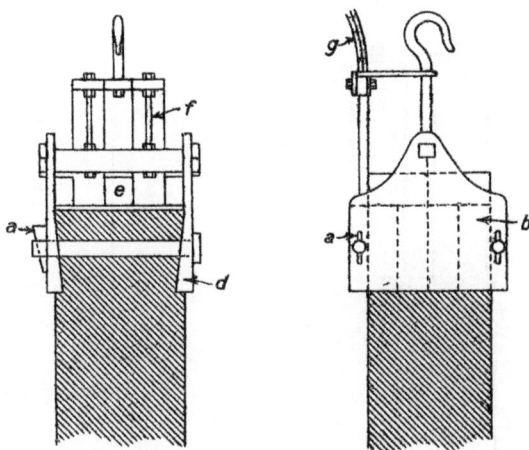

FIG. 224.

ing the ends of electrodes together so as to avoid waste. In the case of graphite, on account of the facility with which it can be machined, the ends of electrodes are generally screw threaded after manufacture so as to enable fitting together as shown in Figs. 193 and 194.

In some cases, with carbon anodes, the ends are provided with dovetail fittings as shown in Fig. 219.

Cast-iron screw connexions are in some cases inserted in the ends of the carbon electrodes during moulding as shown in Figs. 220 to 222. These attachments enable the carbons to be connected to conductors or to be jointed together so as to utilise the ends. This method forms an economical and simple method of connecting together electrodes, and can be used wherever the metal of the connecter, which eventually melts, does not contaminate the charge.

Cowles Furnace Electrodes.—In the Cowles furnace for the manufacture of aluminium alloys (cf. Fig. 6), the electrodes were composed of bundles of carbon rods which were cast into bronze blocks as shown in Fig. 223.

An electrode which is used largely with carbide furnaces is shown in Fig. 224. The head of the electrode is dovetailed and

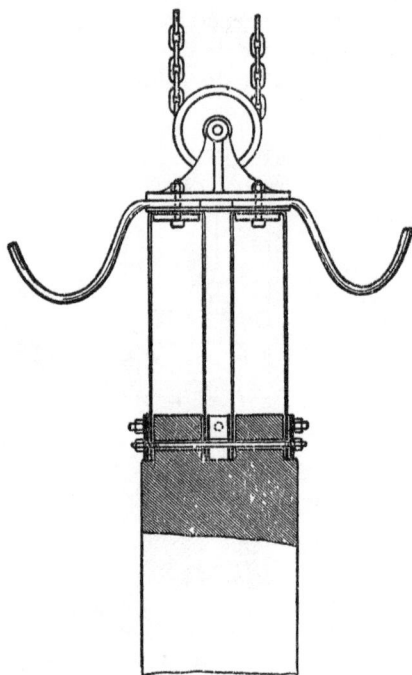

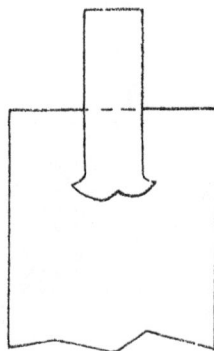

FIG. 226.

FIG. 225.

clamped between two cast-iron plates *b* by means of two wedges *a*. The whole apparatus is supported by an iron structure which itself hangs on an electrically insulated hook. Copper plates *d*, *e*, and *f* serve to make contact with flexible laminated ribbons *g*, carrying the current. The mechanical suspension is independent of the electrical connexions, and the weight of the electrode tends to tighten the contact between the copper plates and the electrode head.

A similar type of electrode terminal to the above was introduced by the Société des Carbures Metalliques, and is shown in Fig. 225.

For the manufacture of aluminium the end of a flat piece of iron is turned in the shape of a fishtail, as shown in Fig. 226, and moulded into the electrode before baking or may be fastened after baking by filling in with molten bronze.

In the Héroult iron-smelting furnace illustrated in Fig. 110, as erected at Sault Ste Marie, the electrode head consists of iron

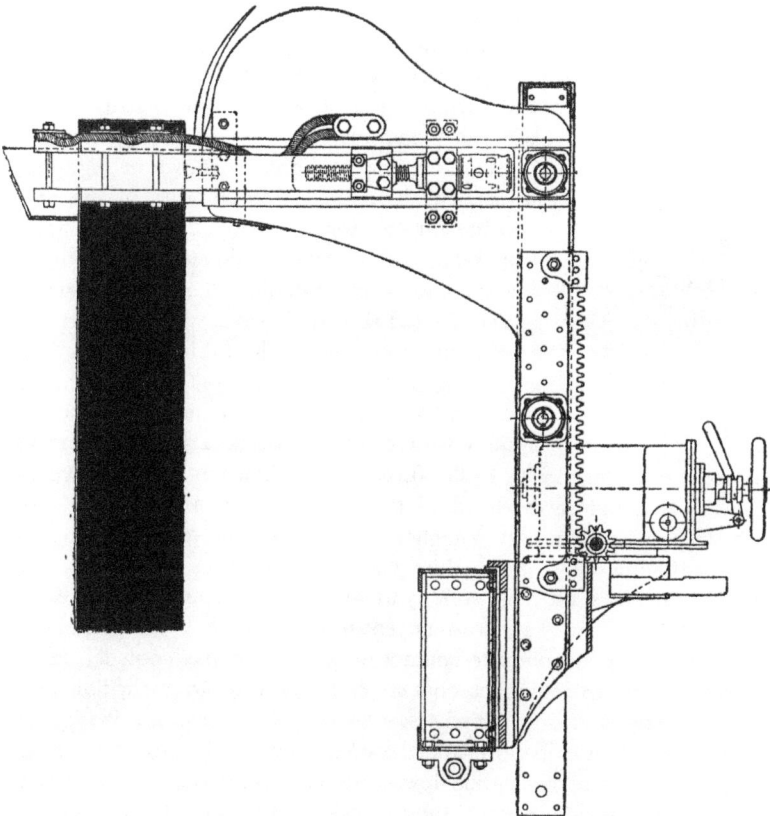

FIG. 227.

wedge-shaped jaws which are secured by two bolts. The electrode is recessed to fit the jaws. The terminal is cooled by a stream of air admitted by the tube which extends vertically above the electrode.

In the present Héroult steel furnaces the terminals consist of four iron plates, one of which is fastened to the horizontal projecting arm of the electrode carrier (cf. Figs. 227 to 229), and the other three, to which the main cables are screwed, are held firmly against the electrode by flexible metal bands which are tensioned by a screw placed in the carrier arm.

Keller Electrode Terminals.—A system of air-cooled electrodes devised by Keller, Leleux & Co., and applied for large electric furnaces, is illustrated in Figs. 230 and 231. Electrode blocks are prepared from a bundle of four rectangular carbon bars of high conductivity while the whole is surrounded by agglomerated carbon made from a paste mixture of higher resistance. The paste is also used as a material for cementing together the bars. Copper plates are, as shown in Fig. 230, pressed against the inside surfaces of the carbon bars at the top by means of wedge-shaped iron blocks which are tensioned by screws on the outside of the bundle. The copper strips are connected with iron plates which extend upwards and are joined to the cables.

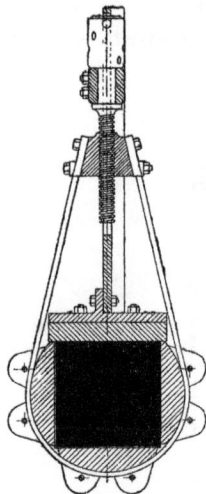

Water-cooled Electrodes.— It is generally found advantageous to use a water-cooling system with electrode terminals. In a system

Fig. 228.

now generally used with Keller furnaces [1] as shown in Fig. 232 a vertical cavity is cut in the head of the electrode and an iron tube with an inner copper lining is attached by inserting the end in the opening and filling in with molten copper or pig-iron. Although the metal shrinks a little on solidifying an intimate contact is ensured. The effect of shrinkage can be entirely overcome by casting about 30 grm. of tin around the contact metal after it has cooled. Heat generated by an imperfect contact causes the tin to melt, flow into the interstices, and to remedy the defect. The costs of this type of contact are said to be low, and the connexion is mechanically strong and its electrical resistance low. An important advantage of this type of connexion is that there are no projecting parts such as brackets or bolts. The electrodes can be used to within less than 1 inch of the metal rods and the necessity of joining electrodes together is thus avoided. The vertical movement of the electrodes requires flexible leads and this is arranged with the Keller electrodes by using thin and very flexible leaf springs of copper. Sheet copper of a thickness of 0·5 mm. is made into two symmetrically arranged

[1] U.S. Pat. 961,139.

bundles and held together at several points by rings. Each side of
the holder has a guiding rod which causes the leaf springs to be

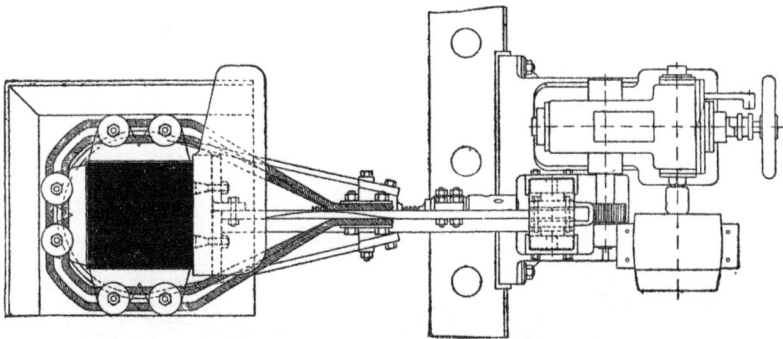

FIG. 229.

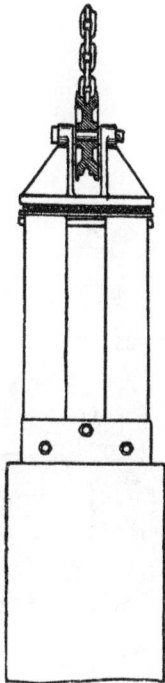

FIG. 230.

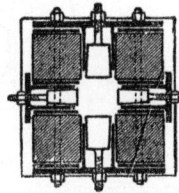

FIG. 231.

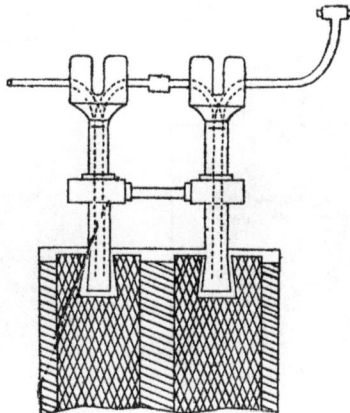

FIG. 232.

compressed or opened in a vertical direction when the electrodes
move up or down. Cf. Fig. 124.

Electrode Terminals in Stassano's Furnace.—In the Stassano furnace the electrodes (cf. Fig. 233) are surrounded by, though not in actual contact with, an annular water-cooled jacket.

Schindler's Water-cooled Holders.[1]—In electrode connexions de-

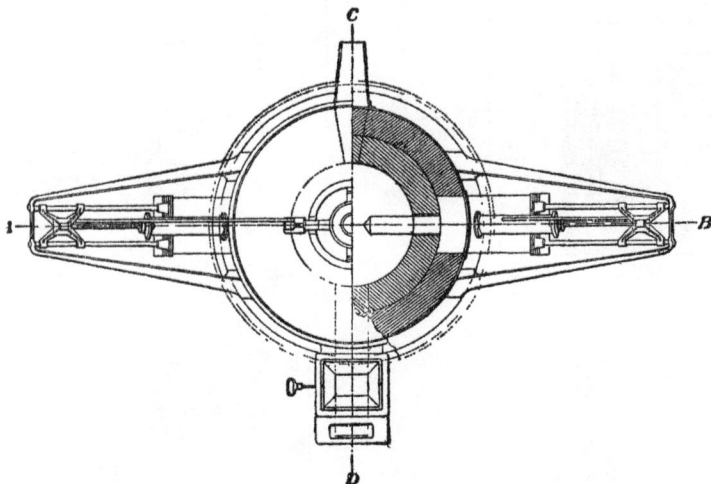

FIG. 233.

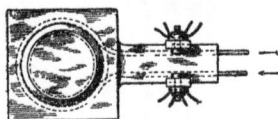

FIG. 235.

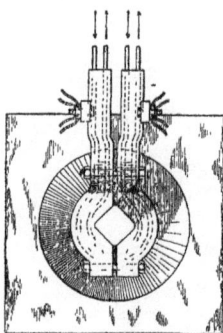

FIG. 234.

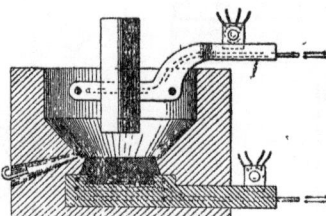

FIG. 236

signed by Schindler, provision is made for the cooling of both top and bottom electrodes by the circulation of water through pipes which are cast into the metal supports. As shown in Figs. 234 to 238,

[1] U.S. Pat. 573,041.

the top electrode holder is made in the form of a clamp tensioned by transverse bolts, while for the bottom electrode, a carbon block is secured by a dovetailed end in a water-cooled metal plate.

Electrodes with Horry Carbide Furnace.—The construction of electrodes and terminals used with the Horry carbide furnace (cf. p. 104) is shown in Figs. 239 and 240. A number of graphite rods are secured by screw connexions to a head consisting of a cast-iron hollow box provided on top with a lug for connecting to the leads. The hollow space is arranged for water circulation. The space between the

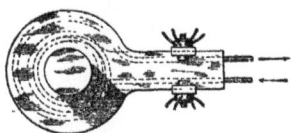

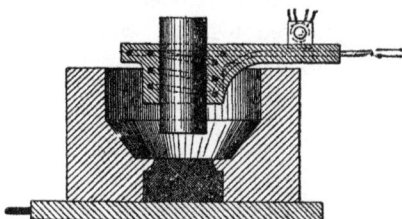

FIG. 237.   FIG. 238.

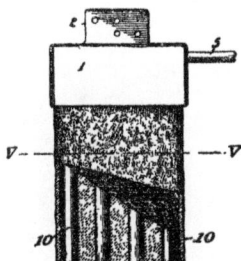

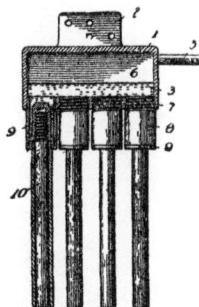

FIG. 239.   FIG. 240.

graphite rods is filled with agglomerated carbon and the whole covered with sheet-iron or wire-gauze to protect from oxidation. These composite electrode plates are then inserted in openings in the furnace roof cover.

Electrode Dimensions.[1]—The current density of electrodes depends on the sectional area. Round electrodes of graphite of from 2·8 to 4 inches diameter, as used, for instance, in the Stassano furnace, can carry 20 to 25 amps. per square cm. (129 to 161 amps. per square inch).

With retort carbon electrodes of a section of 16 by 16 inches,

[1] *Cf.* p. 390.

26

32 amps. per square inch are allowed with a temporary overload of 29 to 48 amps. per square inch. Smaller electrodes have a denser structure, higher conductivity, and greater cooling loss through radiation. The limiting size of electrodes now reached in single blocks is 21·6 by 21·6 inches section, and 25 inches with those of round section. With increasing size the mechanical strength is proportionately decreased, and the current-carrying capacity diminished. The larger electrodes are, in consequence, generally used in bundles or groups. For calcium carbide, ferro-alloy, and similar furnaces, six, eight, or more electrodes each of 10 by 14 inches section are as a rule cemented together with a carbon paste into one bundle and lowered by means of cranes into the furnace.

SECTION XXIII.

POWER EXPENDITURE IN ELECTRIC-FURNACE PROCESSES.

THE importance of the cost of power per unit weight of output in electric-furnace processes varies very much from case to case according to the proportion the power costs bear to other operating expenses and to the value of the product. In the table below the estimated yield and present value of a number of different electro-chemical products for a given power expenditure are given.

In estimating the yield of a particular process per unit of power consumption, regard must be paid to the difference in the amount of energy when measured at its source and at the furnace terminals. In the case, for instance, of hydro-electric power, the turbine losses will enable only from 80 to 90 per cent of the potential energy of the waterfall to be obtained on the shaft of the turbine.

Taking the lower figure of 80 per cent, the following successive losses may be allowed before the power is utilisable at the furnace :—

10 per cent (on 80 per cent) in dynamos,
  balance   72   per cent.
1 per cent (on 72 per cent) in the trans-
  mission lines, balance   71·3 per cent.
3 per cent (on 71·3 per cent) in the trans-
  formers, balance   69·1 per cent.
1 per cent (on 69·1 per cent) in the con-
  ductors, balance   68·5 per cent.

Apart from this actual loss, an expenditure amounting to a further 10 per cent of the above may on an average be allowed for subsidiary operations apart from the furnace process, such as the power needed for crushing or in other ways preparing the raw materials, conveyers and pumps, and the dynamos for lighting. In computing yields based on the output per kw. year, it is necessary, in practice, to define more closely the unit of measurement. Though theoretically the kw. year is equivalent to 8760 kw. hours, yet in practice all plants, though designed for continuous operation, will occasionally be interrupted such as during tapping, changing electrodes, for repairs and inspection, and through accidents. In consequence, 8000 kw. hours may on a general average be taken as representing a kw. year. In the estimates given below[1] the value for the power is that measured at the terminals of the plant in question.

TABLE OF YIELDS AND VALUE OF PRODUCT PER KW. YEAR OF POWER CONSUMPTION.

Product.	Yield per Kw. Year (8000 Hours). Metric Tons.	Value per Metric Ton. £. Feb., 1921.
Calcium carbide, small furnaces	1·0 to 1·2	} 28
large furnaces	1·6 „ 2·2	
Calcium cyanamide (from lime carbon and nitrogen) (20 per cent N_2)	2·3	—
Ferro-silicon (50 per cent Si)	1·0 to 1·1	21
(75 „ „)	0·5 „ 0·6	31
(90 „ „)	0·35 „ 0·4	—
Ferro-chromium (10 per cent Carbon)	1·1	30
(1 „ „)	0·6	108
Ferro-manganese (18-20 per cent Mn)	5·5	—
(70-75 „ „)	2·2	35
Silico manganese and Silico-spiegel	about 10	—
Ferro-nickel (with 50 per cent Ni)	0·3	—
Ferro-tungsten (50 per cent W)	1·0	250
Ferro-molybdenum	0·9	—
Carborundum (crystallised, 0·3 } amorphous, 0·6)	0·9	160
Graphite (bulk)	0·3	280
(formed articles)	0·2	—
Pig-iron (from ore with 50-60 per cent Fe)	2 to 3	} 8
(from mixture of turnings and ore)	8 „ 10	
(from steel turnings)	10 „ 12	
Steel (ordinary grade for casting, from cold turnings)	12 „ 15	} 20
(high-grade from cold charge)	10 „ 12	
(refined in electric furnace from molten charge poured from converter, basic operation)	35 „ 45	
(as above, but with acid operation)	80	
Iron (pure electrolytic)	2	—
Aluminium (of 98-99 per cent Al)	0·25 to 0·27	150
Zinc (by electro-thermic reduction of ore)	1·2	} 25
(Johnson Process with pre-heated charge)	2·5	
(Electrolytic extraction from ore)	2·2	
Copper (Electro-thermic treatment of ore with 25 per cent Cu to give high-grade matter). Yield of metal	2 to 3	} 68
(Electrolytic refining of crude metal)	20 „ 32	
Lead (electrolytic from crude metal)	100	19
Sodium metal	0·55	130
Calcium metal	0·14 to 0·15	—
Magnesium	0·3	2400
Nitric acid (by Arc process)	0·5	70
Carbon bisulphide	7·0	56
Ammonia (as NH_3 by Haber process)	2·0	280

[1] Cf. " J. du Four electrique," 1918, **27**, 49.

SECTION XXIV.

WATER-POWER DEVELOPMENTS AND ELECTRO-CHEMICAL CENTRES.

THE development of a water power is conditioned by topographical circumstances which enable the transfer within a moderately short distance of a supply of water from a high to a lower level. In the rapids of a river or a natural waterfall the energy of this displacement is dissipated as heat and in disintegrating rocks, while by leading the water down inclined or vertical steel tubes, the energy can be applied to the rotation of a turbine and transferred into utilisable mechanical work. The amount of power which can thus be derived is measured by the weight of water passed in unit time and the difference in level between inlet and outlet. In practice a certain loss of energy arises through the resistance offered by the tubes, a value which is proportional to the rate of flow and the length of the pipe and inversely proportional to its diameter or cross section. A further factor which determines the work which can be derived is given by the efficiency of the turbine.

Units.—The actual computation of power is usually expressed in the form of h.p. as unit, which is equivalent to 550 foot-pound per second. In practically all water-power developments the mechanical power is transformed into electrical energy, the yield of which is determined by the efficiency of the electrical generator. The mechanical efficiency of a modern turbine under the most favourable circumstances may be taken as 90 per cent, and that of the electrical generator as 95 per cent, so that the total efficiency in the generation of electrical power will amount to

$$0.9 \times 0.95 = 0.855.$$

It follows from the definition of a h.p. as 550 foot-pounds per second that if 5.50 pounds of water is discharged per second under a head of 100 feet, the energy expended will be at the rate of 1 h.p., or, allowing for the above efficiency, 0.855 electrical h.p.

(e.h.p.) can in practice be generated. Since 1 e.h.p. = 0·746 kilo-watt, the above power will be equivalent to 0·855 × 0·746 = 0·636 kw.

The unit of flow, which is measured in cubic feet per second is generally termed the " cusec ".

It follows from the above efficiency of 85·5 per cent that

$$1 \text{ e.h.p.} = \frac{8·8}{0·855} = 11 \text{ cusecs,}$$

$$\text{or, generally, e.h.p.} = \frac{\text{cusecs flow} \times \text{head in feet}}{10·2}.$$

In practice there is a lower limit for the difference of level below which the water cannot be economically utilised. This limitation is due to the falling efficiency and increasing cost of the large turbines required for developing a flow of large volume of low pressure. A wide swiftly-flowing river may, for instance, possess over a distance of 1 mile a fall of only 1 foot or less, and cannot come into con-sideration for power development.

The different types of waterfalls suitable for power developments can be divided into three categories according to the difference in level or head given. The first class includes falls of low head ranging from about 8 feet, the lowest which can be economically utilised, to about 80 feet. Large volumes of water are required in this class and the turbines used are of the reaction type. In the second class are those of medium head varying between 80 feet and 300 to 400 feet and requiring impulse turbines. The third class contains high heads, varying between 300 to 400 and an upward limit of nearly 5000 feet, for which the Pelton type of jet impulse turbine is used.

PRINCIPLE OF HARNESSING OF WATER POWERS.

The general principles followed in the development of a water power consist in the selection of a feeding-point at some part of the water supply on the higher level where a large uniform flow of water can be diverted to the power conduits. In the case of a stream, some point which is above a natural waterfall or a series of falls or rapids will generally be the most suitable. The water is usually impounded here by means of an artificial reservoir or " forebay " and led to the entrance or intake of the pipe line or " penstock " which slopes downwards to the point of discharge or tail race, the turbines being installed at the lower ends of the pipe lines. To obtain the highest efficiency, it is desirable that the slope of the pipe line should be as steep as possible. The discharge point will in many cases be several miles distant from the intake. Methods which are generally adopted to improve the gradient of the pipe line when the inflow and outflow points are widely separated include the following :—

1. The water is led from the forebay by means of a wide open canal to the intake of the penstocks. Only a comparatively slight loss of head need be involved in the canal, and the length of the pipe line is correspondingly shortened.

2. By excavating the ground at the inflow point or forebay, arranging the penstocks vertically downwards with the turbines at the base at a distance corresponding to the difference in levels and discharging the tail race along a tunnel to the outlet point.

3. In cases where the head of water is obtained by the construction of a dam, the power house may be situated at the base of the dam and a conduit of minimum length obtained. In some cases the power house is constructed in the interior of the dam itself.

Instead of being situated on the site of a natural waterfall or series of rapids, a water power may be developed by discharging water from a lake at a high level to an adjacent stream at lower level or from an inland lake to a loch or sea or by short-circuiting a hair-pin bend in a river.

Minimum Flow.—In the case of a water supply which varies in amount at different periods of the year in accordance with the prevalence of rain, it is only the minimum rate of flow which is of value in large-scale continuous power generation. In order to increase the supply during rainless seasons of the year, it is frequently necessary to arrange for storage either by utilising natural lakes or, by the construction of dams, impounding the water in artificial reservoirs. Storage of this nature must be on a very large scale when a large power is to be developed even when working on a very high head, while with a low head, storage is impracticable. A storage of 31,000,000 cubic feet is required theoretically to give 1 *cusec* throughout the year, or half this amount for a period of six months. This estimate does not take into account losses through evaporation and percolation which increase the necessary storage capacity.

Water Power Available in Different Countries.

Estimates which have been made of the total water power, which is capable of being developed in different countries, are given in the following table, together with the amount at present actually developed :—

Country.	Power Capable of Economic Development. Millions of H.P.	Amount Actually Developed (1920). Millions of H.P.
United States of America . .	60	5·3
Canada	—	1·8
Norway	7·5	1·25
Sweden	6·2	1·1
Finland	2·6	—
Austria-Hungary	6·1	0·5
France	5·9	0·6
Italy	5·6	0·9
Spain	5·0	0·4
Switzerland	1·5	0·3
Germany	1·4	0·6
Great Britain	0·8	0·04
India	—	0·12
British Empire	50 to 70	—

These data emphasise the very backward position of this country in the utilisation of available water-power resources.

Cost of Hydraulic Power.

The cost of hydraulic power is made up mostly of interest and depreciation on capital, sinking fund charges, taxes, and insurance, which are usually much greater than costs of operation, maintenance, and consumable materials. The capital charges vary very widely in different cases; where the available head is great and the supply of water constant they may be comparatively small. Where, on the other hand, much engineering work is required on the construction of dams and culverts and provision for storage, and where the pipe lines and transmission lines are long, the total cost of power may be largely in excess of that generated under favourable circumstances by a steam plant.

In Norway the conditions are particularly favourable for power developments owing to the great height of the falls, often accompanied by easy storage and equable flow. The capital costs of power developments at the most favourable sites in that country has been as low as £6 to £7 per kilowatt developed, while a moderate cost would be £14 to £15. In the U.S.A. a standard cost would be about £26, the cost at Niagara being from £23 to £30. In Scotland, power has been developed at a total capital cost of £26 per kw.[1] These figures relate to pre-war conditions.

[1] Cf. p. 412.

It has been estimated by the Power sub-Committee of the Ministry of Munitions [1] that power can be developed at various places in the British Empire on the basis of pre-war prices at a cost of £10 to £15 per kilowatt.

Some typical estimates of the cost of hydro-electric power developments from a number of widely distributed stations are shown in the following table. [2] The instances marked with an asterisk denote the charges actually made to large consumers for a constant load. The figures are based on pre-war conditions and rates of exchange :—

Station.	Cost.	
	Per E.h.p. Year. £.	Per Kw. Year. £.
Niagara Falls *	3·5 to 4·2	4·7 to 5·7
„ Ontario Power Company to Municipalities	2·1	2·8
„ Power delivered at Toronto . .	2·9	3·9
„ Ontario Power Company to Hydro-Electric Commission of Ontario .	1·8	2·4
Sault Ste Marie, Canada *	2·1	2·8
Cameron Rapids, Ontario	1·9	2·6
Montreal *	3·2	4·3
Kootenay Power Company, British Columbia * .	5·0	6·7
Mexico, El Oro Gold Mines *	10·0	13·5
Kanawha Falls, Va., U.S.A.	2·0	2·7
California	1·8	2·5
Scotland, Kinlochleven	1·7	2·2
Norway, Notodden	0·7	0·9
Sweden, Trollhättan *	1·5 to 2·1	2·1 to 2·8
„ other stations	1·7 „ 3·0	2·3 „ 4·0
Lapland, Porjus Falls *	1·7	2·3
Savoy, Bellegarde, France	2·2	3·0
„ Chedde, France	0·9	1·2
Italy, Tivoli, Power delivered at Rome * . . .	3·0	4·1

POWER AND ELECTRO-CHEMICAL CENTRES.

A centralisation of electro-chemical industries is in many cases taking place in districts, where there is available, at economical rates, large amounts of power to furnish the heavy currents needed in most electro-chemical processes. Electrical energy on a large scale has hitherto been mainly derived from water powers, and recently, to an increasing degree, from steam-power stations.

[1] " Final Report of the Nitrogen Products Committee," p. 28.
[2] J. N. Pring, " Some Electro-chemical Centres ". Manchester University Press, 1908.

410 THE ELECTRIC FURNACE

Load Factor.—The constant steady load taken by electro-chemical processes is specially adapted for obtaining the maximum utilisation from water power, and is also a desirable asset to a steam-power generating station as an increase is thereby obtained in the "load factor," or the average power consumption over any given interval divided by the maximum consumption during that period. A generating station which supplies electricity mainly for lighting and traction purposes will necessarily have a very low-power factor with high peak load, and consequent high cost per unit per current, as a large part of the capacity of the plant will be lying idle during the greater part of the twenty-four-hour day.

Operation During " Off-Peak " Load.—Many schemes have been considered for operating electro-chemical processes intermittently during times of " off-peak " load, so as to serve as " load equalisers," and obtain power at specially favourable rates. However, the disadvantages of intermittent operation have so far prevented these schemes from being applied on any very large scale.

WATER POWER IN GREAT BRITAIN.[1]

The bulk of the power in Britain is situated in the west and north-west of Scotland, in the area north of the latitude of Glasgow, and west of the meridian of Nairn. This area is specially suitable for the development of water power as it has a mean annual rainfall varying from 50 to over 100 inches, and many of the drainage areas lie at a considerable elevation and are situated within a short distance of the sea, while there are numerous lochs which can be made into regulating reservoirs. In the area above mentioned there are at present only two plants of any importance, viz. : those of the British Aluminium Company at Foyers and at Kinlochleven. In addition to these there are within this area twenty-nine sites where the works for regulating the water and developing the power can be economically constructed, giving an aggregate yield of 200,000 h.p.

It is estimated on the basis of a uniform annual rainfall of 42 inches, allowing 28 inches capable of being utilised, that the whole water power available in Scotland from 122 sites amounts to 375,000 h.p. Allowance is not made in this estimate for the additional power available by diverting water from one catchment area to another. In the Loch Ericht basin—which is 25 miles to the east

[1] W. Linton, " Journ. Soc. Chem. Ind.," 1918, **37**, 314 ; Interim Report of the Water-power Resources Committee, Board of Trade, London, 1919. H.M. Stationery Office.

of Fort William—for instance, it is calculated that by bringing into this basin the water from certain adjoining areas, and diverting the whole water to the west coast, it is possible to develop about 34,000 h.p.

The method which has been proposed for utilising most advantageously the water powers in Scotland is to establish works at convenient centres and transmit to these centres the various powers that can be developed within, say, a radius of 20 miles. The most important of these centres would be at Fort William, from which Kinlochleven is distant about 9 miles, and where some 90,000 h.p. could be concentrated. The second is at Inverness, where at least 35,000 h.p. could be supplied. Two centres next in importance are at Poolewe at the head of Loch Ewe in Western Ross-shire, and in the neighbourhood of Lochinver in Sutherlandshire. At these centres 25,000 and 35,000 h.p. respectively could be concentrated. With regard to the cost of development of the above powers, it is estimated by Linton that allowing for working costs, interest on capital, and an adequate allowance for depreciation, etc., the power can be supplied (on the basis of pre-war costs) at less than 0·1d. per unit or £3·7 per kw. year.

Report of Water-power Resources Committee.—The Water-power Resources Committee of the Board of Trade investigated in detail, in 1918, the possibility of water-power development in Scotland. Nine sites which were examined were found to be capable of a total continuous supply of 183,500 e.h.p. It was considered to be practicable to transmit the electrical energy developed at these stations to industrial centres in Scotland, such as Glasgow, Edinburgh, Aberdeen, and Dundee, while as an alternative outlet for a part of this power, industries such as the manufacture of calcium carbide, cyanamide, or other electro-chemical processes could be established close to the generating stations which would be conveniently situated in regard to transport.

The cost of development was estimated on the basis of pre-war contract prices with an addition of 50 per cent, and included all civil engineering and hydraulic works, power house and plant, and compensation for rights and property. The estimates vary from £17 per e.h.p. developed for a power of 9500 e.h.p. at Kilmorack Falls to £45·8 per e.h.p. for 38,000 e.h.p. at Loch Laggan and Loch Treig. The average cost for the development of the total power of 183,500 e.h.p. amounts to £38·5 per e.h.p., while it is calculated that power could be generated at an average cost inclusive of all running expenses and capital charges, of 0·15d. per unit for continuous power at the power station.

Comparison of Water and Steam-power Developments.—It is considered that under pre-war conditions, with coal at 10s. a ton, it would be possible for a large water-power scheme to compete with the best steam-power station, providing the cost of developing the water power does not exceed about £60 per effective e.h.p. With the present largely increased costs of steam-power generation, a much higher capital expenditure would of course be justified in a hydraulic development.

Freight and Transmission of Power.—In comparing the relative merits of steam and water-power stations an additional consideration to be taken into account is that generally with water-power stations, through the isolated position of the site, it will be necessary to transmit the current for a considerable distance to the factories, thus entailing the expense of the transmission lines and a certain loss of energy. On the other hand, steam-power stations can as a rule be situated near the centre where the power is consumed, though in this case the cost of carriage of fuel comes into consideration.

Kinlochleven.—A large power development was completed near Loch Leven for the works of the British Aluminium Company, in extension of an earlier plant which fully utilised the power from the Falls of Foyers.

The Kinlochleven scheme consisted in the construction of a large reservoir at an altitude of 1000 feet above sea-level, and only about 5 miles distant from the coast. By the formation of a dam over half a mile in length, with a maximum height of 80 feet in the middle, a volume of 20,000 million gallons of water is impounded. The water is led along a concrete conduit along the hill face for a distance of about 3½ miles, to the head of the pipe line. The head of water at the turbines is about 900 feet. The following data have been published on this development :—[1]

Drainage area, 55 square miles.

Power generated, 30,000 h.p. (22,800 kw.).

Capital cost of development, £600,000 (£26 per kw.).

The cost of power is estimated as follows : Allowing 5 per cent interest on capital, and depreciation at the rate of 1 to 3 per cent on dam, pipes, buildings, etc., and 7½ per cent on the whole of the power plant, the total cost is estimated at $\frac{1}{16}$d. (0·0625d.) per kw. hour

[1] W. M. Morrison, " Proc. Instit. Civil Eng.," 1912, 187, 93.

or £2·28 per kw. year of 8760 hours. Of this, labour (beginning at the dam and ending at the switchboard), repairs, oil, waste, stores, and other industrial charges amount to 0·00457d. per unit. It is further estimated that when additional units are added, and power developed up to the full capacity of the present pipes, a cost as low as $\frac{1}{20}$d. per kw. hour would be approached.

Falls of Foyers.—The British Aluminium Company began the erection of a works here in 1896 for the preparation of aluminium, after taking over the works of the Cowles Syndicate at Milton, Staffordshire. Use is made here of a water power between Foyers river and Loch Ness, a head of over 300 feet being obtained and a total of 6000 h.p. generated.

Lake Eigiau, near Conway, North Wales.—A development of about 4500 h.p. from this lake was commenced in 1907, and applied mainly for the manufacture of aluminium by the Aluminium Corporation, and for the operation of a light railway in the district.

In the Report of the survey made by the Water-power Resources Committee,[1] it is stated that at one site in North Wales, a power is capable of developing 4400 continuous e.h.p. at an estimated capital cost of £41·8 per effective e.h.p.

Water Powers in England.—Numerous small water powers in many parts of England as the Lake District, Devon, and Cornwall, ranging from 100 to 2000 e.h.p., might be usefully developed for purposes such as lighting, but could not come into consideration for any large scale industrial work.

Hydro-electric Power in America.—It is estimated[2] that in the United States the total hydro-electric force available amounts to 60,000,000 h.p., of which 5,350,000 h.p. have been already harnessed. The cost of installation is generally more expensive in the U.S.A. than in Canada, France, and Switzerland and particularly Norway where the cost is lowest, and amounts to from $100 to $200 per h.p. for all charges.[3] At Niagara Falls the cost of supply to electro-chemical works is from $18 to $22 per h.p. year. In other parts of the States the price is generally higher and probably averages $25 per h.p. year.

Canada.—Within the provinces of the Dominion of Canada and excluding the North-west Territories, practically all of the Yukon, and the Northern and Eastern portion of Quebec, it is estimated[4]

[1] Loc. cit., p. 4.
[2] " Journ. du Four electrique," 1919, **23**, 157. [3] Cf. p. 408.
[4] " Water Powers of Canada," Ottawa, 1916, Department of the Interior.

that 17,764,000 h.p. is available, this amount being inclusive, in the case of Niagara Falls, Fort Frances, and the St. Mary's river at Sault Ste Marie, of only the development permitted by International treaties, and does not include the full possibilities of storage for the improvement of capacities. The developed water powers in Canada in 1919, including all water powers whether for electrical production, pulp grinders, for milling or other uses is reported to be as follows :—

Province.	Horse-power Developed.
Nova Scotia	21,412
New Brunswick	13,390
Prince Edward Island	500
Quebec	586,000
Ontario	831,000
Manitoba, Saskatchewan, and Alberta . . .	120,000
British Columbia	265,345
Yukon	12,000
Total . .	1,849,647

The cost of power in Canada for electro-chemical and metallurgical works taking a steady load averages about $15 per h.p. year ($20·1 per kw. year) ; the capital installation cost representing about $100 per h.p. ($134 per kw.).

The more important hydro-electric centres on the American continent which serve as public supply companies are as follows :—

NIAGARA FALLS.

The Niagara river, in its 25 mile course between Lakes Erie and Ontario along the boundary of the United States and Canada, has a change of level of 330 feet, including a drop of 165 feet at the Falls themselves. The flow of the Niagara river is very even throughout the year and amounts to about 220,000 cubic feet per second. In a treaty of 1910 between the United States and Canadian Governments, the amount of water which may be diverted for power purposes was limited to 56,000 cusecs, or about 25 per cent of the mean flow. Of this total, 36,000 cusecs was assigned to Canada, and 20,000 to the United States. The total power now being generated on both sides of Niagara is 653,000 h.p. Of this amount, 265,000 h.p. is generated in the United States, which also receives 110,000 h.p. transmitted from the Canadian side, making a total of 375,000 h.p. available in the United States. The United

PLATE XVI.—HYDRAULIC POWER COMPANY, NIAGARA FALLS—POWER HOUSE AND GATE HOUSE,

(The Hydraulic Power and Manufacturing Company.)

PLATE XVII.—HYDRAULIC POWER COMPANY, NIAGARA FALLS—GENERATOR ROOM.

(The Hydraulic Power and Manufacturing Company.)

States and Canadian Governments are conferring regarding a proposal to allot another 10,000 cusecs.[1]

At present, the following are the most important of the power companies :—

Approximate power development in 1918.

U.S.A.	E.H.P.
Hydraulic Power and Manufacturing Company	145,000
Niagara Falls Power Company	140,000
Canada.	
Canadian Niagara Power Company	110,000
Ontario Power Company	180,000
Electrical Development Company of Ontario	110,000

Hydraulic Power and Manufacturing Company.—The oldest power project at Niagara Falls is that of the Hydraulic Power and Manufacturing Company which was incorporated as early as 1853, when steps were taken for the construction of a canal 70 feet wide by 10 feet deep, which was later enlarged to a width of 100 feet and a depth of 14 feet. In 1881 power was first supplied for commercial purposes, in 1885, 10,000 h.p. was in use. In 1896 the erection of a second power house was undertaken and situated in the gorge immediately below and facing the Falls. A head of 210 feet is obtained, the intake of the canal being at a point 2 or 3 miles above the Falls. The plant has on several occasions been enlarged and now generates 145,000 h.p. (cf. Plates XVI and XVII). Penstocks are led vertically downwards from the canal to each of the turbines of 8000 h.p. capacity. The turbines are of the horizontal shaft type running at 300 revolutions per minute, the alternators directly coupled to the turbines giving three-phase current at 11,000 volts, 25 cycles.

Niagara Falls Power Company.—The Niagara Falls Power Company scheme was commenced in 1890 as the outcome of a project to construct a large central station and develop an industrial centre. The station is situated on the United State side of the Niagara river at a point about 1 mile above the Falls. A short canal was made to lead the water to the power house which now consists of two buildings on opposite sides of the canal, and the water, after passing through iron gratings to remove any debris or ice, is led through penstocks and thence vertically downwards, a distance of 178 feet to the turbines. The turbines are installed near the bottom of two wheel slots, excavated out of the rock, under

[1] " Metall. and Chem. Engineering," 1918, 18, 439.

the respective power houses.   Each turbine is connected by a vertical shaft which extends upwards to an electric generator installed above on the ground level.   Two-phase alternating current is thus generated at 2200 volts and 25 cycles, operating at a speed of 250 revolutions per minute, each alternator being of 5000 h.p. Three separate systems of transmission from the power house are in use.   For local transmission, two-phase current at 2200 volts is supplied as generated, and conveyed by underground cables. For intermediate distances the current is stepped up to 11,000 volts, three-phase, and transmitted, first by underground cables, and then by overhead conductors to a transformer station about 2 miles from the power house, where it is reconverted to 2200 volts, two-phase.   Long-distance transmission, such as to Buffalo, about 18 miles distant, Tonowanda and Lockport, 35 miles distant, is effected by three-phase current at 22,000 volts along bare overhead conductors.   The Niagara Falls Power Company is now consolidated with the Hydraulic Power Company.

Canadian Niagara Power Company.—After the completion of the second power house of the Niagara Falls Power Company in 1902, a further installation was commenced on the Canadian side by the same company, operating under the title of the Canadian Niagara Power Company.   This plant is situated a short distance above the Horse Shoe Falls, and is in most respects similar in equipment to the older power houses, but the units are here 10,000 h.p. machines, and have a developed capacity of about 110,000 h.p.[1] The two plants of the Niagara Power Company are interconnected by copper cables, so that when necessary, the two plants can be run in parallel for supply either on the American or on the Canadian side.

The Electrical Development Company of Ontario.—This company was formed at Toronto in 1903 for the purpose of generating power from the Niagara river and transmitting to Toronto and other places in Ontario.   Rights were acquired for the generation of power to the extent of 125,000 h.p. and a site secured at Tempest Point, about half a mile above the Falls, for the location of the power plant.   The company also purchased a right of way 78 miles in length between Niagara Falls and Toronto, and an area about 2 miles from the Falls and facing the Chippewa river, which communicates with the Welland Canal, where an industrial district has been established.   The intake to the power house was constructed by projecting a short coffer dam on to the bed of the river, which

[1] Cf. Plate XVIII.

PLATE XVIII.—CANADIAN NIAGARA POWER COMPANY—INTERIOR OF POWER HOUSE.

(The Canadian Niagara Power Company.)

at this point has a rapid fall. The flow of water was thus diverted to the intake, and its level raised about 18 feet. The power house now contains about 11 generators, each of 10,000 h.p. The tail water is led away by a tunnel under the upper rapids which terminates at the base of the Horseshoe Fall, about midway between the Canadian and American banks.

Ontario Power Company.—The third power plant on the Canadian side is that of the Ontario Power Company. This power house is situated in the gorge near the base of the Falls, and the water is taken in at a point above the upper rapids, where a large deep forebay, with smooth water surface, and a series of ice screens have been constructed. The water is led from here to the penstocks, a distance of about $1\frac{1}{4}$ miles, whereby an effective head of 175 feet of water on the turbines is secured. The turbines are arranged horizontally, and coupled direct to the generators. About 18 generators are now in operation, each of 10,000 h.p., which supply power at 12,000 volts, three-phase, 25 cycles. The transforming and distributing station is situated in a building 500 feet distant and 280 feet higher on the cliff. The generators are all controlled from this station and the current is here stepped up to 60,000 volts for long-distance transmission. This company was taken over by the Ontario Hydro-electric Power Commission in 1917, who are arranging to provide payment within twenty-five years out of revenues from contracts with municipalities and with the Niagara Lockport Company. Power contracts have been made at $9 per h.p. year for 100,000 h.p., and at $12·50 with the Niagara Lockport Company for 60,000 h.p.

This Commission has developed the largest system of electric transmission in the world, and now owns altogether ten plants in various parts of the province, aggregating 248,000 h.p., and supplying nearly 200 municipalities, to which Niagara contributes the supply for 118. The transmission lines comprise 455 miles of 110,000 voltage double circuit, and 2100 miles of low tension.

The Chippawa-Queenston Project.[1]—This undertaking is an extension of the system of the Hydro-electric Power Commission of Ontario, and will within a few years' time increase its serviceable capacity by from 200,000 to 300,000 h.p., and ultimately by 1,000,000 h.p. The first instalment of the work is expected to be ready by the spring of 1921, at an estimated cost of £5,000,000.

Instead of the relatively small head of the actual Falls, it is

[1] "The Engineer," 1919, **128**, 427.

proposed tó utilise 307 feet of the difference in level, amounting
to 330 feet, between the surfaces of Lakes Erie and Ontario.   For
this purpose water will be taken from the Welland river which at
present enters the Niagara river at a point 2 miles above the Falls.
The Welland river itself will be deepened and widened for a distance
of 4½ miles from its mouth, and along this stretch the direction of
the flow of water will be reversed, and by means of a canal leading
from the Welland river the water will be conveyed for a distance
of 9 miles to the power station at Queenston, and after passing
down the short length of penstocks will be discharged into the
Niagara river.   The head of 307 feet which will be given will
provide the generation of 30 h.p. per cusec instead of the 14 h.p.
per cusec given by the station of the Ontario Power Company.
The mean velocity of flow will be 2 feet per second in the Welland
river and 6 feet to 7 feet per second in the canal, resulting in a
discharge of 10,000 cusecs at low water level.   At the delivery end
of the canal a forebay will be excavated.   Penstocks of riveted
steel plates, 14 feet in diameter and about 450 feet in length, will
extend down to the power house at the foot of the river bank.   The
initial generating plant will consist of 4 to 6 units of 50,000 h.p.
each.   On each turbine shaft there will be a three-phase 25 cycle,
12,000 volt internal revolving field generator of 43,900 kilovolt-
amperes at a power factor of 85 per cent.

Industries at Niagara Falls.—The large bulk of the power
developed from the Niagara river is utilised in the immediate neighbourhood for industrial and mainly electro-chemical purposes.   Projects have been suggested at different times for the transmission of
power from here to distant places such as New York and Chicago,
but the demand in the immediate neighbourhood has increased more
rapidly than the productive capacity.   The result has been that the
power produced has been absorbed locally, or within such moderate
distances as Buffalo (18 miles), Toronto (80 miles), and Syracuse
(160 miles).   The main industrial district at Niagara Falls is situated
on the American side.   A list of these works which are supplied by
the two American Power Companies, together with the amount of
power being employed in April, 1918, is given in the following
table :—[1]

[1] " Met. and Chem. Engineering," 1918, **19,** 120.

(1) *Hydraulic Power Company.*		(2) *Niagara Falls Power Company.*	
	H.P.		H.P.
Aluminium Company of America	59,886	Buffalo General Electric Company	26,182
Ice Companies	250	Acheson Graphite Company	7,908
General Abrasives Company	1,706	Hooker Electro-Chemical Company	9,165
National Electrolytic Company	2,892	Isco Chemical Company	709
Union Carbide Company	42,500	Mathieson Alkali Company	11,750
Hooker Electro-Chemical Company	6,534	Niagara Electro-chemical Company	13,506
Isco Chemical Company	1,191	Niagara Alkali Company	1,363
National Carbon Company	4,761	Oldbury Electro-Chemical Company	2,654
Niagara Alkali Company	6,014	Star Electrode Works	5,683
Oldbury Electro-chemical Company	6,000	Carborundum Company	12,700
United States Light and Heat Company	1,319	Union Carbide Company	17,789
Titanium Alloys Manufacturing Company	7,074	Other Smaller Companies	23,935
Other smaller Companies	4,514		
Total	144,641	Total	133,344

WATER POWERS IN TENNESSEE AND SOUTHERN STATES.[1]

Muscle Shoals.—It is estimated that a total power of 660,000 h.p. is available from the Tennessee river in the neighbourhood of Muscle Shoals. A development is now taking place, and it is estimated that power will be generated at a cost of \$7 to \$9 per h.p. year. This district is well supplied with raw materials and industries which have already been installed here and operated by steam power include the manufacture of ferro-alloys and ferro-phosphorus, while a large factory for the production of nitrates has been erected by the United States Government[2] for war requirements, the process being operated by steam power until the water power is developed. Other electro-chemical centres in Tennessee are as follows :—

Maryville, near Knoxville.—An aluminium plant is in operation here, the raw materials being obtained from Georgia. A large hydro-electric development to give eventually 400,000 h.p. is being planned on the Tennessee river in this neighbourhood.

Chattanoga.—The manufacture of ferro-alloys has recently been commenced here. High-grade manganese deposits are available in East Tennessee and the adjacent portion of Georgia.

Kingsport, Tennessee.—At Kingsport there is a large steam-power plant which furnishes current for several chemical and electro-chemical operations.

[1] W. G. Waldo, " Met. and Chem. Engineering," 1918, **18**, 444.
[2] Cf. p. 138.

27 *

Caxton, Maryland.—An electrolytic copper refinery is in operation here with an output of 360,000 tons per annum.

In the Southern States there are good prospects for the smelting and refining of zinc, the ore of which is mined by the American Zinc Company. Bauxite is very extensively mined, the output in 1917 being 400,000 tons.

It is considered that the electric smelting of iron ores can be advantageously undertaken in this district. On account of the high sulphur and phosphorus content of the ore, this cannot be successfully treated by the Bessemer process.

In the Southern States very cheap fuel is available, and it has been shown that steam power plants can compete with water power in this district where specially favourable hydro-electric developments are possible.

Natural Gas.—A promising field for economic power development in this district is from natural gas by utilising directly in gas engines. A large supply containing 95 per cent methane is available at Caddo in Northern Louisiana.

Power Stations in the Southern States.

The Tennessee Power Company operates a number of hydro-electric power stations which are interconnected and, in 1918, had a combined capacity of 85,000 kw. in addition to steam-power stations with a total capacity of 24,000 kw. This company supplies power to the Aluminium Company of America at Maryville, Tennessee, and to the American Zinc Company at Mascot, Tennessee.

The Georgia Railway and Power Company at Tallulah Falls in 1918 were generating 100,000 h.p., which is being extended to 200,000 h.p.

The Southern Power Company in North Carolina has a number of interconnected stations, having an aggregate capacity of 160,000 h.p. The transmission lines are constructed for a voltage of 100,000, and the circuits of this company are connected with those of the Carolina Light and Power Company.

Pacific Coast.[1]

Large amounts of water power are available on the Pacific Coast much of which can be developed at a low cost, and large mineral resources and good transport facilities are also available.

---

[1] J. W. Beckman, " Met. and Chem. Engineering," 1918, **19**, 30.

Of the extensive iron deposits California possesses an ore of high quality while abundant supplies of timber are available for the production of charcoal. The electrical production of iron and steel has been successfully established in California.[1] Another promising field for the operation of electro-chemical processes in this district is in the manufacture of artificial fertilisers for which there is a large demand in California. A large supply of limestone would be available for the manufacture of cyanamide and calcium nitrate. A large possible field for minor industries also exists in the manufacture of ferro-alloys and chemicals for which a large market is offered by China, Japan, and Australia.

CANADA.

Shawinigan Falls.[2]—Shawinigan Falls is situated about half-way between Quebec and Montreal, and at a distance of about 85 miles from the latter. The town is served by two railways, the Canadian Northern and the Canadian Pacific Railways. A large power development has been carried out here by the Shawinigan Water and Power Company. Very wide concessions were originally conferred by the Government on this Company in the expropriation of land required for transmission lines and in other ways. At the present time there has been installed at the generating stations at Shawinigan Falls and Grand' Mere, Quebec, 330,000 h.p., of which 265,000 is used for the generation of electric power, and the balance of 65,000 is utilised directly as mechanical power by certain of the local industries at Shawinigan Falls. The Shawinigan Company further owns the Grès Falls, 6 miles below Shawinigan Falls on the same river, at which point 75,000 h.p. can be developed as soon as required. Moreover, at Grand' Mere, 9 miles above Shawinigan Falls, provision has been made for the addition of 60,000 h.p.

The principal industries at present in operation are as follows :—

The Northern Aluminium Company.—This is a branch of the Aluminium Company of America and is situated outside the main electro-chemical district, but adjacent to the power developments of the Shawinigan Company. The Aluminium Company has its own generating station, though in addition, power is now purchased by them. The works cover about 15 acres of ground and contain, besides the reduction rooms for the production of aluminium, an ingot room, a rolling mill, and a wire-drawing and cable plant.

[1] Cf. p. 186.
[2] H. C. Randall, " Met. and Chem. Engineering," 1918, **19**, 561.

The Canada Carbide Company occupies some 15 acres of land in the industrial district. The company manufactures calcium carbide and acetylene gas and utilises about 50,000 h.p.

The Shawinigan Electro-metals Company, whose plant was constructed and brought into operation in 1915, produces metallic magnesium in many forms and occupies about 5 acres with its various processes. About 2500 h.p. is consumed in both alternating and direct-current furnaces.

The Canadian Electrode Company first began operations in 1915, and subsequently extended. This company manufactures the larger sizes of carbon electrodes for electric furnaces, and has an output of about 3 tons per day.

The Canadian Electro-products Company manufactures acetic acid, acetaldehyde, paraldehyde, acetone, and other similar products from acetylene gas supplied by the Canada Carbide Company. This plant was primarily constructed to supply acetone and acetic acid to the British Government, but now continues as one of the most important industries at this centre (cf. p. 110).

The Prestolite Company constructed in 1917 a plant for compressing acetylene gas for charging Prestolite cylinders. The works are situated near the Canada Carbide Company, from where the calcium carbide is purchased.

Fraser, Bruce, & Company in 1917 erected a small electric furnace plant of about 1000 h.p. for the manufacture of low-phosphorus pig-iron. This plant has been very successful and is now being extended.

The Canadian Aloxite Company, a branch of the Carbide Company, erected in 1917 a plant covering about 15 acres of land near the upper end of the electro-chemical district. In these works the Company utilises over 20,000 h.p. for the production of such electric furnace products as aloxite, carborundum and ferro-silicon.

The American Electro-products Company.—In 1917 the requirements of the U.S. Government for acetic acid led to the establishment of a plant adjacent to and practically duplicating the Canadian Electro-products Company's plant.

The Canadian Ferro-alloys, Limited.—This is the latest addition to the Shawinigan electro-chemical district, with a plant erected near the Canadian Carbide Company for the manufacture of ferro-silicon, and utilising some 10,000 h.p. While not electro-chemical, two industries in the Shawinigan district are worthy of mention on account of the large amount of power they consume—the Laurentide Company at Grand' Mere utilising 28,000 h.p., and the Belgo-

Canadian Company at Shawinigan Falls utilising 18,000 h.p. The companies are engaged in the manufacture of paper and pulp.

SAULT STE MARIE.

A water power has been developed at Sault Ste Marie by utilising the difference in level—amounting to about 20 feet—between Lakes Superior and Huron. About 18 feet of this difference in level is concentrated at the St. Mary's Rapids. The total energy available from these rapids during times of minimum flow, is estimated at about 110,000 h.p., while about 60,000 h.p. is at present utilised.

The method by which this water power is controlled is by the construction of three parallel canals through the narrow neck of land which separates Lake Superior from Lake Huron. The largest of the canals is on the American side of the boundary, and is 2½ miles long, 200 feet wide and 22 feet deep. The water, to the amount of 108,000,000 cubic feet per hour, passes into a wide forebay, through and across which is the power house owned by the Michigan Lake Superior Power Company, which extends over 1368 feet in length. The turbines, on account of the low available head, number 320, each of 125 h.p. The Union Carbide Company, which adjoins this power house, takes 10,000 h.p. for the manufacture of calcium carbide. The two smaller canals are on the Canadian side, where about 20,000 h.p. is developed and employed by the Algoma Steel Corporation and its allied industries.

PROVINCE OF QUEBEC, MONTREAL DISTRICT.

The Cedars Rapids Manufacturing and Power Company produces 90,000 h.p., of which 60,000 h.p. is transmitted to aluminium works at Massena, and the remainder to Montreal.

The Canadian Light and Power Company.—The plant of this company is at St. Timothee, on the opposite bank of the St. Lawrence from the Cedars Rapids Company. A power of 20,000 h.p. is developed here and transmitted to Montreal, where it is mainly used for the operation of the tramways.

The Lachine Rapids, Montreal.—The total power capacity of these rapids is estimated at 400,000 h.p. A total development has so far been made of 30,000 h.p.

At Chambly, a hydro-electric plant on the Richlieu river provides Montreal with 20,000 h.p. for light and power purposes.

BRITISH COLUMBIA.

West Kootenay Power and Light Company.—This company develops power at the Bonnington Falls on the Kootenay river, which is a tributary of the Columbia river, and situated in the centre of the mining district of British Columbia. A natural head of 70 feet is obtained at these falls, and by the construction of a dam, the head has been increased to 80 feet.

The power house is of reinforced concrete built actually in the old river bed, a coffer dam having been constructed from the bank to an island in the river, thus removing the water from the whole site. The water is led to the turbines down a tube formed in concrete without any steel lining. Each 8000 h.p. turbine consists of three inward flow Francis runners. The electrical equipment consists of two 5625 k.v.a. 3-phase, 60-cycle, Canadian General Electric "Umbrella" type generators. A second power station at the lower Bonnington Falls, 1 mile lower down the river, develops 4000 h.p. Power is also developed by this company at a station on the Kettle river, where there are installed three 750 k.v.a., 3-phase, 60-cycle generators.

British Columbia Electric Railway Company, Ltd., develops at Lake Buntzen, near Vancouver, a total power of 84,500 h.p. The development has consisted in raising the surface of Coquitlam Lake by the construction of a dam, and delivering the water thereby stored through a tunnel $2\frac{1}{4}$ miles long to Lake Buntzen, a small lake due west of Coquitlam Lake, and thence through steel pipes to a power house on the shore of Burrard Inlet, an arm of the sea. This same company has also developed a power of 25,000 h.p. at Jordan river, Vancouver Island. In both cases the power is utilised for lighting, industrial purposes, and the operation of tramways.

The Powell River Company, Ltd., has a development of 24,000 h.p. at the Powell river.

HYDRO-ELECTRIC STATIONS IN SWEDEN.

Power Stations for Notodden Factories.—The Notodden factories, which now use about 60,000 h.p., secure their power from two neighbouring waterfalls, Lienfos and Svaelgfos. At Lienfos, about 2 miles from Notodden, the power station is equipped with four units of 5000 h.p. each. The Svaelgfos power house is situated above and about 1 mile from Lienfos. Four sets of turbo-generators are installed and generate 10,000 h.p. each. The power of these two stations is transmitted to Notodden by six separate transmission lines. The three-phase alternating current of fifty periods is transmitted with a voltage of 10,000.

A second power station at Svaelgfos is now under construction in which by additional regulation of distant lakes in the same watershed, the operation of two units, each of 10,000 h.p., will be possible.

Rjukan Hydro-electric Power Station.[1]—The supply of water for the power station comes from Lake Mösvand, 3000 feet above sea-level. The surface of the lake has been raised, and the flow regulated by means of a dam. The Rjukan power works commence at a distance of 5 miles below the Mösvand dam. An intake reservoir of about $\frac{3}{4}$ square mile area is formed by means of a dam built across the river Maana on the top of the Skarsfos. The water is conducted from this reservoir to the power station of Vemork (Rjukan 1) by a tunnel excavated in the rock. The tunnel has a length of 12,907 feet and a fall of 29 feet, corresponding to an average gradient of 1 : 480. From the tunnel the water is conducted into a distributing basin, cut entirely from the rock, the sides being lined with concrete. From the distributing basin the water is led to the power station in ten large steel pipes. Each pipe is 2296 feet long, and the total height of the fall 941 feet. The diameter of the pipes decreases from 79 inches at the top to 49 inches at the bottom. The upper part of the line is of riveted, and the lower of welded, pipes. The pipes traverse a valve house placed below the distributing basin. In order to reduce the danger of injuries arising from breakage of the pipes, a valve device is introduced in the pipe line so that the flow of water can be cut off automatically in case the speed at any time is too great. The cut-off valves can also be manipulated from the power station electrically.

The power station has an interior length of 360 feet and breadth of 67 feet. Ten main turbines are installed, each developing 14,500 h.p. at 250 revs. The water running from the turbines is collected in a common lower race from where it is led directly into the tunnel leading to the power works at Saaheim (Rjukan 2). The turbines at the Vemork plant are directly coupled to three-phase alternating current generators of a normal output of 17,000 k.v.a. each. Nine of these are constructed as double generators, consisting of two generators of 8500 k.v.a. each, built together on a common shaft, whereas the tenth generator is built as a single unit. The normal working voltage is 10,500 and the power factor 0·6. The total energy developed at this station is 145,000 h.p.

The whole of the power from these stations is used in connexion with the nitric acid process (cf. p. 122).

[1] " Engineering," 1914, **97**, 141.

The factory at Saaheim was opened in 1915, and is operated from its own power house, which contains 9 turbo-generators, each of 18,000 h.p., or a total of 162,000 h.p., and utilises a head of water of 800 feet (cf. p. 122). The water used is led by means of a canal from the tail race of the Vemork station, 3 miles higher up the valley.

A view of the Vemork Power Station is given in Plate XIX.

THE PORJUS WATER-POWER DEVELOPMENT, LAPLAND.

Considerable electrical developments have been made from the extensive water-power resources in the North of Sweden.

This district extends northward for some distance in the Arctic Circle and is served by a railway which connects ports on the Gulf of Bothnia, and crossing the Scandinavian peninsular in a north-easterly direction terminates in Norway at Narvik, a port on the North Sea. The largest water course in this region is the Lule river, on which, from a series of prominent Falls, a total of 300,000 e.h.p. is capable of being developed. The principal Fall is at Porjus near the source of the river from the Lule lake, where a total head of about 160 feet is available. A dam 1350 yards long has now been constructed above the Falls, to regulate the flow from the lake, and the water is led by means of a canal 650 yards long to the forebay of the power station. The danger from ice formation is reduced by leading the canal at a short distance from its source through a tunnel in the rock, whence it emerges in a small forebay in front of the power station. From here the water passes down vertical penstocks to the turbines 150 feet below, and from there the tail race is led away by a common tunnel for a distance of 1300 yards to the river basin below the Porjus rapids. The turbo-generators are thus in an underground hall which has been excavated out of rock at a depth of 150 feet below the surface. The switchgear building is situated on the ground level directly over the power house, and communication between the two is provided by a cable shaft. The warm air leaving the generators is used to heat and keep dry the buildings. In 1915 there were five turbines in operation, each of 12,500 h.p., while the general lay-out was arranged to allow of the duplication of this installation. Current is generated at 4000 volts single-phase, 15 cycles, with some of the generators and three-phase at 10,000 volts, 25 cycles with others.

The switchgear occupies two floors of the surface building. The

PLATE XIX.—VEMORK (RJUKAN I) POWER STATION.

(The Norwegian Hydro-Electric Nitrogen Company.)

transformers are placed on the lower floor, receive the current from the generators, and step up to 80,000 volts with the single-phase and 70,000 volts with the three-phase current. Power at this voltage is transmitted to intermediate transforming stations, the farthest of which is near Riksgrans, a distance of 160 miles from Porjus. At the transforming stations the power is transformed down to 15,000 volts for the operation of the railway. This railway, known as the Riksgräns Railway, is the one connecting Sweden with the Norwegian port of Narvik above referred to, and is now operated electrically over its central section. The chief importance of the railway is in the transport of iron-ore from the deposits at Kiruna. The maximum power consumption on the electrified section amounts to 22,000 e.h.p. Power from the Porjus Falls is also supplied to the Luossavaara-Kirunavaara Mining Company and to a recently erected electric smelting works at Porjus.

THE PORJUS SMELTING WORKS.[1]

The Aktiebolaget Porjus Smältverk contains at present the following installation :—

(*a*) Three ferro-alloy furnaces, two of which are operated with 650 kw. single-phase current, and one two-phase furnace of 2000 kw. The transformer for this furnace is Scott-connected, the voltage between the common return and the other electrodes being about 80 volts.

(*b*) Two electric shaft furnaces for the smelting of iron-ore. These are each of 3000 k.v.a. capacity, constructed with six electrodes and equipped with three single-phase transformers with a high-tension of 10,000 volts, 25 periods, and a low-tension varying between 110 and 60 volts. Two electrodes are connected to each transformer.

The iron-ore used at these works is from the Lapland Iron Ore Mines and has the following average composition :—

	Per cent.
Si	0·3 to 1·0
Mn	0·3
P	0·020
S	0·010

The pig-iron produced is mainly used for steel production in the open-hearth and crucible furnaces.

[1] From data kindly supplied by the Aktiebolaget Porjus Smältverk, Stockholm.

HYDRO-ELECTRIC STATIONS IN INDIA AND BURMA.[1]

The position in the interior of India is characterised by the occurrence of enormous water power supplies during the limited monsoon season and the comparative drought during the larger part of the year. Many of the rivers fall to a very low ebb or dry up altogether in their higher reaches before the end of the dry season. On this account the impounding of water supplies in lakes or river valleys is generally necessary for utilisation in power development. Storage of water has for a long time been carried out for irrigation purposes. Irrigation will always make the first demands on water supplies, yet after all requirements for this purpose have been met, there will still remain large supplies available for power generation.

Many of the rivers in Northern India are perennial on account of their origin from the snows of the Himalayas. However, though the world's greatest water powers are to be expected in these districts, the disadvantages attending these instances is that they are situated almost exclusively in the trans-frontier regions and exhaust most of their fall to the plains in Kashmir, Tibet, Nepal, Sikkim, and Bhutan. The boundary lies for the most part near the foot-hills, and while the water can be used thereafter for irrigation, it has reached the inert stage from the power point of view. In the province of Sikkim, however, which is under British administration, development of water power is now taking place, while in Burma there are large perennial rivers and streams which offer great possibilities.

The amount of hydro-electric power which has so far been developed in India amounts to a total of 83,000 kw. (111,860 h.p.). The principal developments are detailed in the following list :—[2]

1. *The Tata Hydro-electric Station at Khapoli, Western Ghats, Bombay.*—This installation is operated entirely from water stored in lakes during the few monsoon months. From the lakes the water is led by means of canals and a tunnel over a distance of 4·6 miles through the watershed to the forebay from which the pressure pipes, of a total length of 13,000 feet, lead down to the power house.

The head obtained is equal to a vertical height of 1725 feet, 67,000 e.h.p. is developed continuously. The total cost of hydraulic development, power house, and plant, excluding transmission, is estimated at £27 per kw.

---

[1] "Preliminary Report on the Water Power Resources in India," J. W. Meares, Calcutta, 1919, Superintendent Government Printing.

[2] In the estimates of the cost of development, the figures given are converted from rupees at the pre-war rate of Rs.15 = £1.

The power from this station is all utilised for miscellaneous industrial purposes in Bombay.

2. *Cauvery Power Scheme, Mysore.*—A power is now developed on the Cauvery river of 22,650 e.h.p., which by the construction of a bigger dam, is now being increased to give 40,000 e.h.p. The water is led from the supply point by an open channel for a distance of 17,000 feet to the pipe line which is 1000 feet in length and provides a vertical head of 420 feet. The cost of development inclusive of all items except transmission, amounted to £37 per kw. The power from this development is utilised at the Kolar Gold Mines and for general supply in Mysore and Bangalore.

3. *Jhelum Power Installation, Kashmir.*—This development is derived from rapids in the Jhelum river. No storage is provided. The forebay is situated at the point of supply from the river and the pipe line of 750 feet length gives a vertical head of 395 feet. The total power now generated amounts to 5360 e.h.p. which is capable of extension to 20,000 e.h.p. The cost of development, inclusive of all items except transmission, amounted to £38 per kw.

Smaller water-power installations in operation include those at Darjeeling in Bengal; the Gokak Water Power Company in Bombay; Jammu in Kashmir; the Government Cordite Factory in the Nilgiri Hills, Madras; the Simla Municipality plant on the Sutlej river; and the Mussoorie-Dehra Scheme in the United Provinces. In the Mussoorie-Dehra scheme, a development of 2400 e.h.p. has been made at an inclusive capital cost of £16 per kw.

In Burma, there are two water-power developments each of 500 e.h.p. installed capacity at the Burma Ruby Mines, Mogok, and the Kanbauk Wolfram Mine.

In addition to the above installations, other schemes at which actual construction work is now in progress are the following :—

1. *The Andhra Valley Power Supply Company.*—This is an extension of the existing development of the Tata Hydro-electric Power Supply Company in the Western Ghats, Bombay. Storage is provided in an artificial lake during the monsoon season. A pipe line of a total length of 4700 feet leads the water down directly from the lake to the power house whereby a vertical head of 1743 feet is obtained. A plant is being installed with a capacity of 68,000 e.h.p., which will ultimately be extended to 90,000 e.h.p.

2. *Sikkim Project.*—This project is situated at Lagyap-la, East of Gangtok, and is being developed by Messrs. Burn & Co. The source of water is from a lake fed by snows.

3. *Mansan Falls, Burma.*—This installation is situated in the

Northern Shan States near Hsipaw on the Lashio branch line from Mandalay. The power is being developed from the Mansan Falls whereby a head of 270 feet is obtained with a length of pipe line of 996 feet. The capacity of the plant being installed amounts to 9750 e.h.p., while the ultimate capacity of the site during minimum flow will amount to 13,400 e.h.p. The total capital cost of the development, plant, and buildings is estimated at £22 per kw.

The power from this installation is being transmitted to the smelting works of the Burma Mines at Namtu, some 20 miles distant.

Koyna Valley Power Project.[1]—A project is being undertaken by Messrs. Tata, Sons, & Co. to develop a large water power in the Koyna Valley in an area in the Western Ghats, beginning at a point about 90 miles S.E. of Bombay. By the construction of a dam across the valley, it is intended eventually to form a reservoir extending from here in a south-westerly direction to a distance of about 45 miles. An intermediate dam at half this distance, which it is intended to construct as a first measure, will enable a portion of the scheme to be developed first, and make it possible to take advantage eventually of the somewhat higher head of water which will be offered by the northern section of the reservoir. This power scheme was preceded by a Government project for impounding on a smaller scale, water for irrigation purposes. The catchment area above the final dam site is 346 square miles, lying lengthwise along the Western Ghats. The minimum annual run-off is estimated at 100,000,000,000 cubic feet of water. After providing 15,000,000,000 cubic feet Deccan irrigation, there will be a balance left of 85,000,000,000 cubic feet minimum gross annual supply for power purposes. The height of the river bed at the dam site is 1906 feet above sea-level. It is estimated that a dam raised to 290 feet in height above the river bed would give the reservoir about 112,600,000,000 cubic feet available storage capacity. It is proposed to draw off the water by means of a tunnel leading at a slope of 1 in 500 from near the dam for a distance of 4380 yards westward and emerging at the entrance to the pipe line, where the water will descend to the power house, a further 1880 yards to the west. The mean static head of water obtained on the turbines will be 1606 feet, so that on a basis of 8000 hours per annum continuous supply, a power of 300,000 e.h.p. will be generated continuously with the consumption of 77,184,000,000 cubic feet of water per annum. The turbo-

[1] A. T. Arnall, " Koyna Valley Power Project, Bombay ".

generators will be about 8700 kw. normal output each, generating a three-phase supply at 25 cycles. The maximum length of the dam at the site selected will be only 2900 feet. The estimated total cost of construction (at Rs. 15 = £1) will be £5,000,000. Allowing £200,000 for annual overall expenses, and 10 per cent interest on the capital, a total power of 250,000 h.p. on this basis can be supplied at £2·8 per h.p. year, or if 300,000 h.p. is supplied at £2·3 per h.p. year. A modification of the project on which this estimate is based, has, as mentioned above, been introduced by the decision to develop the scheme in two stages by the initial construction of a dam just below the junction of the Kandati and Koyna rivers, giving a reservoir of about half the capacity. The cost of partial development will thus be reduced, and the ultimate output of the whole project will be increased, because a northern half development may have a greater head of water on the turbines than has a whole development.

The success of this project will, of course, depend on alliance with large industries as there is no existing market for the power. However, the advantages offered by the low cost of the power, transport facilities, and favourable situation should not make this district an exception to the conditions which have existed at Niagara Falls and other large power centres where a demand for power has always been created largely in excess of the supply.

ELECTRO-CHEMICAL AND METALLURGICAL INDUSTRIES IN INDIA AND BURMA.[1]

Metallurgical enterprise in India on modern lines dates from 1903, when the Hutti Gold Mines began operations in Hyderabad State. The production of pig-iron by the Tata Iron and Steel Company at Sakchi first took place in 1911, and of steel in 1912, while ferro-manganese was first manufactured by this company in 1915. In 1918 the Bengal Iron and Steel Company at the instance of the Indian Munitions Board, applied one of their blast furnaces to the manufacture of ferro-manganese. In 1909, after several years of preliminary work, the Burma Mines, Ltd., commenced the treatment of old Chinese slags at the Bawdwin Mines for lead and silver and are now smelting ores for these metals. Finally, The Cape Copper Company, after several years of development work, commenced in 1918 the regular production

[1] L. L. Fermor, " Indian Munitions Board, Industrial Handbook, 1919 " ; A. T. Arnall, loc. cit.

of blister copper. The production on a commercial scale of the metals lead, silver, steel, ferro-manganese, and copper has thus been initiated in India within the decade 1909 to 1918, and very large developments may be expected in the future.

The aim of the Indian Government is to render the country more self-supporting in all essential materials, the practice followed in the past of importing rather than manufacturing having been found particularly undesirable during the recent war, especially where articles required as munitions were concerned.

Prospects of Electric Smelting in India.—Considering first those processes which belong to the domain of ordinary metallurgical practice there are several directions in which electric smelting promises to find application in processes where the current is used solely to replace carbonaceous fuel. The position with regard to the availability of coke in India is that, although the supplies of coal are large, only a small proportion is suitable for the preparation of dense hard coke as required for blast furnaces. Moreover, the coking coals of India are typically high in phosphorus and moderately high in ash so that the resultant coke is always much higher in these impurities than good English cokes. Though the high ash content can be neutralised by suitable fluxing, the phosphorus enters the metallic product. For this reason, Indian pig-iron is phosphoric and necessitates the adoption of the basic process in the production of steel. The supplies of charcoal would not suffice for its continued use as fuel in blast furnaces. Electric smelting which entails the use of carbon only for the reduction may therefore find a successful application for the production of low-phosphorus pig-iron, using either coke, which might be imported, or charcoal, and employing cheaply developed water power. A development may also be expected in certain districts where suitable ore and cheap water power are available, while fuel is scarce.

Other materials associated with the use of electric furnace processes which promise to find a successful development in India are enumerated below :—

Aluminium.—The conditions for the manufacture of this metal in India appear to be very favourable. Large supplies of bauxite equal in quality to the best French bauxite exist in several districts in the country. Laterite with a composition closely resembling bauxite, occurs in large amounts in India and Burma, and though bauxite, mostly obtained from South France, has hitherto been exclusively used in the preparation of aluminium, the Indian laterite should be utilisable for this manufacture.

A method of Serpek (*vide* p. 143) yields pure alumina as a by-product in a process for the preparation of synthetic ammonia, so that by working the two processes in conjunction, the alumina could be obtained from this source.

Copper.—Apart from the smelting of copper ores in the blast furnace as carried out by the Cape Copper Company in the Singhbhum district, there are important ore deposits in the state of Sikkim where conditions appear to be favourable for the application of electric smelting. In experiments which have been conducted in France at Livet (Isère) and Ugine (Savoy), with the Keller and Girod furnaces (*vide* p. 321), it has been demonstrated that copper ores can be efficiently smelted with the production of a high grade matte. The circumstances in Sikkim which should favour this method are the large supplies of hydro-electric power which can be cheaply developed and on which work has already been commenced (*vide* p. 429) while the high price of coke detracts from the blast furnace.

Ferro-manganese.—There are numerous deposits of manganese ores, chrome-ores, and wolfram in India and Burma, and, by the application of hydro-electric power, the possibility of producing ferro-manganese, ferro-chromium, and ferro-tungsten in electric furnaces is very promising. India is practically the greatest producer of manganese ore of any country, the output in 1913 amounting to 828,000 metric tons,[1] the bulk of which is at present exported. Apart from the demand for local consumption, a great inducement to smelting in the country would be the saving on freight of the exported material, the value of manganese ore in 1913 being only Rs.22 per ton as compared with ferro-manganese at Rs.120 to Rs.180 per ton. Meanwhile the production of this alloy containing 65 to 75 per cent Mn by the more customary blast furnace methods has already been successfully inaugurated at the Tata Iron and Steel Company at Jamshedpur and at the Bengal Iron and Steel Company at Kulti. The phosphorus content of these alloys, however, is from 0·5 to 0·8 per cent, which is considerably above the upper limit—i.e. 0·30 per cent—of phosphorus acceptable abroad in normal times. In view of the high content of phosphorus in both the Indian coke and manganese ore, the amount of this element in the resulting ferro-manganese can only be kept low by electric furnace operation, where the bulk of the coke used in the blast furnace is dispensed with.

[1] "Mineral Industry," 1915, **24**, 494.
28

It is estimated by W. F. Smeeth[1] that the cost of electrical production of ferro-manganese (77 per cent Mn), manufactured in Mysore, assuming the electric power to cost Rs.7.5s. per h.p. year (8000 hours) and that it takes 6500 units per ton, would be as follows :—

	Rs.	as.
2 tons ore at Rs.7	14	0
7 cwt. charcoal at Rs.25	8	12
7 cwt. limestone at Rs.5	1	12
6500 kw. hours at 0·2 as.	81	4
Electrodes	6	0
Repairs and relining	5	0
Management, labour, and sundries	9	0
Interest and depreciation	7	4
Total . . .	133	0

Ferro-Chromium.—Most of the Indian chromite comes from Baluchistan and is exported from Karachi. The content of Cr_2O_3 is from 35 to 57 per cent. Dr. Smeeth gives the following estimate for the manufacture in Mysore of 60 per cent ferro-chrome, with 6 to 8 per cent carbon :—

	Rs.
2 tons selected ore, delivered at furnace	40
Electric energy, 8000 units at 0·2 as.	100
Electrodes	10
Charcoal, labour, repairs, and fixed charges . . .	40
Total	190

Ferro-tungsten.[2]—Burma is the largest producer of wolframite in the world, and the year 1915 marked the passing of the Burma tungsten industry out of German into British hands. The fact that practically the whole supply of tungsten and ferro-tungsten used by Great Britain before the war was obtained from Germany led to a great scarcity of these materials during the year 1915, and steps were then taken to establish works at home for the manufacture of ferro-tungsten steel. The Company known as the High-speed Steel Alloys was formed for this purpose and works erected in Lancashire. This Company's agents are in Burma and control a great part of the mining operations there. The production of the ore in Burma in 1915 amounted to 2645 tons the value per ton being Rs.1680. Though this value would justify transport to European smelters, favourable power centres in India, such as the projected

---

[1] " Mineral Resources of Mysore," 1916, p. 99, also A. T. Arnall, loc. cit.
[2] A. T. Arnall, loc. cit.

development in the Koyna valley, should provide inducements for the production of ferro-tungsten as favourable as at any centre.

Electric Smelting of Iron-ore.—Though no installation is yet actually in operation, it is considered [1] that the electric smelting of iron-ore to produce high-grade steel, and merchant steel, can be profitably conducted in Mysore. Assuming that electric energy from a proposed hydro-electric scheme near the West Coast would cost £3 per h.p. year, or 0·05d. per unit; ore, 2s. per ton; charcoal, £1 per ton; limestone, 10s. per ton; electrodes, 6s. per ton of steel; and, with other items of cost, it is estimated the cost of steel at the works would be about £5·3 per ton with 65 per cent ore, and £5·6 per ton with 60 per cent ore.

Zinc.—The main source of zinc is at present from the Bawdwin Mines in Burma, obtained in the concentration of the complex lead, silver, zinc-ores. When present extensions are completed, the annual output of zinc concentrates will be about 25,000 tons, while further extensions are proposed for the future. A plant is now being constructed at Jamshedpur for the smelting of these concentrates which contain 48 per cent zinc and 30 per cent sulphur by the ordinary retort process, and the recovery of the sulphur as acid. In view of the water-power resources in Burma, it is possible that zinc might be profitably extracted there by an electrolytic or electro-thermal process.

Magnesium.—One of the chief processes for the extraction of magnesium consists in the electrolysis of the dissolved oxide, and with cheap electric power, this manufacture should find a profitable application in India. Indian magnesite is at present obtained mainly in Madras and Mysore, the estimated cost of production being £2·5 per ton. Very large deposits of dolomite are also found near Belgium.

Calcium Carbide.—The manufacture of calcium carbide, both for export as such and for the further preparation of cyanamide, is one of the most promising undertakings which is being considered for installation at the projected Koyna water-power development near Bombay. At present considerable imports are made into India of calcium carbide and nitrogen compounds. Unlike the newer dominions, cultivation of nearly all suitable land in India has been carried out intensively, and under the encouragement of low prices and better education of the natives in farming, the application of artificial fertilisers should find a very extensive development.

---

[1] W. F. Smeeth, Bulletin No. 5, " Mysore Geological Dept.," 1909, p. 134.

With regard to the supply of raw materials for the manufacture of calcium carbide, the limestone required for this process should be of high grade, containing 97 to 98 per cent of carbonate of lime. Most of the deposits so far examined have been of low grade. Many of these would be suitable if used together with charcoal as the reducing agent, though some very high-grade deposits which have so far been found in small quantities, such as calcite with 99·6 per cent carbonate of lime, in Katiawar, may later prove to be obtainable in sufficient amounts to meet all the requirements of the industry.

The situation with regard to the supply of carbon is not yet satisfactorily settled. In view of the high phosphorus content of cokes so far produced in India, the choice of material at the present rests between charcoal and imported anthracite.

Refractory Materials.—The natural resources of India provide for nearly all classes of refractory materials needed in metallurgical and electric-furnace construction. Highly pure silica is available, and is used for the manufacture of silica bricks by Messrs. Burn & Company, and by the Kumardhubi Fireclay and Silica Works, Limited. Good firebricks are manufactured from fireclay by these firms and others. As a basic lining for furnaces, magnesite and dolomite are both used, the former, of excellent quality, being obtained from Salem and Mysore. Chromite and bauxite are available for neutral linings.

POWER DEVELOPMENT IN TASMANIA.[1]

A hydro-electric development has been in operation in Tasmania since May, 1915, utilising the Great Lake as the main water-storage basin. The output in 1916 amounted to about 10,000 e.h.p., while the development is capable of being increased to 70,000 h.p. by extending the water storage at the Great Lake, and to 150,000 h.p. by utilising two other lakes as water storage basins. The minimum charge for current at Hobart has been fixed at £2 per h.p. year for a continuous consumption of 10,000 h.p. and upwards.

It is considered that hydro-electric power in Tasmania will within the next few years be applied to the following processes :—

Zinc-ore reduction processes, 40,000 h.p. (*vide* p. 318).

Manufacture of caustic soda and bleaching powder, 1500 h.p.

Calcium carbide and possibly cyanamide, 4500 h.p.

Steel and ferro-alloys, 1000 h.p.

[1] F. H. Campbell, " J. Soc. Chem. Ind., " 1916, **35,** 1265.

Other industries proposed are the manufacture of aluminium, the production of nitric acid by oxidation of ammonia or by direct synthesis, electrical smelting of iron-ore, preparation of per-salts, use of chlorine prepared electrolytically for de-tinning scrap, and the manufacture of permanganate and aniline.

TIDAL POWER.[1]

The possibility of developing very large amounts of power from estuaries and river basins by utilising the differences of level of water produced by the rise and fall of tides, has for long been recognised.

Up to the present, however, no power development of this kind, of any appreciable size, has been carried out, though several projects are now under serious consideration. The main disadvantages of deriving power from this source are the intermittent nature of the supply and the relatively low head obtainable, involving the use of very large turbines of low speed and limited energy output.

In the case of all tidal projects, the engineering work involved includes the construction of a barrage for the impounding of water in a tidal reservoir. The turbines are installed in tubes or passageways leading horizontally through the structure and controlled by sluice-gates. During the intervals when a certain difference of level occurs in the water on either side of the barrage, the turbines are rotated by the inflowing or outflowing water. The power obtainable from a given area of tidal basin varies with the square of the tidal range, and since the cost per h.p. of the turbine installation increases rapidly as the working-head is diminished, a high tidal range is of first importance.

In Great Britain, the highest tides are found in the estuary of the Severn, the spring tides at Chepstow being 42 feet, and the neap tides 21 feet. In France, the maximum range occurs at St. Malo, where there is a spring tide of 42·5 feet, and a neap tide of about 18 feet.

A preliminary technical investigation of the feasibility of power development from the Severn estuary has been made on behalf of the Water-power Resources Committee of the Board of Trade[2] by Sir Philip Dawson and Prof. A. H. Gibson. The project under consideration involves the construction of a dam across the Severn estuary, and thus provides for the combination with it the improvement of navigation facilities in the river, and the provision of a

---

[1] "Nature," 1920, 105, 427; Third Interim Report, Board of Trade, Water-power Resources Committee. London, 1920, H.M. Stationery Office.
[2] Ibid.

roadway and railway line across the estuary. It is considered that the most favourable system will consist in operating the turbines on a falling tide only. The water level in the enclosure would thereby never sink below a predetermined level, fixed probably at about half-tide, and the effect would be to provide a large dock.

The procedure recommended consists in filling the estuary through sluices during the rising tide. At high tide the sluices are closed, and when the tide has fallen through a definite height the turbine gates are opened, and the turbines operate on a more or less constant head until low tide. The maximum output would be obtained when the constant working head is approximately one-half the tidal range. With the Severn estuary, the working head during neap tides would thus amount to about 8 feet.

The preliminary estimates indicate that energy could be generated and utilised at a favourable rate, and that the power obtainable throughout a daily interval of ten hours would be of the order of 260,000 kw. (or 348,000 h.p.). To obtain continuity of power supply where a single barrage is employed, it is necessary to provide auxiliary means for storing a portion of the energy during the periods of maximum output. For this purpose it is considered that the most promising method consists in the use of a supplementary high-level reservoir into which water is pumped, and afterwards used in secondary turbines to develop energy as required. The following alternative systems have also been proposed: (1) To employ the tidal power scheme in conjunction with one or several steam-power stations; (2) the combination of a tidal development on the Severn with a similar one on the Dee estuary, the tidal periods of which are several hours different from those of the former; (3) the application of the power to electro-chemical and electro-metallurgical processes worked intermittently. It is estimated that, with an intermittent supply, tidal power can be developed in this country more cheaply than from any other source.

In France, the Ministry for Public Works has made provision for the carrying out of experiments in connexion with the development of tidal power at two sites on the French coast.

SECTION XXV.

STEAM POWER STATIONS AND ELECTRO-CHEMICAL CENTRES.

Electrical Power Generation from Coal.—It has been frequently assumed that electro-chemical industries, on account of their large consumption of energy, can only be operated with economy from water powers. This view has probably arisen from the earlier developments of these industries having mainly taken place in centres near specially favourable water powers as in Norway and the Alps. Power is obtained at these selected places at a cost far below that possible with steam power, but it must be remembered that the most favourable of these sites in Europe have now been utilised, and, with further extensions, the cost of development may be expected always to increase. In some instances developments of water power have already been made at a cost which, on a basis of pre-war values, made the price of power higher than that now attainable in large-scale steam-power stations.

No further appreciable increase in efficiency can be realised with hydraulic turbines, a mechanical efficiency on the turbines approaching 100 per cent having already been obtained, whereas, with fuel-power stations, a continual improvement in this respect is being attained with steam turbines, though the best at present still give at the generators an over-all thermal efficiency from the heat value of the coal consumed of only 18 to 20 per cent.

A further considerable increase in economy may be expected by combining power generation with the recovery of by-products by low-temperature carbonisation of the coal. For these reasons the tendency for the continual increase in the price of coal may be expected to be counterbalanced.

With regard to the prospects for the extension of electro-chemical industries in centres where cheap water power is not available, apart from the progressive increase in the economy of power generation from fuel, the item of power cost is generally by no means the chief determining factor in electro-chemical processes. The question of

439

the availability of raw materials and the vicinity of subsidiary processes are frequently of greater economic importance.

Estimates which have been made in the past of the cost of power generated from fuel have shown very wide variations, partly on account of the difference in efficiencies allowed for, but more particularly through the arbitrary manner in which it is generally necessary to allocate many of the items contributing to the total costs. The items of cost can be classified in the following three categories :—

1. *Working Charges*, which includes items such as fuel, oil, water, and their conveyance and the expenses of labour, management, and accessory duties.

2. *Capital Charges*, which comprise the interest on the outlay on machinery and buildings, rent on land, rates, insurance, etc.

3. *Depreciation*, which involves a sinking fund to repair and replace plant and buildings as necessary.

In spite of this classification, however, many of the items of expenditure cannot be definitely allotted to any particular class, and the estimate of the items in classes 2 and 3 is of a somewhat arbitrary nature. Further, practically all the elements constituting the cost are variables with time, and in the last few years have undergone a progressive rapid increase.

Coal Consumption and Thermal Efficiency.—One of the main factors determining the cost of power from coal, and one which in different cases shows the widest variations, is the thermal efficiency of the steam producers and power generators. This efficiency determines the consumption of coal per unit of electrical power generated.

The heat value of coal is, in practice, expressed by the standard of the British Thermal Unit which is defined as the quantity of heat required to raise the temperature of 1 lb. of water by 1° Fahrenheit. The unit of electrical power is expressed by the Board of Trade unit which is equivalent to 1 kw. hour. Accordingly we have the following relationship :—

$$1 \text{ watt-second} = 1 \text{ coulomb} = 0.239 \text{ calorie.}$$
$$1000 \text{ cal.} = 1 \text{ kg. cal.} = 3.968 \text{ B.T.U.}$$
$$1 \text{ kw. hour} = 860 \text{ kg. cal.} = 3412 \text{ B.T.U.}$$

Hence the amount of coal expended per kw. hour of electrical power will amount to

$$\frac{3412}{C \times E} \text{ lb.,}$$

where C is the calorific value per lb. in B.T.U., and E represents the over-all efficiency in the power generation. To take a typical example, with a total generator efficiency of 18 per cent which may be realised with large steam turbines and with coal of an average quality giving 12,500 B.T.U., the coal consumption will amount to

$$\frac{3412}{12,500 \times 0\cdot18} = 1\cdot54 \text{ lb.}$$

With regard to the coal consumption actually obtained at large power stations in this country, the summarised monthly returns of the Coal Mines Department for 1918 and 1919, show an average fuel consumption from 438 stations for the year ended March, 1919, of 3·47 lb. per unit. The lowest of any station having an output of over 50,000,000 units per annum was 1·8 lb. per unit. In a census taken by the Lancashire and Cheshire Interconnection Committee, the fuel consumption varied from 2·2 lb. to 8·5 lb. per unit.[1]

Load Factor.—The cost of production of electrical power is largely determined by the load factor or the ratio of the average power consumption taken over the whole year to that of the maximum or peak demand. The importance of this factor is that it is necessary to design the machinery on a scale to supply the maximum demand, so that interest on capital and upkeep, and other standing charges, are largely determined by the fluctuations which take place in the load.

In a power station supplying miscellaneous industries, a power factor of 50 per cent will rarely be reached. An improvement in the load factor may be effected by combining industrial with railway supply, a method which may be expected to be applied more extensively in future through the development in the electrification of railways.

In a power station supplying only a continuous electro-chemical process and provided with spare complete generating sets and transformers—which, however, do not count in the total capacity, but enable stoppages of individual units for repairs without interruption of supply—a load factor of 90 per cent may be obtained, but this figure can rarely be exceeded.

Estimates which have hitherto been published of the cost of power relate to the less unsettled and varying conditions existing before the war. At the present time, factors of cost are too conditional to enable a definite computation of any value to be made. A few instances are given below of careful estimates which have been made at different times. By adjusting the different factors in these items in accordance with the present increased costs, these data can be revised to apply to the present fluctuating conditions at different times and in different localities.

[1] Cf. J. A. Robertson, " The Electrician," 1919, **83**, 630.

(a) *Estimate by R. S. Hutton.*[1]—Cost of Power Generation in 1906 from Steam. Working costs per B.T.U.

Station.	Coal.		Wages.	Water Oil Stores.	Repairs, Etc.	Total per Unit.
	Per Ton.	Per Unit.				
	s. d.	d.	d.	d.	d.	d.
Newcastle (Carville) .	5 6	0·078	0·022	0·004	0·016	0·121
Sheffield (Neepsend) .	5 8	0·096	0·072	0·003	0·038	0·209
Messrs. Watson (Linwood)	8 0	0·148	0·022	0·013	0·022	0·205

CAPITAL COSTS.

		Per Unit. d.
10 per cent. depreciation on £15 per kw.	. . . .	0·042
5 per cent. interest ,, ,,	. . . .	0·021
		0·063

(b) *Data from Steam-power Plant at Connor's Creek, U.S.A., erected for the operation of electro-chemical factories for war requirements.*[2]

Costs per kw. hour of net output for twelve months ending 30 June, 1917 :—

	Cost per kw. hour. Cents.
1. Working costs:—	
Superintendence	0·010
Wages	0·047
Fuel	0·240
Lubricants	0·001
Station supplies and expense	0·005
2. Maintenance :—	
Buildings	0·011
Steam equipment	0·019
Electrical equipment	0·001
Total	0·334 cents.

Kw. hour output	210,039,700 kw. hour
Maximum demand (30 minutes) . . .	50,000 kw.
Average load	23,900
Load factor	0·478
Coal per kw. hour—pounds	1·52
B.T.U. per kw. hour	20,040
Over-all thermal efficiency	17 per cent
3. Capital cost of installation	$100·00 per kw.
Interest and depreciation at 15 per cent = $15 per kw. year . . .	0·358 c. per kw. hour consumed.

The total cost including interest on capital and depreciation thus amounts to 0·792c. or 0·4d. per unit.

Modern Steam-power Generating Stations in England.—Data on the cost of power generation at two modern stations in this

---

[1] " Engineering," 1906, **82,** 779.

[2] A. Dow, " Trans. Amer. Electrochemical Soc.," 1918, **32,** 65.

country, viz. The Stuart Street Generating Station at Manchester, and one at Birchills, Walsall, are given by Sir John Snell [1] as follows :—
The estimate is based on wages and working costs prevailing in 1916, with coal at 12s. per ton. The load factor at Stuart Street Station is taken at 37 per cent and that at Birchills at 30 per cent.

	Stuart St., Manchester.	Birchills, Walsall.
Capacity, kw.	120,000	12,000
Total capital costs per kw.	£8·9	£6·6
Annual load factor	37 per cent.	30 per cent.
Total working charges, per kw. hour	0·281d.	0·209d.
Capital charges at 8½ per cent.	0·156	0·077
Total	0·437d.	0·286d.

The higher cost at Stuart Street compared with Birchills is mainly accounted for by the whole plant at Birchills being of modern construction, while the Manchester plant is composite and includes old equipment.

It is estimated by E. M. Lacey [2] that in the case of the Birchills Station, power at a load factor of 80 per cent can be supplied under the above conditions at 0·17d. per kw. hour (£5·0 per kw. year or £3·7 per h.p. year).

In a typical case of a municipal electrical supply for miscellaneous services under present-day conditions, statistics of the Manchester Corporation Electricity Department for the year ending March, 1920, give the following returns :—

Quantity of power generated, 241 million units (kw. hour). Power sold, 182 million units. Total load factor, 25·2 per cent. Plant factor $\left(\frac{\text{Max. demand} \times 100}{\text{Plant installed}}\right)$, 71·0 per cent. Coal consumed per unit sold, 3·44 lb. Cost of coal, 26s. 8d. per ton.

Average cost per kw. hour sold.	d.
Fuel, etc.	0·492
Oil, waste, water, and engine-room stores	0·024
Wages and salaries	0·169
Repairs and maintenance	0·181
Rent and rates	0·082
Management expenses	0·059
Capital charges, interest, debt redemption, etc.	0·401
Income tax	0·052
Total	1·460

[1] " Proc. Instit. Civil Engineers," 1916 to 1917, **204**, 226.
[2] *Ibid.*, p. 194.

Estimate by Government Committee, 1916 to 1917.[1]—This estimate of power costs was compiled by a Power Sub-Committee of the Ministry of Munitions in 1916. A detailed investigation was made of the position and possibilities of the generation of electrical power on a large scale including various methods of carbonisation of fuel and the recovery of by-products such as ammonia. The object at the time in view was the supply for the manufacture of synthetic nitrogen compounds on a large scale for war requirements and subsequently for general industrial supply.

The results, in the case of coal-fired power stations, are summarised as below :—

The estimate was prepared for a proposed large power station having an installed capacity of 125,000 kw., using five turbo-alternators of 25,000 kw. each, one being held in reserve, so as to provide an effective load of 100,000 kw. The calculation is based on the latest pre-war figures.

A. *Capital Cost.*

Items.	Per Kw. Installed.	Per Kw. Maximum Load.
	£.	£.
(a) Land for complete station	0·16	0·200
(b) Buildings and foundations, coal silos and trans-porters, railway sidings, roads, etc. . . .	1·50	1·875
(c) Coal and ash handling plant	0·24	0·300
(d) Boilers, superheaters, reheaters, feed heaters, mechanical stokers, induced draught plant, chimneys, etc.	1·80	2·250
(e) Turbo-alternators and exciters, surface condensers, air pumps and auxiliaries . . .	2·50	3·125
(f) Steam and water-piping, circulating and feed pumps, air pumps and strainers, etc. . .	0·60	0·750
(g) High and low-tension switchgear, reactances, etc.	0·80	1·000
(h) Engineering supervision, inspection, contingencies, etc.	0·60	0·760
Total	8·20	10·260

Capital Charges.—The capital charges are made up as follows : (1) 4½ per cent interest on capital, (2) 2½ per cent depreciation on buildings, etc. (item *b*), and on the corresponding proportion of contingencies and engineering fees ; and (3) 5 per cent depreciation

[1] " Final Report of the Nitrogen Products Committee, Ministry of Munitions," p. 27. London, 1920, H.M. Stationery Office.

on the remaining capital (items *c* to *g*) and on the corresponding proportion of contingencies and engineering fees. No sums are included for insurance, imperial taxes, or local rates.

Working Charges.—The working expenses are estimated on the basis of

Maximum load	.	.	.	.	.	100,000 kw.
Calorific value as fired		.	.	.	.	12,500 B.T.U. per lb.
B.Th.U. per unit output		.	.	.	20,000
Cost of coal	.	.	.	.	.	10s. per ton.

The estimated costs are as given in the following table :—

Items.	Load Factor = 95 per cent. Kw. Year = 8322 hours. Units = 832,200.	
	Per Unit.	Per Kw. Year.
	d.	£.
(a) Salaries and wages	0·00403	0·140
(b) Oil, stores, and sundries . . .	0·00288	0·100
(c) Repairs and maintenance . . .	0·00864	0·300
(d) Coal : 594,430 tons at 10s. . . .	0·08572	2·972
(e) Capital charges	0·02633	0·913
Total	0·12760	4·425

It is of interest to note from this estimate that under pre-war conditions the possibility is shown of generating power from coal at a cost which is no higher than that which, in some cases (cf. p. 409), is charged for water power supplied in large amounts. The introduction of further economies which may be expected in the generation of power from fuel would indicate that the coal areas of this country should be placed in a position as favourable for the location of future developments in all branches of the electro-chemical industry as districts supplied with large water powers.

The above estimate of the cost of power generated from steam may be revised in accordance with the present (1921) costs, in the following approximate estimate of the writer. The values apply, as in the above estimate, to a plant of 125,000 kw. installed capacity and 100,000 kw. maximum load, and the load factor (0·91) is taken as the highest that could be expected with the most favourable electro-thermal processes when operated continuously.

Capital Charges.—The total capital costs are taken at £20 per kw. installed or a total of £2,500,000.

Interest and Depreciation on capital is taken at 12 per cent on the total.

Working Charges and Maintenance.—Coal is taken at 30s. per ton. Salaries, wages, and the remaining items in the above estimate are taken at double the value allowed in the above estimate.

The calorific value of the coal is taken as above at 12,500 B.Th.U. per ton.

Estimate for 1920.	Load Factor = 91·3 per cent. Kw. Year = 8000 hours. Units Output = 800,000,000.	
	Per Unit.	Per Kw. Year.
	d.	£.
(a) Salaries and wages	0·008	0·28
(b) Oil, stores, and sundries	0·006	0·20
(c) Repairs and maintenance	0·018	0·60
(d) Coal: 571,400 tons @ 30s. per ton . .	0·256	8·56
(e) Capital charges	0·090	3·00
Total	0·378	12·64

Power Production by Coal Carbonisation and Gasification.—After a detailed inquiry into the costs of generation of steam power from fuel with recovery of by-products, it was concluded by the Power Sub-Committee of the Ministry of Munitions[1] that complete gasification of coal by Mond and similar recovery plants, combined with the use of gas-fired boilers, will not as a rule be economical with large-scale plants. The increased consumption of fuel per unit of power produced and the greatly increased capital costs of the plant compared with direct-fired methods will not usually be counterbalanced by the value of the by-products obtained. However, the position is more promising with low-temperature carbonisation, followed by the gasification of the coke in Mond or similar recovery producers. Such a process yields the maximum of by-products, but at present is hampered by thermal inefficiency and by heavy capital expenditure.

During the low-temperature carbonisation there is yielded a gas of about 700 B.Th.U. per cubic foot, and soft coke, together with by-products consisting of ammonium sulphate and tar; the gasification of the coke yields gas of about 140 B.Th.U. per cubic foot and a further amount of ammonium sulphate as a by-product. The coal would thus be treated as a chemical from which saleable by-products would be recovered and the resulting gas would be consumed for steam raising.

[1] " Report of the Nitrogen Products Committee," p. 171.

On the basis of gas-fired boilers having a thermal efficiency of 75 per cent, the combination process involves the consumption of 14·4 tons of coal (12,000 B.Th.U. as fired) per kw. year generated (8760 hours) as compared with 6·5 tons in the case of a direct-fired power plant.

The capital cost of the carbonisation and producer plant is estimated at £15·78 per kw. installed, and the working costs, with interest and depreciation on capital at £4·44 per kw. year of 8760 hours, while coal at 10s. a ton would represent a yearly cost of £7·2 making a total gross cost of £11·64 per kw. year.

The yield and value of by-products per kw. year generated would amount to

Ammonium sulphate . . .	0·55 ton
Tar	1·31 tons

The net receipts from these, at a market value of £13·83 and £2·75 per ton respectively, and after deducting £2·75 per ton for the cost of fixing the ammonia and handling, would amount to £9·83, thus leaving for the net cost of the gas fuel, to be supplied to the boilers £1·8 per kw. year of 8760 hours, in place of £3·12, the cost of direct fuel firing, with coal at 10s. per ton. This, in a station of 100,000 kw. capacity would furnish power at £3·25 per kw. year, as against £4·58 for a direct coal-fired station.

It can be calculated from the above data that the economy of by-product recovery processes when employed in power generation becomes more unfavourable with increasing cost of coal. It would appear, therefore, that the ability of by-product power schemes to compete financially with direct firing is essentially a question of the availability of coal at a low or moderate price.

It is of interest to note that in the case of a large-scale nitrogen installation with a steam-driven power station operated on the carbonisation and gasification system, the nitrogen recovered from the coal itself would be approximately equal to the amount obtained from the air by the arc process, and to about one-fourth of the nitrogen fixed by the cyanamide process.

Situation of Large-power Stations.—It has often been advocated that large-power stations should be situated at the collieries, and the energy transmitted therefrom to the areas where power is required. However, the first and principal requirement of a cheaply operated power station is an abundant natural supply of circulating water for condensing purposes which can be utilised without undue cost for pumping. A 100,000 kw. load will require at least 6,000,000 gallons of water per hour, or on a 95 per cent load factor, some 136,000,000 gallons a day. There are no localities in Great Britain where this amount of water can be obtained in proximity to collieries, and it is impracticable to transmit so large an amount of water over any considerable length of pipe or culvert. There are, however, several districts where a water-side station could be built at no great distance from a coalfield, and the only practical compromise is to place the power station site by the water-side, and to carry the coal from the pit to the station. The site would generally also enable both rail or water-borne coal to be delivered, and would afford facilities for the transport of raw and finished materials by both systems.

Gas Engines for Small and Large-power Stations.—For large-power stations, gas engines at the present time are impracticable and uneconomical. The largest units hitherto installed commercially are under 4000 kw. and consist in effect of two engines, each of under 2000 kw. The space occupied by large gas engines is very great, their speed is necessarily low, and the consequent cost is very high. For small blocks of power the operating costs are about the same as, or even less than, for steam engines. With large amounts of power, however, the cost of labour and repairs, owing to the multiplication of units, becomes excessive as compared with that for modern large-scale steam-turbo units.

Waste Heat from Blast Furnaces and Coke-Ovens.—Evidence given before the Royal Commission on Coal Supplies in 1905 showed that the amount of waste gas from blast furnaces in Great Britain at that date amounted to 1000 h.p. for every 100 tons of pig-iron made, and that insufficient use was made of the gas. It was estimated that the potential power of the waste gas was equivalent to that from 2,000,000 to 3,000,000 tons of coal annually. The Power Sub-Committee of the Ministry of Munitions in 1917 concludes that the only economical and proper way to utilise such surplus power from blast furnaces and coke ovens is to collect electrical energy generated therefrom in a general distribution system. The energy can then be regulated and used so as to supply the needs of the iron and steel works themselves and any surplus power can be distributed for general industrial purposes. Skilful co-ordination will be required too in order to obtain the best results. Such schemes have already been accomplished effectively by companies like the Newcastle-on-Tyne Electric Supply Corporation, and a large annual saving of coal has been effected by the combination of a widely extended high-pressure electrical transmission system with a number of waste-heat stations at isolated blast furnaces and coke ovens.

Steam-generated Electrical Power Stations.

A number of large electrical power generating stations have for some time been operated in this country by public supply companies. Steam power mainly from coal is utilised, and these stations are of special importance to the prospects of home electro-chemical enterprises on account of the low figure to which the cost of power generation has now been reduced. These power stations are now promising to form serious rivals to the more favourable water-power centres in the attraction of industries consuming large amounts of electrical power. A significant instance of this is that under pre-war conditions, steam-generated power has in some cases in this country been supplied to large consumers of steady loads at a rate below that obtaining at Niagara Falls.

During recent years a great stimulus has been given to these public electric supply stations by the adoption of electric furnace processes. These processes have hitherto related mainly to steel furnaces of which a total of 117 have been installed in the United Kingdom, the majority being at Sheffield, during the years 1914 to 1919.[1] At Birchills, Walsall,[2] a power plant was completed about 1916 at an estimated total capital outlay of £6·6 per kw. installed, enabling power to be supplied at a load factor of 80 per cent at 0·17d. per kw. hour or £5·0 per kw. year.

Newcastle-upon-Tyne Electric Supply Company.[3] — The main steam-operated electrical power station in this country is that of the Newcastle-upon-Tyne Electrical Supply Company which commenced about thirty years ago as a public supply company with a generating plant of 200 kw. capacity, and now ranks as one of the largest electrical stations of any type in the world.

The generating stations of this company are situated at several centres where special facilities exist for the supply of fuel, water, and transport, and in some cases where use can be made of waste heat and gases from blast furnaces and exhaust steam. The stations are inter-connected and transmit power over a large area comprising the industrial district on the North-Eastern Coast and extending over Northumberland, Durham, and North Yorkshire. The main power stations are situated at Carville, Dunston, Philadelphia, and

---

[1] Cf. " J. Instit. Elec. Eng.," 1919, **57**, 406.   [2] Cf. p. 443.
[3] From data kindly supplied by Messrs. The Newcastle-upon-Tyne Electric Supply Company.

Grangetown and generate a total supply of 245,000 h.p. On completion of an extension system now in course of erection at Haverton Hill in the Tees district, this capacity will be increased to 370,000 h.p. At this new station turbo-generator units of 27,000 h.p. each are being installed and, by the use of higher steam pressures and temperatures, will represent the latest advance which has been made in the development of steam turbines.

The main waste-heat station is at Weardale on the N.E. coast, where 6650 h.p. is utilised from coke-ovens, the gas from which is burnt in tubular boilers. Stations at Newport and Tees Bridge utilise exhaust steam from blowing-engines. The steam is led through a superheater to the power-house, and employed in exhaust steam turbines of some 3000 kw. output.

Distribution.—The company's main distribution is by a network of high-tension underground cables and by pole lines at pressures up to 20,000 volts. All power stations and distributing centres are linked together in order to ensure continuity of supply in the event of any one of the supplies failing.

Consumers.—The main consumers of this supply are the tramways and railways, the dockyards, shipyards, and engineering works on the Tyne and a number of electro-chemical works including the Castner-Kellner Company.

An important advantage which has been gained by a wide distribution of the supply and variety of consumers is that a high load factor is obtained as intermittent demands for large amounts of power tend to equalise the total load.

ELECTRO-CHEMICAL INDUSTRIES IN SOUTH AFRICA.

(a) *Power Generation.*[1]—The water-power resources of South Africa are considered to be in the main, in inaccessible districts, or, as in the case of Rhodesia and the Victoria Falls, at large distances from industrial centres. An important asset in South Africa, however, in the production of power is the availability of cheap coal enabling the favourable operation of steam power stations.

The principal power stations in the Union are those on the Witwatersrand which, in 1914, had a total capacity of approximately 220,000 kw. At some of the larger stations power is said to be generated at a general cost price of 0·3d., and at one station at 0·2d. per kw. hour.

---

[1] " Final Report of the Nitrogen Products Committee," p. 117.

(b) *Electro-chemical Processes.*—A Report[1] issued by the Development of Resources Committee dealing with possibilities of electro-chemical manufacture in South Africa concludes that calcium carbide, cyanamide, and its derivatives such as sodium cyanide and dicyandiamide could be manufactured economically. The cost of electricity is estimated at 0·35d. per unit with a modern steam generating plant, assuming a capacity of about 5000 kw. and a load factor of about 83 per cent. With still cheaper electric energy, the arc process of nitrogen fixation would become practicable.

The manufacture of iron and steel in the electric furnace also offers a wide field in South Africa.

PERIODICITY OF POWER SUPPLY.

The most usual frequency of current formerly employed in public supply services was that of 25-cycles. The tendency, however, is now to employ higher frequencies. In Great Britain, 50-cycle current has now been adopted as a general standard, while in the United States, 60-cycles has been applied as the standard.[2] These higher frequencies are disadvantageous for the operation of large electric furnaces, and involve special construction in order to obtain a reasonably high power factor.

For example, in the case of a furnace operated with three-phase current of 60-cycles, delta-connected with a voltage of 110, if a power factor of 0·9 is assumed, then a limit will be imposed in the power consumption of the furnace at about 2750 kw. If, however, this furnace is given six phases in place of three, it will take nearly double the power under the same conditions. If, further, the current be resolved into twelve phases, applied to twelve electrodes arranged around a circle, the power capacity will be increased to nearly four times. The six-phase secondary can be acquired by a mere change of the secondary connexions from a three-phase transformer, and twelve phases can be similarly developed from two such transformers, one with its primary delta-connected, and the second star-connected, the secondary voltages of the two being the same.

---

[1] " Journ. S. African Instit. Elec. Eng.," 1916, 896 ; " Journ. Soc. Chem. Ind.," 1916, 35, 897.
[2] W. S. Horry, " Trans. Amer. Electrochem. Soc.," 1914, 25, 59.

APPENDIX A.

THERMAL RESISTIVITIES.[1]

Material.	Temperature of Measurement. Centigrade.	Thermal ohms per 1-inch cube.	Per 1 cm. cube.
Silver	0° to 100°	0·094	0·24
Copper	0° ,, 100°	0·11	0·27
Aluminium	0° ,, 100°	0·27	0·69
Graphite (Acheson)	100° to 900°	0·32	0·82
Brass	0° ,, 100°	0·36	0·92
Iron	100° ,, 1245°	0·49	1·3
Carbon (electrode)	100° ,, 942°	0·72	1·9
Carborundum (brick)	1000°	4·1	10·3
Quartz	0°	5·9	15
Retort carbon	0°	9·1	23
Magnesia brick	1000°	13	34
Chromite brick	1000°	16	42
Firebrick	1000°	22	57
Porcelain	95°	38	96
Cement (Portland)	90°	132	336
Silica brick	1000°	47	120
Kieselguhr brick	1000°	52	133
Glass	28°	87	220
Plaster of Paris	20° to 155°	221	562
Asbestos	20° ,, 155°	139 to 416	353 to 1060
Ebonite	48°	251	637
Gas-works "breeze"	—	440	1120
Infusorial earth	—	263 to 745	675 to 1890
Magnesia (calcined) pure	20° to 155°	572	1450
Charcoal	—	490 to 720	1260 to 1840
Cotton	—	590 ,, 2810	1520 ,, 7120
Lampblack	20° to 155°	697	1770

[1] C. Hering, "Met. and Chem. Engineering," 1911, 9, 653.

MELTING-POINTS.

Material.	Melting-Point, Degrees Centigrade.	Authority.
Tin	232	
Zinc	419	
Antimony	630	
Aluminium	658	Waidner and Burgess
Silver (in air)	955	
Silver (in reducing atmosphere)	962	
Copper (in air)	1063	Waidner and Burgess
Gold	1064	
Copper (in reducing atmosphere)	1083	Waidner and Burgess
Manganese (98 per cent Mn) .	1207	Burgess
Nickel	1452	Day and Sosman
Steel (0·5 per cent C) . .	1475	Hadfield
Iron (electrolytic) . . .	1507	Burgess
Chromium	1510	Hadfield
Palladium	1549	Day and Sosman
Platinum	1755	,, ,,
Kaolin (Al_2O_3, $2SiO_2$) . .	1800	Le Chatelier
Silica	1800	,,
Chromic Oxide . . .	1990	Kanolt
Alumina	2050	,,
Molybdenum	2500	
Lime	2572	,,
Zirconia	2700	,,
Magnesia	2800	,,
Tungsten	3000	

BOILING-POINTS OF METALS AT 760 MM. PRESSURE.

Metal.	Boiling-Point.	Authority.
Magnesium	1120°	Greenwood
Antimony	1440°	,,
Lead	1525°	,,
Aluminium	1800°	,,
Manganese	1900°	,,
Silver	1955°	,,
Chromium	2200°	,,
Tin	2270°	,,
Copper	2310°	,,
Iron	2450°	,,

SPECIFIC HEATS AT 0° C.

Material.	Specific Heat.
Aluminium	0·22
Iron	0·11
Copper	0·093
Nickel	0·11
Platinum	0·032
Silver	0·056
Mercury	0·0332
Glass . . - . . .	0·19
Graphite	0·155
Retort Carbon . . . ;	0·165

ELECTRICAL RESISTANCES, see p. 21.

UNITS OF ENERGY.

1 Board of Trade electrical unit = 1 kw. hour.
1 kilowatt $\qquad$ = 1·34 horse-power.
1 horse-power $\left\{ \begin{array}{l} = 746 \text{ watts.} \\ = 33,000 \text{ ft. lb. per minute.} \end{array} \right.$
1 joule = 1 watt-second = 10,000,000 ergs = 0·239 calorie.
1 B.T.U. = 1058 joules.
1 kg. calorie = 3080·9 ft. lb.

COST OF ELECTRIC POWER; CONVERSION TABLE, PENCE PER
B.T.U. AND £ PER E.H.P. YEAR AND KW. YEAR OF
8000 HOURS.

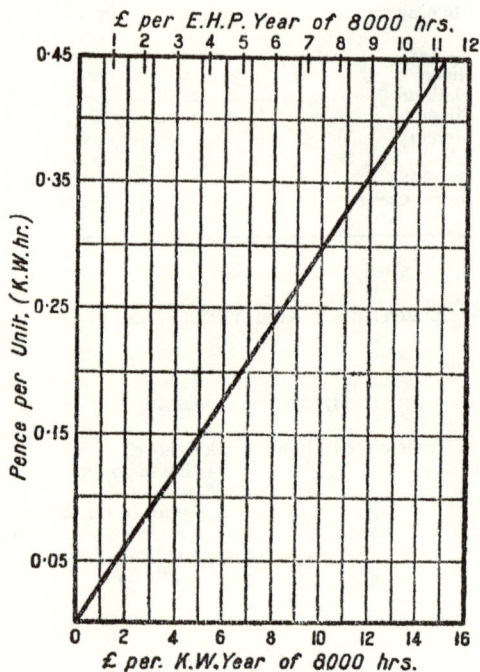

FIG. 241.

APPENDIX B.

BIBLIOGRAPHY.

(This list is intended to supplement the references given in the foot-notes throughout the text.)

CONTENTS.

BIBLIOGRAPHY.

BOOKS ON ELECTRIC FURNACES.

BRONN. " Der elektrische Ofen im Dienste der keramischen Gewerbe und der Glas-und Quarz-glaserzeugung," 1910 (Wilhelm Knapp, Halle-a-S.).

KERSHAW. " Electro-thermal methods of iron and steel production," 1913 (Constable). Review in 'Trans. Faraday Soc.,' 1914, 10, 300.

MOISSAN. " The electric furnace," trans. by Mouilpied, 1904 (Edward Arnold).

RIDEAL. " Industrial electrometallurgy," 1919 (Balliere, Tindall & Cox).

RODENHAUSER AND SCHONAWA. " Electric furnaces in the iron and steel industry," trans. by Baur, 1913 (John Wiley). Review, 'Trans. Faraday Soc.,' 1913, 9, 335.

STANSFIELD. " The electric furnace, its construction, operation and uses," 2nd ed., 1914 (McGraw Hill Book Company).

WRIGHT. " Electric furnaces and their industrial application," 1910 (Constable).

ELECTRODES.[1]

BARHAN, G. B. " Electrodes for electric furnaces." 'Elec. Rev.' (London), Apr. 18, 1913.

BARNETT, C. A. " Electric furnace electrodes, their manufacture and uses." 'Sibley J. Eng.,' 1915, 30, 27.

BAY, I. " Evolution of the electrode." 'L'Electricien,' 1909, 37, 385 et seq.

CLOCHER, WM. " Manufacture of carbons." 'Elec. Rev.,' Jan. 20, 1911; 'Met. Chem. Eng.,' 1911, 9, 137.

COLLINS, C. L. " Graphite electrodes in electrometallurgical processes." 'Trans. Am. Electrochem. Soc.,' 1902, 1, 53.

---

[1] Cf. A. D. Little, " The electric furnace as applied to metallurgy." 'Trans. Amer. Electrochem. Soc.,' 1920, vol. 37.

"Design of electric furnace electrodes." 'Electrochem. Met. Eng.,' 1909, **7**, 502.

"Effect of heat leakage on the electrode loss." 'Met. Chem. Eng.,' 1910, **8**, 59.

"Electrode holders construction for electric furnaces." 'Stahl u. Eisen,' 1913, **33**, 472, 555; 'Met. Chem. Ind.,' 1913, **11**, 321.

"Electrode manufacture." 'Elec. Times,' 1919, **55**, 1311.

"Electrode situation (in France)." 'J. four eléc.,' 1917, 17-18.

ESCARD, J. "Electrodes for electric furnaces. Their manufacture, properties and utilisation." 'Gén. Civil,' Aug. 4, 1917; 'Elec. Rev.,' Sept. 14, 1917; 'Gen. Elec. Rev.,' 1918, 21, 664, 781.

FAVARNEY, O. "Electrodes of pure graphite and their use in electric metallurgy." 'Rev. Industriel,' 1912, **43**, 464.

FITZGERALD, F. A. J. "On carbons for electric metallurgy." 'Trans. Am. Electrochem. Soc.,' 1907, **11**, 317.

—— "On testing carbon electrodes." 'Trans. Am. Electrochem. Soc.,' 1902, **2**, 43.

FITZGERALD, F. A. J., AND HINCKLEY, A. I. "Experiments with furnace electrodes." 'Trans. Am. Electrochem. Soc.,' 1913, **23**, 333.

FITZGERALD, F. A. J. "Electrode suspension in electric furnaces." 'Stahl. u. Eisen.,' 1913, **33**, 472.

FORRSELL, J. "Current densities and energy losses in electrodes." 'Met. Chem. Eng.,' 1910, **8**, 26.

HANSEN, C. A. "Furnace electrode losses." 'Electrochem. Met. Ind.,' 1909, **7**, 358.

HERING, C. "Design of furnace electrodes." 'Elec. World,' 1910, **55**, 1508.

—— "Chilling or heating action of furnace electrodes *versus* least electrode loss." 'Met. Chem. Eng.,' 1910, **8**, 188.

—— "Determination of the constants of materials for furnace electrodes." 'Trans. Am. Electrochem. Soc.,' 1910, **17**, 151.

—— "Electrode construction for furnaces." 'Met. Chem. Eng.,' 1911, **9**, 67.

—— Electrode efficiency of furnaces." 'Electrochem. Met. Ind.,' 1909, **7**, 473.

—— "Empirical laws of furnace electrodes." 'Trans. Am. Electrochem. Soc.,' 1910, **17**, 171.

—— "Furnace electrodes." 'Met. Chem. Eng.,' 1910, **8**, 391.

—— "Furnace electrode losses." 'Electrochem. Met. Ind.,' 1909, **7**, 400.

—— "Properties and behaviour of furnace electrodes." 'Met. Chem. Eng.,' 1910, **8**, 128.

KENNELLY, A. E. "Modification in Hering's law of furnace electrodes." 'Proc. Am. Inst. Elec. Eng.,' 1910, **29**, 267.

KUNZE, W. "Automatic electrode regulation for arc furnaces." 'Stahl. u. Eisen.,' 1918, **38**, 125, 152, 184, 212.

MAHLKE, A. "New form of electrode." 'Met. Chem. Eng.,' 1911, **9**, 42.

"Manufacture of electric furnace electrodes." 'L'Elecn.' 1909, **38**, 339.

"Production of carbon electrodes for metallurgical purposes." 'Stahl. u. Eisen.,' 1912, **32**, 1857.

"Properties and uses of furnace electrodes." 'Elec. World,' 1917, **70**, 963.

ROEBER, E. F. "Electrode losses in electric furnaces." 'Trans. Am. Electrochem.,' 1909, **16**, 363.

ROUSH, G. A. "Manufacture of carbons for steel furnaces." 'J. Ind. Eng. Chem.,' 1909, **1**, 286.

"Soderburg continuous self-baking electrodes." 'Iron Age,' Oct. 7, 1920; 'Trans. Am. Electrochem. Soc.,' 1921.

"Standard electrodes for electric furnaces." 'Iron Age,' 1916, **98**, 1369.

STYRI, H. "Electrode cooling." 'Met. Chem. Eng.,' 1917, **17**, 233.

TURNBULL, R. T. "Furnace electrodes practically considered." 'Trans. Am. Electrochem. Soc.,' 1912, **21**, 397; 'Chem. Eng.,' 1912, **15**, 68.

FERRO-ALLOYS.[1]

General.

BARDELL, E. S. "Manufacture of ferro-alloys in the electric furnace." 'Mining Journ.,' 1918, **123**, 708.

EASTON, W. H. "Electric furnace for melting alloys." 'Elec. World,' 1918, **72**, 295.

"Energy consumption in the manufacture of ferro-alloys." 'Elec. World,' July 24, 1920; 'Tech. Rev.,' 1920, **7**, 175.

GIBSON, C. B. "Manufacture of ferro-alloys in the electric furnace." 'Elec. J.,' 1919, **16**, 366.

KEENEY, R. M. "Electric smelting of chromium, molybdenum and vanadium ores." 'Trans. Am. Electrochem. Soc.,' 1913, **24**, 167; 'Met. Chem. Eng.,' 1913, **11**, 585.

—— "Manufacture of ferro-alloys in the electric furnace." 'Bull. Am. Inst. Min. Eng.,' 1918, **140**, 1321; **143**, 1651; 'Iron Age,' **102**, 624 *et seq.* ; 'Eng. Min. J.,' **106**, 405; 'Chem. Met. Eng.,' **19**, 281.

LYON, D. A., KEENEY, R. M., AND CULLEN, J. F. "Electric smelting of ferro-alloys." 'Iron Trade Rev.,' 1915, **56**, 717 *et seq.*

[1] A. D. Little, loc. cit.

"New electric furnace for making ferro-alloys." 'Elec. World,' 1918, **71**, 566.

NORTHRUP, E. F. "Electrolytic production of carbon free alloys." 'Chem. Met. Eng.,' 1919, **21**, 258.

SEEDE, J. O. "Electric furnaces in the production of steel and ferro-alloys." 'Gen. Elec. Rev.,' 1918, **21**, 767.

WIDMER, G. "France and the electrometallurgy of ferro-alloys." 'J. four eléc.,' 1918, **27**, 65.

YENSEN, T. D. "Preparation of pure alloys for magnetic purposes." 'Trans. Am. Electrochem. Soc.,' 1917, **32**, 269.

FERRO-SILICON.

ANDERSON, R. J. "Metallurgy of ferro-silicon." 'Iron Trade Rev.,' 1917, **60**, 1025; 'Eng. Min. J.,' **103**, 1095.

ESCARD, M. "Ferro-silicon." 'La Lumière eléct.,' Mar. 6, 1919.

"Ferro-silicon and aluminium in Norway." 'J. four eléc.,' 1917, **26**, 211.

FERROMANGANESE.

BUCK, E. C. "Bibliography on the manufacture of ferromanganese." 'Met. Chem. Eng.,' 1917, **17**, 638.

"Electric smelting of ferromanganese ore in California." 'Elec. Rev. West. Elec.,' 1918, **73**, 940.

HÄRDÈN, J. "Electric furnace for melting ferromanganese." 'Elec. Rev.' (London), 1918, **82**, 116.

JAKOKI, J. "Ferromanganese in the blast furnace." 'Stahl. u. Eisen.,' 1909, **29**, 1191.

FERROTUNGSTEN.

"Electric production of ferrotungsten." 'Eng. Min. J.,' 1912, **93**, 173.

KEENEY, R. M. "Electric furnace data for ferrotungsten." 'Blast Furnace,' 1918, **6**, 486.

FERROMOLYBDENUM.

DITTUS, E. F., AND BOWMAN, R. G. "Direct production of molybdenum steel in the electric furnace." 'J. Ind. Eng. Chem.,' 1911, **3**, 717; 'Trans. Am. Electrochem. Soc.,' 1911, **20**, 355.

EVANS, J. W. "Ferromolybdenum manufacture in Canada." 'Can. Chem. J.,' 1918, **2**, 208.

GUICHARD, M. "Production of molybdenum from molybdenite." 'Compt. rend.,' **122**, 1270.

LENHER, W. "Preparation of molybdenum." 'Metallurgie,' 1906, **3**, 549.

MATHEWS, J. A. "Molybdenum steels." 'Chem. and Met. Eng.,' 1921, **24**, 395.

IRON.[1]

ALLEN, H. "Electric furnace in the iron and steel industries." 'Cassier's Mag.,' 1905, **27**, 358.

—— "Application of electricity to the smelting of iron ores." 'Mech. Eng.,' 1907, **19**, 39.

—— "Application of the electric furnace to the metallurgy of iron and steel." 'Electrochem. Ind.,' 1904, **2**, 307.

ARMSTRONG, D. E. "Electric low-phosphorus pig iron." 'Can. Chem. J.,' 1918, **2**, 190.

ARNOU, G. "Direct reduction of iron ores in the electric furnace.' 'Rev. métal.,' 1910, **7**, 1054.

—— "Present state of the electrometallurgy of iron." 'Lumière eléc.,' **12**, 4.

—— "Progress realised in the reduction of iron ores in the electric furnace." 'Lumière eléc.,' **16**, 269.

BAILY, T. "Data on the operation of the electric furnace." 'Elec. World,' 1918, **71**, 780.

BEILSTEIN, A. "Electric pig iron production in Scandinavia." 'Stahl u. Eisen,' **33**, 1270.

BENNIE, P. McN. "Application of the electric furnace to the metallurgy of iron and steel." 'Electrochem. Ind.,' 1904, **2**, 307.

—— "The electric furnace." 'Iron Age,' **85**, 216.

—— "The electric furnace for iron and steel." 'Iron Trade Rev.,' 1904, **27**, 63.

—— "Electric furnace in the iron and steel industry." 'Electrochem. Met. Ind.,' **7**, 322.

—— "Electric furnace pig iron in California." 'Trans. Am. Electrochem. Soc.,' **15**, 35; 'Electrochem. Met. Ind.,' 1909, **7**, 251.

BIBBY, J. "Development in electric iron and steel furnaces." 'Iron Coal Trades Rev.,' **98**, 611; 'Engineer,' **127**, 513; 'Elec. World,' 1919, **74**, 84, 712; 'Foundry Trade J.,' 1919, **21**, 311; 'Electrician,' **83**, 214; 'Elec. Rev.' (London), 1919, **84**, 136, 166, 176.

---

[1] Cf. A. D. Little, loc. cit.

464   THE ELECTRIC FURNACE

BORCHERS, W. "Reduction of oxide ores in the electric furnace." 'Stahl u. Eisen,' 1911, **31**, 706.

BOVING, J. O. "Electric iron smelting." 'Electrician,' 1917, **79**, 613.

—— "New data on electric smelting in Sweden." 'Iron Age,' **93**, 1268.

BRIDGE. "Electric furnaces." 'L'Electricien,' **38**, 187.

CAIN, J. R., SCHRANN, E., AND CLEAVER, H. E. "Preparation of pure iron and iron-carbon alloys." 'J. Ind. Eng. Chem.,' **8**, 217; 'U.S. Bur. Standards, Bull.,' 1916, **13**.

CAMPBELL, D. F. "Progress in the electrometallurgy of iron and steel." 'Trans. Faraday Soc.,' **7**, 198; 'Electrician,' 1912, **68**, 149; 'Chem. News,' **104**, 192.

CARCANO, F. E. "Production of pig iron in the electric furnace and the industrial utilisation of pyrite residues." 'Electrochem. Met. Ind.,' **7**, 155.

CATANI, REMO. "Large electric furnace in the electrometallurgy of iron and steel." 'Trans. Am. Electrochem. Soc.,' 1909, **15**, 159; 'Electrochem. Met. Ind.,' **7**, 268.

—— "Reduction of iron ore in the electric furnace." 'Electrochem. Met. Ind.,' 1909, **2**, 153.

CIRKEL, F. "Preparation of pig iron in electric furnace." 'Stahl u. Eisen,' 1906, **26**, 868, 1369.

CONE, E. F. "High grade pig iron from scrap steel." 'Iron Age,' 1917, **100**, 485, 497, 629.

"Cost of electric pig iron production in North Sweden." 'Engineering,' 1917, **104**, 621.

CRAWFORD, J. "Progress of electric smelting at Héroult, California." 'Met. Chem. Eng.,' 1913, **11**, 383.

DE GEER, D. "Electric smelting of pig-iron at Domnarfvet." 'Chem. Met. Eng.,' 1921, **24**, 429.

DOUBS, F. "Production of white cast iron in the electric furnace from cold and molten charge." 'Stahl u. Eisen,' 1911, **31**, 589.

ECKMANN, S. H. "Electricity in iron and steel works and allied metal industries." 'Electrician,' **70**, 389.

"Electric furnace for pig iron." 'Iron Trade Rev.,' 1914, **55**, 521.

"Electric furnace pig iron at Trollhattan." 'Met. Chem. Eng.,' 1912, **10**, 413.

"Electric pig iron in Norway—a new type of furnace using coke successfully." 'Iron Age,' 1915, **95**, 1120.

"Electric smelting of iron ore in California." 'Iron Age,' 1913, **92**, 124.

"Electric smelting of iron ores in northern Sweden." 'Iron Age,' 1917, **100**, 605.

"Electric smelting in Sweden." 'Electrician,' 1918, **80**, 800.

"Electrometallurgical industry of Sweden." 'J. four eléc.,' Dec., 1916.

"Electrothermic iron ore smelting in Scandinavia." 'Eng. Min. J.,' 1915, **100**, 351.

ELLIOTT, G. K. "Improving the quality of gray iron by the electric furnace." 'Trans. Am. Electrochem. Soc.,' 1919, **35**, 175 ; 'Iron Age,' 1919, **103**, 939.

ELWELL, E. F. "Refining of iron and steel in induction type furnaces." 'Trans. Am. Inst. E. E.,' **30**, 621.

ESCARD. "Electrolytic iron." 'Le Gen. Civ.,' Aug. 23, 1919 *et seq.* ; 'Tech. Rev.,' 1921, **8**, 233.

ETCHELLS, H. "Application of the electric furnace to the metallurgy of iron and its alloys." 'Electrician,' 1918, **81**, 734.

FRANK, K. G. "Progress in the iron and steel industry and the electric furnace." 'Proc. Am. Inst. E. E.,' 1915, **34**, 2547.

FRICK, O. "Electric reduction of iron ores." 'Met. Chem. Eng.,' 1912, **10**, 71 ; 'Elec. World,' **58**, 1432.

—— "Results with a Rennerfelt furnace." 'Iron Age,' 1918, **101**, 563.

GORROW, R. C. "Electric furnace in the foundry." 'Met. Chem. Eng.,' 1915, **13**, 882.

"Greaves-Etchells electric furnace." 'Elec. Rev.' (London), 1917, **80**, 395.

GREENE, A. E. "Electric heating and the removal of phosphorus from iron." 'Trans. Am. Inst. Min. Eng.,' **74**, 269 ; 'Trans. Am. Electrochem. Soc.,' 1912, **22**, 123.

HANSON, H. J. "Smelting iron electrically with coke as fuel." 'Iron Trade Rev.,' 1913, **53**, 1003.

HÄRDÈN, J. "Electric iron ore smelting at Hardanger, in Norway." 'Electrician,' **52**, 766 ; 'Met. and Chem. Eng.,' 1914, **12**, 82, 223, 280, 444.

—— "Electric smelting and reduction of iron ores in England." 'Electrician,' 1911, **58**, 467.

—— "Induction furnace notes." 'Met. Chem. Eng.,' 1913, **11**, 558.

—— "Utilisation of manganese ores in Sweden." 'Met. Chem. Eng.,' 1917, **17**, 701.

HERLENIUS, J. "Swedish electric pig-iron furnace." 'Chem. and Met. Eng.,' 1921, **24**, 108.

IRRESBERGER, CARL. "The electric smelting furnace of Gronwall-Dixon." 'Stahl u. Eisen,' 1918, **38**, 90.

JOHNSON, W. M.   "Electrometallurgy of iron and steel."   'Met.
    Chem. Eng.,' 1914, **12**, 165.
—— "Two stage electric smelting of iron ore."   'Iron Age,' 1912,
    90, 450.
KALMUS, H. T.   "Recent developments in the electrothermic
    production of iron and steel, 1911-1912."   'Canada, Dept.
    of Mines, Mines Branch,' 1912, pp. 107-120.
KERSHAW, J. B. C.   "Electric furnace methods of iron production."
    'Iron Trade Rev.,' 1912, **50**, 41.
KNUDSEN.   "The electrometallurgical industries of Scandinavia."
    'J. four eléc.,' 1919, **28**, 17.
LEFFLER, J. A.   "Electric iron ore smelting in Sweden."   'Elec-
    trician,' 1915, **75**, 729; 'Engineering,' 1915, **100**, 131.
—— "Electric iron smelting at Trollhattan, Sweden."   'Iron Coal
    Trades Rev.,' June 9, 1911.
—— "Electric pig iron and steel plant at Trollhàttan, Sweden."
    'Engineering,' **92**, 374; 'Met. Chem. Eng.,' **9**, 505.
LOUDEN, T. R.   "The electric smelting of iron ores in Canada."
    'Appl. Sci.,' 1914, **8**, 219.
LYON, D. A.   "Use of electric furnace pig iron in the open-hearth
    furnace."   'Met. Chem. Eng.,' 1912, **10**, 539.
MCKNIGHT, W. M.   "Faults of the small electric arc furnace."
    'Iron Age,' 1916, **97**, 1008.
MARSHALL, A. H.   "Use of electricity and its bearing on fuel saving
    in the iron and steel trades."   'Electrician,' 1918, **80**, 550.
MERCER, R. G.   "Electric furnaces in the United Kingdom."
    'Electrician,' 1919, **82**, 694.
MORRISON, W. L.   "Electric smelting on the Pacific Coast."
    'Journ. of Electricity,' 1919, **42**, 67.
NEUMANN, B.   "Materials and thermal balance of the electric pig
    iron furnace."   'Stahl u. Eisen,' 1915, **35**, 1152; 'Iron
    Age,' 1916, **97**, 834.
—— "New results from the electric smelting of iron in the experi-
    mental plant at Trollhàttan."   'Stahl u. Eisen,' **32**, 1409.
"New installations of Héroult electric furnace."   'Iron Age,' 1915,
    96, 337.
OESTERREICH, MAX.   "Large Helfenstein electric furnace."   'Iron
    Age,' **9¹**, 1482; 'Stahl u. Eisen,' **33**, 305.
OSANN, B.   "Ingot iron from the electric furnace."   'Iron Coal
    Trades Rev.,' Nov. 6, 1918.
PETINON, F.   "Electric furnace in the foundry."   'Met. Chem.
    Eng.,' 1915, **13**, 650.
—— "Production of pig iron in the electric furnace."   'Elec. Rev.'
    (London), 1912, **71**, 44.

PRENTISS, F. L. "Uses of electricity in malleable foundry." 'Iron Age,' 1919, **103**, 537.

RICHARDS, J. W. "Gas circulation in electric reduction furnaces." 'Trans. Am. Electrochem. Soc.,' 1912, **21**, 403.

ROBERTSON, T. D. "Iron and steel melting in electro-metal furnaces." 'Electrician,' 1912, **70**, 501.

RODENHAUSER, W. "Improvements in the electric furnace and new fields of application in iron smelting." 'Chem. Ztg.,' **36**, 1294.

SCOTT, E. K. "Electric furnace in iron and brass foundries." 'Foundry,' 1913, **41**, 379.

SEBILLOT, A. "Water-jacketed electric blast furnace." 'J. four. eléc.,' 1918, **27**, 215.

SIFTON, CLIFFORD. "Electric furnace in Canada. 'Electrician,' 1918, **80**, 674.

SIMPSON, I. "Reduction of iron ores by the electrothermic process." 'Bull. Can. Min. Inst.,' 1919, **87**, 709.

STANSFIELD, A. "Electric smelting of iron ores." 'Bull. Can. Min. Inst.,' 1919, **87**, 706.

—— "Electric smelting of iron ores in British Columbia." 'Eng. Min. J.,' 1919, **107**, 224; 'Met. and Chem. Eng.,' 1919, **20**, 630.

—— "Electric smelting possibilities in British Columbia." 'Elec. Rev.,' 1919, **74**, 805.

TURNBULL, R. "Electric pig iron in war times." 'Iron Age,' **100**, 886, 870; 'Iron Trade Rev.,' **61**, 828; 'Met. Chem. Eng.,' 1917, **17**, 459; 'Trans. Am. Electrochem. Soc.,' 1917, **32**, 119.

—— "Electric pig iron after the war." 'Trans. Am. Electrochem. Soc.,' 1919, **34**, 143; 'Met. and Chem. Eng.,' 1919, **20**, 178.

—— "Electric pig iron from steel scrap." 'Iron Age,' 1918, **102**, 1026.

VOM BAUR, C. H. "Rennerfelt electric furnace operations." 'Trans. Am. Electrochem. Soc.,' 1917, **31**, 111; 'Iron Age.,' **99**, 1206.

LABORATORY FURNACES.[1]

ARNDT. "Pressure and vacuum furnaces." 'Electrician,' 1916, **77**, 637, also 'Elektrotechnische Zeitschrift,' 1916, **37**, 119.

ARSEM. "Vacuum furnace." 'Trans. Am. Electrochem. Soc., 1906, **9**, 153; 'Journ. Am. Chem. Soc.,' 1906, **28**, 921.

[1] Cf. E. and E. A. Griffiths, "A carbon tube furnace for testing softening points and compressive strengths," p. 25. London, 1920, H.M. Stationery Office.

30 *

CALHANE ¹AND BARD. "An efficient electric furnace for high temperature." 'Met. and Chem. Eng.,' 1912, 10, 461.

FINK. "Vacuum furnace metallurgy." 'Trans. Am. Electrochem. Soc.,' 1912, 21, 445.

GOECKE. "The vacuum electric furnace and its use." 'Metallurgie,' 1911, 8, 667. (Carbon Tube Furnace.)

HANSEN. "Experimental arc furnace." 'Electrochem. and Met. Ind.,' 1909, 7, 206. (Laboratory form of arc furnace for metallurgical operations.)

HÄRDÈN. "Recent development of the Kjellin and Ròchling-Rodenhauser electric induction furnaces." 'Trans. Faraday Soc.,' 1908, 4, 120.

HARKER. "Small carbon tube furnace." 'Engineering,' Mar. 17, 1916.

HUTTON AND PETAVEL. 'Journ. Inst. Elec. Eng.,' 1902-3, 32, 222. Describes electric furnace equipment of Manchester University, with details of the electrochemical industry at that time. *Ibid.*, p. 236. Electrode holders for arc furnaces.

LUMMER AND PRINGSHEIM. "The radiation scale of temperature and its realisation up to 2300° abs." 'Ber. Deutsch. Phys. Gesell.,' 1903, 3; 'Electrician,' 1903, 51, 673.

MENDENHALL AND FORSTYTHE. "High temperature measurements with the Stefan Boltzmann Law." 'Physical Review,' 1914, 4, 62.

MALM. "High temperature experimental furnace." 'Met. and Chem. Eng.,' 1915, 13, 70. (Small arc furnace.)

NORTHRUP. 'Trans. Am. Electrochem. Soc.,' 1911, 1914; 'Journ. Franklin Institute,' 1913-16; 'English Mechanic and World of Science,' June 18, 1916 and subsequent numbers.

—— "A new high-temperature furnace." 'Met. and Chem. Eng.,' 1914, 12, 31, 305. Describes a graphite resister furnace with conical electrodes. Furnace designed for experiments on resistivity of molten metals.

RUFF. "Carbon tube furnace." 'Berichte d. Deutsch. Chem. Ges.,' 1910, 43, 1564.

RUFF AND GOECKE. "Carbon tube furnace." 'Zeits. f. Angew. Chem.,' 1911, 24, 1459. Describes vacuum furnaces.

RUHSTRAT AND GRIMMER. Brit. Pat. 24472 (1903). Describes method of cutting spiral on carbon tube.

STAHLER AND ELBERT. 'Berichte d. Deutsch. Chem. Ges.,' 1913, 46, 2070. Pressure furnace.

TUCKER. 'Electrochem. and Met. Ind.,' 1907, 5, 227.

WOLF AND MÜLLER. "Pressure and vacuum furnaces." 'Zeit. f. Electrochem.,' 1914, 20, 1, 177.

Yensen. "Vacuum furnace," for melting pure iron and its alloys. Arsem type. 'Electrician,' 1916, **77**, 363; see also 1915, **75**, 119.

NITROGEN FIXATION—ARC PROCESSES.

Andriessens, H. "Nitrogen fixation, new process for." 'Chem. Trade Journ.,' 1918, **62**, 116.

Bodenstein. "Nitric acid from air." 'Z. Angew. Chem.,' 1906, **19**, 14.

Buhler, F. A. "Die Ammonsalpeterfabrik in Notodden." 'Chemische Ind.,' 1911, **34**, 210.

Eyde. 'Journ. Ind. Eng. Chem.,' 1912, **4**, 771.

Fischer and Braehmer. "Oxidation of nitrogen at high temperatures." 'Ber.,' 1906, **39**, 940.

Fischer and Mark. "Thermal relation between ozone, nitric oxide and hydrogen peroxide." 'Centralblatt,' 1907, **1**, 85; 'Berichte,' 1906, **39**, 3631; 'Chem. Zeit.,' **30**, 1291; 'Journ. Soc. Chem. Ind.,' 1907, **26**, 200.

Friderich, L. "Electrochemical production of nitric acid." 'Cen.,' 1906, **1**, 1765.

Gros. "Formation of oxides of nitrogen in electric furnace." 'Compt. Rend.,' 1920, **170**, 811.

Guye. "Combustion of nitrogen, theory of." 'Chem. Ind.,' 1906, **29**, 85; 'Journ. Soc. Chem. Ind.,' 1906, **25**, 270.

Jellinek. "Rate of decomposition of nitric oxide at high temperatures." 'Zeit. Anorg. Chem.,' 1906, **49**, 229.

Kilburn-Scott, E. "Oxidation of nitrogen, descriptions of furnaces." 'Journ. Soc. Chem. Ind.,' 1915, **34**, 113.

Lamy, E. "Met. and Chem. Engineering,' 1911, **9**, 100.

Nernst. "Formation of nitric oxide at high temperatures." 'Zeit. Anorg. Chem.,' 1906, **49**, 213.

Pennock, J. D. "Birkeland-Eyde process for nitric acid." 'Journ. Am. Chem. Soc.,' 1906, **28**, 1251.

Rossi. "Oxidation of nitrogen in contact with heated Nernst filament." 'Bull. Soc. Chim. de France,' 1906, **36**, 756.

Steinmetz. "Theoretical study of nitrogen fixation by an electric arc." 'Chem. Met. Eng.," 1920, **22**, 299.

Witt. "Notodden process." 'Dingler's Polytech. Journ.,' 1906, 143.

NITROGEN FIXATION—CYANIDE PROCESSES.

Briner and Baerfuss. "Fixation in form of HCN by the electric arc." 'Journ. Soc. Chem. Ind.,' 1920, **39**, 105 A.

FERGUSON AND MANNING. "Fixation equilibrium studies in the Bucher process." 'Journ. Ind. Eng. Chem.,' 1919, 946.

LANDIS. "Canadian cyanide." 'Tech. Rev.,' 1920, **7**, 326.

MORIMOTO. "Cyanide process." 'Journ. Soc. Chem. Ind.,' 1919, **38**, 10 A.

ROBINE AND LENGLEN. "The cyanide industry," 1906.

THOSSELL. "Nitrogen fixation as cyanide." 'Z. Angew. Chem.,' 1920, **33** (1), 239.

NITROGEN FIXATION—CYANAMIDE PROCESSES.

BREDIG, FRAENKEL AND WILKE. 'Zentralblatt,' 1907 (11), 1219. 'Zeit. f. Elektrochem.,' 1907, **13**, 69.

CARO, N. 'Zeit. Angew. Chem.,' 1906, 1569.

CREIGHTON. "Nitrogen absorption by carbides. Cyanamide process." '*Chem. News,' 1919, **119**, 109.

FRANK, A. 'Journ. Soc. Chem. Ind.,' 1908, **27**, 1093.

LAMY. 'Met. and Chem. Engineering,' 1911, **9**, 99.

LANDIS. "American Cyanamide Company's Plant." 'Met. and Chem. Engineering,' 1914, **12**, 265 ; 1915, **13**, 213 ; 'Journ. Ind. Eng. Chem.,' 1915, **7**, 433 ; 1916, **8**, 156.

PRANKE, J. 'Journ. Ind. Eng. Chem.,' 1914, **6**, 415 (Agricultural Uses.)

SIEBNER. "Cyanamide industry." 'Chem. Zeit.,' 1913, 1087. "Cyanamide industry." 'Dingler's Polytech. Journ.,' 1906, 43 ; 'Engineering,' 1914, **98**, 267.

"Storage of cyanamide." 'Chem. Trade Journ.,' 1915, **56**, 1475. Tofani, Ger. Pat., 246077 (1910).

WASHBURN. 'Trans. Am. Electrochem. Soc.,' 1915, **27**, 385.

NITROGEN FIXATION—SYNTHETIC AMMONIA PROCESSES.

CLAUDE. "Ammonia, synthetic." 'Tech. Rev.,' 1920, **7**, 326.

—— "Ammonia synthesis at high pressures." 'Chim. et Ind.,' 1920, **4**, 5 ; 'Chem. Age,' 1920, **3**, 56.

—— "Ammonia, synthesis of, at very high pressures." 'Bull. Soc. Chim. de France,' 1920, **27**, 705.

—— "Claude process, its installation in England." 'Chem. Age,' 1920, **2**, 466.

—— "Manufacture of hydrogen for ammonia synthesis." 'Compt. rend.,' 1921, **172**, 974.

CREIGHTON. "Ammonia synthesis, Haber process." 'Chem. News,' 1919, **119**, 98.

HABER. 'Zeit. Angew. Chem.,' 1913, **3**, 323; 1914, **I**, 473; 'Journ. Soc. Chem. Ind.,' 1914, **33**, 52.

—— "Ammonia, synthetic, at Oppau, Germany " (Haber process). 'Chem. and Met. Eng.,' 1921, **24**, 305, 347, 391. 'La technique moderne,' Nov. 1920.

—— "Synthetic ammonia." 'Chem. Trade Journ.,' 1914, **54**, 70, 155 ; 'Met. and Chem. Eng.,' 1913, **II**, 211.

JONES, C. H. "Nitrogen Fixation by the Haber process." 'Met. and Chem. Eng.,' 1920, **22**, 1071.

MAXTED. "Ammonia synthesis, catalysts and plant." 'Chem. Age,' 1919, **I**, 515.

—— " Ammonia synthesis direct." 'Rev. de. Prod. Chim.,' 1919, 658. •

" Nitrogen fixation, British development of Haber process." 'Chem. Trade Journ.,' 1918, **62**, 380.

TOUR. "Ammonia, synthetic." 'Journ. Ind. Eng. Chem.,' 1920, **12**, 844.

—— " Some interpretations of ammonia synthesis equilibrium." *Ibid.*, 1921, **13**, 298.

NITROGEN FIXATION—HAÜSSER PROCESS.

BENDER. Ger. Pats., Nos. 277435, 279007, 280966.

HAÜSSER. 'Zeit. Verein. Deutsch. Ingen.,' 1912, 1157; 'Chem. Trade Journ.,' 1914, **55**, 46.

HERMAN. Ger. Pats., Nos. 281984, 283535.

MATIGNON. 'Chem. Trade Journ.,' 1914, **34**, 69.

NITROGEN FIXATION—SERPEK PROCESS.

SERPEK. "Inorganic synthesis of ammonia." 'Chem. Zeit.,' 1913, 1196.

TUCKER. "Serpek process." 'Journ. Soc. Chem. Ind.,' 1913, **32**, 1143.

VON HERRE. "Serpek process." 'Chem. Zeit.,' 1914, **38**, 317.

WASER. 'Chem. Zeit.,' 1915, 903.

NITROGEN FIXATION—GENERAL.

BERNTHSEN. "Synthetic formation of ammonia." 'Zeit. Angew. Chem.,' 1913, **26**, 10.

BERTHELOT. "Synthesis of nitric acid by silent discharge." 'Comp. Rend.,' 1906, **142**, 1367.

472   *THE ELECTRIC FURNACE*

BUCHER. "Fixation of nitrogen." 'Journ. Soc. Chem. Ind.,' 1917, **36**, 451; 'Journ. Ind. Eng. Chem.,' 1917, **9**, 233.

CREIGHTON. "Absorption of nitrogen by carbides." 'Chem. News,' 1919, **119**, 109.

—— "Methods of nitrogen fixation." 'Chem. News,' 1919, **119**, 82.

CROOKES. "The wheat problem," 2nd edition. London, 1905.

CROSSLEY. "The fixation of atmospheric nitrogen." 'Thorpe's dictionary of applied chemistry,' Vol. III., 698.

ERLWEIN. "Fixation of atmospheric nitrogen." 'Zeit. Angew. Chem.,' 1907, **20**, 351.

FRANKLAND. "Utilisation of atmospheric nitrogen for industrial purposes." 'Journ. Soc. Chem. Ind.,' 1907, **26**, 175.

"German industry and the war." 'Nature,' 1918, **102**, 66.

GILBERT. "Sources of nitrogen compounds in the U.S." 'Nature,' 1917, **98**, 431.

GRAU AND RUSS. "Combustion of air in electric arc." 'Centralblatt,' 1907, **1**, 1171.

HABER, F. "Modern chemical industry." 'Journ. Soc. Chem. Ind.,' 1914, **33**, 49.

HARDING. "Nitrogen fixation by silent electric discharge." 'Tech. Rev.,' 1920, **6**, 469.

HARDING AND McEACHRON. 'Am. Inst. Elect. Eng.,' 1921; 'Journ. Soc. Chem. Ind.,' 1921, **40**, 213 A.

HOSMER. "Literature of the nitrogen industries." Reprint from 'Journ. Ind. Eng. Chem.,' Vol. IX., No. 4, Apr., 1917.

JURISCH. "Saltpetre and sein Ersatz," 1908.

KELER. "Summary." 'Zeit. Angew. Chem.,' 1914, **27**, 244.

KILBURN-SCOTT. "Cyanamide process compared with the electric arc." 'Met. and Chem. Eng.,' 1918, **19**, 411.

"Kitzinger process." 'Chem. Trade Journ.,' 1918, **63**, 88.

LANDIS. "Nitrogen fixation." 'Journ. Ind. Eng. Chem.,' **7**, 433.

LOWRY. "Oxidation of nitrogen in presence of ozone." 'Trans. Chem. Soc.,' 1912, **101**, 1152.

LUNGE. "Coal tar and ammonia," 1916.

MAILLOUX. "Electrochemical process for nitric acid." 'Met. and Chem. Eng,' 1917, 325.

MANNING. "Nitrogen fixation." 'Bur. of Mines, Bull., Washington,' 178 B., 1919.

MARSHALL. "Explosives," 1916, Vol. I., chap. iv. (Nitre in India).

MAXTED. "Economic aspect of methods." 'Chem. Age,' 1919, 590.

"Nitrogen fixation in England." 'Times Eng. Supp.,' 1919, 25.

PARSONS. "Fixation of nitrogen." 'Chem. Trade Journ.,' 1917, **61**, 299, 319.

RIDEAL AND TAYLOR. "Nitrogen fixation: catalysts in different processes." 'Chem. Age,' 1919, **1**, 488.

SKAUPY. "Chemical reactions by electrical discharge." 'Engineering,' 1916, **102**, 618; 'Berichte,' 1916, 2005.

TOLMAN, R. C. "Government fixed nitrogen research—Recommendations for future investigations for U.S. Government." 'Chem. Met. Eng.,' 1921, **24**, 595.

WARBURG AND LEITHAUSER. "Oxidation of nitrogen by silent discharge." 'Centralblatt,' 1907, **1**, 1173.

WASER. "Summary." 'Chem. Zeit.,' 1913, 1286.

WHITE. "Nitrogen fixation processes." 'Chem. Trade Journ.,' 1919, **44**, 341.

—— "Nitrogen fixation in U.S." 'Journ. Ind. Eng. Chem.,' 1919, **11**, 231.

NITROGEN FIXATION—AMMONIA OXIDATION PROCESSES.

CREIGHTON. "The technical oxidation of ammonia to nitric oxide." 'Chem. News,' 1919, **119**, 146.

OSTWALD. "Catalytic process." 'Met. and Chem. Eng.,' 1913, **11**, 438.

PERLEY. "Ammonia, catalysts used in oxidation of." 'Met. and Chem. Eng.,' 1920, **22**, 127.

"The oxidation of ammonia applied to vitriol chamber plants." (Nitrogen Products Committee) H.M. Stationery Office, 1919.

NON-FERROUS METALS.[1]

General.

BAILY, T. F. "Electric furnace progress." 'Metal Ind.,' 1919, **17**, 316.

COLLINS, E. F. "Electric furnace for melting non-ferrous metals," 'Elec. World,' **73**, 1110; 'Iron Trade Rev.,' **64**, 1023; 'Iron Age,' **103**, 1288; 'Met. Ind.,' **17**, 221.

—— "Melting of some non-ferrous metals and their alloys in the electric furnace." 'Met. and Chem. Eng.,' 1919, **21**, 673; 1920, **22**, 148.

EASTON, W. H. "Electric furnace for melting alloys." 'Elec. World,' 1918, **72**, 295.

---

[1] Cf. A. L. Little, loc. cit.

"Electric furnace development for non-ferrous metals." 'Metal Ind.,' 1919, **14**, 444.

HERING, CARL. "Ideals and limitations in the melting of non-ferrous metals." 'Proc. Inst. Metals' (London), 1917, **17**, 243.

MILLER, D. D. "Electrical melting of non-ferrous metals." 'Foundry,' 1918, **46**, 110.

—— "Present progress and development of the electric furnace in the non-ferrous metallurgy." 'Met. and Chem. Eng.,' 1917, **17**, 537; 'Iron Age, **100**, 1122.

STANSFIELD, A. "Electric furnaces as applied to non-ferrous metallurgy." 'Chem. Trade Journ.,' 1916, **58**, 317, 339; 'Cassier's Mag.,' June, 1916.

BRASS.

BAILY, T. E. "Baily electric metal-melting furnace." 'Metal Ind.,' 1917, **15**, 399.

—— "Electric furnace for brass melting." 'Met. and Chem. Eng.,' 1917, **17**, 461; 'Elec. World,' 1917, **70**, 820.

—— "Resistance furnace for melting brass." 'Trans. Am. Electrochem. Soc.,' 1917, **32**, 155.

—— "Baily electric brass-melting furnace." 'Elec. Rev.', 1919, **74**, 438.

BLAKESLEE, R. N. "Operating brass making induction furnaces." 'Elec. World,' 1919, **74**, 642.

BOOTH, C. H. "Booth electric rotating brass furnace." 'Metal Ind.,' 1919, **17**, 317.

—— "New electric rotating electric furnace." 'Iron Age,' 1919, **103**, 1699; 'Metal Ind.,' 1919, **17**, 317.

GILLET, H. W. "Symposium on electric furnaces. Utilisation of electric brass furnaces." 'Journ. Ind. Eng. Chem.,' 1919, **11**, 664; 'Met. and Chem. Eng.,' 1919, **21**, 6; 'Metal Ind.,' 1919, **17**, 316.

—— H. W., AND RHOADS, A. E. "Melting brass in a rocking electric furnace." 'Bur. of Mines Bull.,' 1918, 171; 'Journ., Ind. Eng. Chem.,' 1918, **10**, 459; 'Metal Ind.,' **16**, 265; 'Elec. World,' **72**, 360.

MANNING, V. H. "Electrical furnaces for the melting of brass." 'Elec. World,' 1919, **73**, 493.

Melting of non-ferrous metals and alloys." 'Elec. World,' 1919, **73**, 1110.

"Melting silver, nickel, and bronze alloys by electricity." 'Eng. Min. Journ.,' 1919, **107**, 323.

St. John, H. M. "Electric brass melting—its progress and present importance." 'Elec. Journ.,' 1919, 16, 373.

—— "Present status of brass melting." 'Elec. World,' 1918, 71, 1129; 'Iron Age,' 1918, 101, 1281.

"Trend towards electric brass melting." 'Met. and Chem. Eng.,' 1918, 18, 434.

COPPER.

Dewar, W. "Electric smelting of copper ores." 'Min. Mag.,' London, 16, 288.

Lyons, D. A., and Keeney, R. M. "The smelting of copper ores in the electric furnace." 'U.S. Bur. of Mines, Bulletin,' 1915, 81.

MANGANESE.

Carothers, J. N. "Electric smelting of manganese ores." 'Elec. Eng.,' 1918, 52, No. 5, 26.

Härdèn, J. "Utilisation of manganese ores in Sweden." 'Met. and Chem. Eng.,' 1917, 17, 701; 'Iron Age,' 1918, 101, 938.

NICKEL.

"Electrometallurgical treatment of nickel minerals in New Caledonia." 'J. four. eléc.,' 1919, 24, 3.

Zeerleder, A. von. "Smelting of copper-nickel sulphide ores in the electric furnace." 'Metall. u. Erz.,' 1916, 13, 453, 473, 494.

TIN.

"Electric tin smelting." 'Eng. Min. Journ.,' 97, 167.

Mattonet, F. "Electric reduction of tin ore." 'Metallurgie,' 5, 186.

"Recent experiments in electrometallurgy of tin." 'J. four eléc.,' Jan., 1912; 'Min. Eng. World,' 36, 868.

"Smelting of tin ore in the electric furnace." 'Mining Journ.,' 1913, 103, 1002.

ZINC.

Clerc, F. L. "Electric zinc smelting." 'Met. and Chem. Eng.,' 1913, 11, 668.

"Electric zinc furnaces." 'Met. and Chem. Eng.,' 1913, 11, 583.

"Electric zinc smelting." 'Eng. Min. Journ.,' 1912, **94**, 1109.

"Electric zinc smelting process." 'Met. and Chem. Eng.,' 1917, **16**, 158.

"Electrolytic zinc—Tainton-Pring process." 'Eng. and Min. Journ.,' 1921, **III**, 341, 456.

"Electrolytic zinc refining." 'Chem. Trade Journ.,' 1916, **58**, 363.

"Electrometallurgy of zinc." 'Elec. Rev.' (London), 1913, **73**, 408.

FERRARIS, E. "Zinc problem in Italy." 'J. four eléc.,' 1918, **27**, 182.

JOHANSSON, E. A. "Electric furnaces for zinc reduction." 'Met. and Chem. Eng.,' 1918, **18**, 654.

JOHNSON, W. M. "Electric zinc smelting." 'Trans. Am. Electrochem. Soc.,' 1913, **24**, 191; 'Met. and Chem. Eng.,' 1913, **II**, 582.

LAIST, F. "Electrolytic zinc plant of the Anaconda Copper Mining Company at Great Falls, Mont." 'Chem. and Met. Eng.,' 1921, **24**, 245.

NATHUSIUS, H. "Electrochemical production of zinc." 'Metall. u. Erz.,' 1918, **15**, 87, 108.

PETERS, F. "Recent developments in the electrometallurgy of zinc." 'Eng. Min. Journ.,' **89**, 1017.

SNYDER, F. T. "Condensation of zinc vapours from electric furnaces." 'Trans. Am. Electrochem. Soc.,' 1911, **19**, 317.

THOMSON, J. "Condensation in electric zinc smelting." 'Met. and Chem. Eng.,' 1918, **19**, 62.

REFRACTORIES AND ABRASIVES.[1]

ACHESON. 'Journ. Frankl. Inst.,' 1893, **136**, 194, 297.

—— "Graphite, its formation and manufacture." 'Journ. Frankl. Inst.,' 1899, **147**, 475.

—— "Researches on electric furnace products." 'Trans. Faraday Soc.,' 1911, **7**, 217.

ARSEM. "Transformation of other forms of carbon into graphite." 'Trans. Am. Electrochem. Soc.,' 1911, **20**, 105.

BANCROFT, WALKER AND MILLER. "Study of a small carborundum furnace," 8th International Congress of Applied Chemistry, Electrochemistry Section, 1912. 'Trans. Am. Electrochem. Soc.,' 1912, **22**, 73.

BLEININGER AND BROWN. "The testing of clay refractories, with special reference to their load-carrying capacity at furnace temperatures." 'Bur. Standards Technologic Paper,' No. 7, 1911.

—— 'Trans. Am. Cer. Soc.,' 1910, **12**, 336; 1911, **13**, 210; 1912, **14**, 391.

[1] Cf. E. and E. A. Griffiths, loc. cit.

Bohm. "The technical application of crude zirconia." 'Chem. Zeit.,' 1911, **35**, 1961.

—— "Alundum and graphite." 'Scientific Amer. Suppl.,' No. 1499, Sept. 24, 1904.

Bölling. "Silundum." 'Electrochem. and Met. Ind.,' 1909, **7**, 25.

Fitzgerald. "Carborundum." Engelhardt Monographs on Applied Electrochemistry (in German). 'Journ. Frankl. Inst.,' 1897, **143**, 81.

Fitzgerald. "Carborundum furnace." 'Electrochem. and Met. Ind.,' **4**, 53.

—— "Acheson graphite furnace." 'Electrochem. and Met. Ind.,' **3**, 416; see also 1909, **7**, 187.

—— "Artificial graphite," vol. **15** of the Engelhardt Monographs.

Gary. 'Mitt. K. tech. Versuchsanstalten,' 1896, **14**, 63.

Goodwin and Mailey. "Physical properties of fused magnesia." 'Trans. Am. Electrochem. Soc.,' 1906, **9**.

Gillett. "Carborundum." 'Journ. Phys. Chem.,' 1911, **15**, 213.

Heraeus. 'Zeit. Angew. Chem.,' 1905, **18**, 49.

Hofman and Demond. 'Trans. Am. Inst. Mining Eng.,' 1894, **24**, 42.

Hofman and Stroughton. 'Trans. Am. Inst. Mining Eng.,' 1898, **28**, 440.

Hofman. 'Trans. Am. Inst. Mining Eng.,' 1895, **25**, 3; 1898, **28**, 435.

Jacobs. "Alundum," U.S.A. Patent, No. 659,926 (1900).

Jochum. 'Thonindustrie Zeit.,' 1903, **27**, 764.

Kanolt. "Melting-points of some refractory oxides." 'Bull. Bur. Stds.,' 1914, **10**.

—— "Melting-points of fire-bricks." 'Trans. Am. Electrochem. Soc.,' 1912, **22**, 95; and 'Bur. Standards Technologic Paper,' No. 10, 1912.

Mellor and Moore. "The effects of loads on the refractoriness of fire-clays." 'Trans. Eng. Cer. Soc.,' 1915-16, **15**, 117.

Meyer. "Refractory properties of Zirconia." 'Met. and Chem. Eng.,' 1914, **12**, 791; 1915, **13**, 263.

Nesbitt and Bell. "Practical methods for testing refractory fire-brick." 'Met. and Chem. Eng.,' 1915-16, **15**, 205.

Northrup. "Properties of fibrox." 'Met. and Chem. Eng.,' 1916, **15**, 409.

Parker and Hensel. 'Trans. Am. Cer. Soc.,' 1905, **7**, 185.

Richards. "Silicon and boron." 'Scientific Amer. Suppl.,' Nos. 2058, 2059, June 12 and 19, 1915.

RICHARDS. "Graphite manufacture." 'Electrochem. and Met. Ind.,' 1902, **I**, 52.

——— "Carborundum manufacture." 'Electrochem. and Met. Ind.,' 1902, **I**, 50; 1909, **7**, 189.

RODD. "Zirconia as a refractory." 'J. S. Chem. Ind. Rev.,' 1918, **37**, 213.

RUFF, SEIFERHELD AND SUDA. "Fusion and volatilisation of refractory oxides in the electric vacuum furnace." 'Zeit. Anorg. Chem.,' 1913, **82**, 373.

RUFF, SEIFERHELD AND BRUSCHKE. "Preparation of refractory objects of zirconium dioxide." 'Zeit Anorg. Chem.,' 1914, **86**, 389.

SAUNDERS. "Carborundum." 'Trans. Am. Electrochem. Soc.,' 1912, **21**, 425.

SOSMAN. "The common refractory oxides." 'Journ. Ind. Eng. Chem.,' 1916, **8**, 985.

TONE. "Silicon." 'Electrochem. and Met. Ind.,' 1909, **7**, 192; **2**, 111; **4**, 464; **5**, 141. Preparation in arc furnaces using sand and coke.

——— "Silicidised carbon-silfrax." 'Trans. Am. Electrochem. Soc.,' 1914, **26**, 181.

TUCKER AND LAMPEN. "Carborundum." 'Journ. Am. Chem. Soc.,' 1906, **28**, 853.

TUCKER. "Manufacture of silicon in a small laboratory furnace." 'Met. and Chem. Eng.,' 1910, **8**, 19.

VOGEL. "Quartz fusion furnace." 'Electrochem. Zeit. (Berlin),' 1911, 121, 181, 218.

WATTS AND MENDENHALL. "Softening of carbon at high temperature." 'Phys. Rev.,' 1911, **33**, 65.

WEBER. 'Trans. Am. Inst. Mining Eng.,' 1904, **35**, 637.

WEDEKIND. "Synthesis of borites in the vacuum electric furnace," 'Ber. Deutsch. Chem. Ges.,' 1913, 1198.

STEEL.[1]

AMBERG, R. "The function of slag in electric steel refining." 'Trans. Am. Electrochem. Soc.,' 1912, **22**, 133; 'Met. and Chem. Eng.,' 1912, **10**, 601.

"Arc furnaces for steel making." 'Iron Coal Trade's Rev.,' Jan. 20, 1917.

ARNOU, G. "Electric steel direct from the ore." 'Lumière eléc.,' **13**, 304; 'Rev. métal.,' **7**, 1190; 'Engineer,' **91**, 834.

——— "Notes on electric steel." 'Rev. metal.,' **7**, 1054.

---

[1] Cf. A. D. Little, loc. cit.

BAILY, T. F. "An electric furnace for heating bars and billets." 'Trans. Am. Electrochem. Soc.,' 1911, **19**, 285; 1912, **21**, 419.

—— "The electric furnace as a soaking pit in the steel mill." 'Iron Age,' 1916, **97**, 311.

BIBBY, J. "Electric steel refining furnace." 'Trans. Faraday Soc.,' 1919, **14**, 78; 'Engineering,' **107**, 649.

BJORKSTEDT, WM. "The induction furnace, its efficiency and refining capabilities." 'Met. and Chem. Eng.,' 1914, **12**, 146.

BOOTH, W. K. "The Booth-Hall electric furnace." 'Electrician,' 1919, **82**, 588; 'Met. and Chem. Eng.,' 1918, **18**, 211; 'Iron Age,' **101**, 45; 'Iron Trade Rev.,' **62**, 162.

BUCHANAN, W. "Electric steel melting plant." 'J. Chem. Met. Soc.,' South Africa, 1917, **18**, 83.

CAMPBELL, D. F. "Canada's electric steel plant at Toronto." 'Iron Age,' 1918, **101**, 1053; 'Elec. Rev.,' 1918, **73**, 177.

"Changing the voltage of electric steel furnace by variable transformer connection." 'Elec. Rev. West. Elec.,' 1918, **72**, 636.

"Chetwynd electrical purification process." 'Engineering,' 1915, **99**, 283.

CHURCHILL, F. A. "Seattle electric steel foundry." 'Iron Trade Rev.,' 1914, **55**, 1043, 1050.

CLARK, E. B. "Electric furnaces for steel making." 'Trans. Am. Electrochem. Soc.,' 1914, **25**, 139; 'Chem. Eng.,' **19**, 157; 'Iron Age,' 1914, **93**, 1007; 'Met. and Chem. Eng.,' 1914, **12**, 336.

CLARK, GEO. T., AND PHILLIP, F. "World's largest electric steel plant in Toronto." 'Can. Eng.,' 1919, **36**, 327.

COGSWELL, W. H. "Electric furnace steel." 'Elec. Journ.,' 1917, **14**, 142.

CONE, E. F. "Status of the electric steel industry." 'Iron Age,' 1919, **103**, 60.

CORNELL, SIDNEY. "Open hearth *versus* the electric furnace in the manufacture of commercial steels." 'Met. and Chem. Eng.,' 1915, **13**, 630.

CRAFTS, W. N. "Producing steel in electric furnace." 'Iron Age,' 1914, **93**, 1066.

CROWLEY, J. A. "Gronwall-Dixon electric furnace." 'Mech. Eng.,' 1917, **38**, 306.

DALTON, A. C. "Electric steel direct from ore mines." 'Iron Age,' 1914, **94**, 877; 1915, **96**, 1184.

DARLING, CHAS. R. "Electric furnaces." 'Nature,' 1919, **103**, 235.

DE FRIES, H. A., AND HERLENIUS, J. "Developments in the Rennerfelt furnace." 'Iron Age,' 1919, **103**, 190.

DIXON, J. L. "Notes on electric steel melting." 'Trans. Am. Electrochem. Soc.,' 1917, **31**, 53; 'Met. and Chem. Eng.,' **16**, 577.

DUFRESNE, A. O. "Electric steel furnaces in the Province of Quebec." 'J. four eléc.,' 1917, **26**, 273.

"Electric furnace operation at Buffalo." 'Elec. World,' 1919, **73**, 1378.

"Electric furnaces in the steel industry and their relation to the central station business." 'Met. and Chem. Eng.,' 1919, **20**, 73.

"Electric furnace steels and alloy steels." 'J. Ind. Eng. Chem.,' 1916, **8**, 947; 'Met. and Chem. Eng.,' 1916, **15**, 448.

"Electric furnaces for heat treatment of steel." 'Electrician," 1919, **83**, 375.

"Electric steel direct from titaniferous iron ores." 'Iron Age, 1915, **96**, 1416.

"Electric steel furnaces in California." 'Metal Trades,' 1919, **10**, 245.

"Electric steel furnaces in Great Britain." 'Iron Age,' 1917, **100**, 519, 1251.

"Electric steel furnaces in Japan." 'Electrician,' **80**, 139.

"Electric steel furnaces in Sheffield." 'Elec. Rev.,' 1913, **73**, 154.

"Electric steel in Germany and Austria." 'Met. and Chem. Eng.,' 1915, **13**, 398.

ETCHELLS, H. "Application of electric furnace methods to industrial processes." 'Trans. Faraday Soc.,' 1919, **14**, 71.

—— "Electric steel at Sheffield." 'Iron Age,' **97**, 1379.

—— "Modern steel making by electricity." 'Journ. Elec.,' 1919, **43**, 310.

FLINTERMAN, R. F. "Electric steel castings." 'Trans. Am. Electrochem. Soc.,' 1918, **33**, 263; 'Met. and Chem. Eng.,' 1918, **18**, 511, 610; 'Iron Age,' **101**, 1398.

—— "The electric furnace in the steel casting plant." 'Met. and Chem. Eng.,' 1917, **16**, 574; 'Trans. Am. Electrochem. Soc.,' 1917, **31**, 69.

—— "Electric process for small steel castings." 'Iron Age,' 1917, **99**, 1144.

FRESCH, O., RENNERFELT AND VON ECKERMANN. "Experiments with the Rennerfelt furnace." 'J. four. eléc.,' 1918, **27**, 101.

FRICK, O. "Frick furnace for electric refining of steel." 'Iron Age,' 1913, **92**, 1113.

—— "Frick electric steel induction furnace." 'Iron Age,' 1913, **92**, 670, 744.

GIROD, P. "The electric steel furnace in foundry practice." 'Met. and Chem. Eng.,' 1912, **10**, 663.

—— "Girod electric steel furnace." 'Engineering,' 1917, **104**, 519.

Gosron, R. C. "Producing electric steel castings." 'Iron Trade Rev.,' 1919, 65, 1706.

Gray, J. H. "Steel making in electric furnaces." 'Iron Age,' 1915, 96, 1238; 'Met. and Chem. Eng.,' 1915, 13, 656.

Greene, A. E., and Amberg, R. "Function of the slag in steel refining." 'Met. and Chem. Eng.,' 1912, 10, 656.

Hadfield, R. "Electric furnace in steel work." 'Electrician,' 1918, 80, 903.

Hess, H. L. "Electric furnaces as applied to steel making." 'Mech. Eng.,' 1919, 41, 245.

Holden, J. A. "Economic production of electric steel." 'Iron Age,' 1919, 104, 440.

—— "Electric carbon tool steel—details of British production in a 3-ton Heroult furnace." 'Iron Age,' 1918, 101, 1222.

Keeney, R. M. "Pig steel from ore in the electric furnace." 'Bull. Am. Inst. Min. Eng.,' 1914, 86, 349; 90, 1289; 'Iron Age,' 1914, 93, 810.

Kershaw, J. B. E. "Electric furnace method of steel production." 'Iron Trade Rev.,' 1913, 51, 865 *et seq.*; 52, 197 *et seq.*

—— "Induction furnace for steel refining." 'Engineer' (London), 1912, 114, 643.

Lellis, A. de, and Rines, C. "New resistance type electric steel furnace" (the Lellis furnace). 'J. four eléc.,' 1917, 26, 325.

Lipin, W. "Nathusius electric steel furnace." 'Met. and Chem. Eng.,' 1912, 10, 227.

"Ludlum electric steel furnace." 'Electrician,' 1917, 80, 215.

Lyon, D. A. "Noble Electric Steel Company's plant." 'Trans. Am. Electrochem. Soc.,' 1909, 15, 39; 'Electrochem. and Met. Ind.,' 1909, 7, 252.

Mathews, J. A. "Electric furnace in steel manufacture." 'Iron Age,' 1916, 97, 1327; 'Iron Trade Rev.,' 58, 1264.

Moffet, F. J. "Electric steel furnaces in England. 'Elec. Rev.,' 1918, 73, 1000; 'Iron Age,' 1919, 103, 120.

—— "English electric furnace development." 'Blast Furnace and Steel Plant.,' 1919, 7, 169.

Müller, A. "Metallurgy of the acid electric steel process." 'Stahl u. Eisen,' 34, 89; 'Iron Age,' 93, 670; 'Chem. Eng.,' 19, 154.

Nathusius, Hans. "Refining of steel in the Nathusius electric furnace." 'J. Iron Steel Inst.,' 1912, 85, 41.

Neumann, B. "New electric steel furnace." 'Met. and Chem. Eng.,' 1918, 18, 211.

"New electric steel plant and rolling mill." 'Iron Age,' 1917, 99, 1003.

31

"New sidelights on electric steel making." 'Iron Age,' 1917, **99**, 1132.

"Operating records of electric steel furnaces." 'Elec. World,' 1919, **74**, 125.

"Progress in electric steel." 'Iron Age,' 1919, **103**, 571.

RENNERFELT, I. "Rennerfelt electric steel furnace." 'Iron Age,' 1914, **93**, 200.

RICHARDS, J. W. "The passing of crucible steel." 'Met. and Chem. Eng.,' 1916, **8**, 563.

—— "Pig steel made directly from ore in the electric furnace." 'Met. and Chem. Eng.,' 1912, **10**, 397.

ROBERTSON, T. D. "Electric steel-making furnaces." 'Engineering,' 1915, **99**, 176 ; 'Electrician,' 1915, **74**, 630.

ROBINSON, T. W. "Electric steel furnace experiments at South Chicago." 'Met. and Chem. Eng.,' 1912, **10**, 373.

—— "Triplex process of making electric steel." 'Iron Age,' **101**, 1471 ; 'Iron Trade Rev.,' **62**, 1432 ; 'Iron Age,' **101**, 1476.

SCOTT, E. K. "Electric steel furnaces." 'Elec. Rev.' (London), 1917, **81**, 45. 'Iron Coal Trades Rev.,' July 6, 1917.

SCOTT, WIRT S. "Development of an electric furnace for annealing treatment and forging of steel." 'Chem. and Met. Eng.,' 1918, **19**, 86.

SNYDER, F. T. "Cost of electric furnace steel." 'Iron Age,' **96**, 926 ; 'Iron Trade Rev.,' 1915, **57**, 1091.

—— "Electric steel." 'Iron Trade Rev.,' 1914, **55**, 1077.

—— "Electric steel costs." 'Trans. Am. Electrochem. Soc.,' 1916, **28**, 221.

STANLEY, G. H., AND BUCHANAN, W. "Design and operation of a small Kjellin furnace (South Africa)." 'Met. and Chem. Eng.,' 1918, **18**, 349.

STOBIE, V. "Large electric steel melting furnaces." 'Iron Coal Trades Rev.,' 1919, **98**, 618 ; 'Foundry Trade J.,' 1919, **21**, 304 ; 'Electrician,' 1919, **83**, 526 ; 'Engineer,' **107**, 749.

—— "Manufacture of electric steel in the Stobie furnace." 'Mech. Eng.,' 1916, **35**, 502 ; 'Engineer,' 1915, **119**, 616. 'Iron Age,' 1915, **95**, 1171.

"Webb electric steel furnace." 'Iron Age,' 1918, **102**, 257.

"Wild-Barfield electric muffle." 'Engineering,' 1918, **106**, 143.

WILE, R. S. "Electric steel furnace of new design." 'Iron Age,' 1915, **96**, 966.

WILLS, W. H., AND SCHUYLER, A. H. "Heat losses from an electric steel furnace." 'Trans. Am. Electrochem. Soc.,' 1915, **28**, 207.

INDEX.

PRINTED IN GREAT BRITAIN BY THE UNIVERSITY PRESS, ABERDEEN